BLACK&DECKER®

THE COMPLETE PHOTO GUIDE TO

SHEDS, BARNS & OUTBUILDINGS

Creative Publishing international

MINNEAPOLIS, MINNESOTA
www.creativepub.com

Creative Publishing international

400 First Avenue North, Suite 300
Minneapolis, Minnesota 55401
1-800-328-0590
www.creativepub.com

Printed in China

10 9 8 7 6 5 4 3 2 1

Library of Congress Cataloging-in-Publication Data

The complete photo guide to sheds, barns & outbuildings.
p. cm.
"Includes Garages, Gazebos, Shelters and More."
"Black & Decker."
Includes index.
Summary: "Provides more than 25 detailed plans and instructions for building a wide variety of outdoor storage buildings"--Provided by publisher.
ISBN-13: 978-1-58923-522-9 (soft cover)
ISBN-10: 1-58923-522-3 (soft cover)
1. Sheds. 2. Barns. I. Black & Decker Corporation (Towson, Md.) II. Title.

TH4955.C66 2010
690'.89--dc22

2010004485

President/CEO: Ken Fund

Home Improvement Group

Publisher: Bryan Trandem
Managing Editor: Tracy Stanley
Senior Editor: Mark Johanson
Editor: Jennifer Gehlhar

Creative Director: Michele Lanci-Altomare
Art Direction/Design: Jon Simpson, Brad Springer, James Kegley

Lead Photographer: Joel Schnell
Set Builder: James Parmeter
Production Managers: Linda Halls, Laura Hokkanen

Page Layout Artist: Danielle Smith
Shop Help: Charles Boldt
Edition Editor: Betsy Matheson Symanietz
Technical Editor: Philip Schmidt
Proofreader: Jane Hilken
Illustrator: Trevor Burks, Robert Leonnatt (p. 216, 217, 218 top, 219)

The Complete Photo Guide to Sheds, Barns & Outbuildings
Created by: The Editors of Creative Publishing international, Inc., in cooperation with Black & Decker.

NOTICE TO READERS

For safety, use caution, care, and good judgment when following the procedures described in this book. The publisher and Black & Decker cannot assume responsibility for any damage to property or injury to persons as a result of misuse of the information provided.

The techniques shown in this book are general techniques for various applications. In some instances, additional techniques not shown in this book may be required. Always follow manufacturers' instructions included with products, since deviating from the directions may void warranties. The projects in this book vary widely as to skill levels required: some may not be appropriate for all do-it-yourselfers, and some may require professional help.

Consult your local building department for information on building permits, codes, and other laws as they apply to your project.

Contents

The Complete Photo Guide to Sheds, Barns & Outbuildings

Introduction 7

OUTBUILDINGS BASICS

Getting Started 11

Choosing a Site for Your Building. 12

Building Codes & Zoning Laws. 14

Working with Plans 15

Power & Rental Tools. 16

Fasteners & Hardware. 17

Selecting Lumber. 18

Working with Concrete 20

Safety Considerations 21

Environmentally Friendly Building Materials. 22

Construction Techniques. 25

Constructing an Outbuilding Overview 26

Building the Foundation. 28

Framing & Raising Walls. 34

Framing the Roof 44

Sheathing Walls 52

Installing Fascia & Soffits. 56

Completing the Roof 60

Installing Windows & Service Doors 68

Installing Overhead Doors 76

Installing Siding & Trim. 82

Contents (Cont.)

PLANS & PROJECTS

Barns & Sheds 93

Pole Barn 94

Mini Gambrel Barn 114

Simple Storage Shed 130

Metal & Wood Kit Sheds 142

Clerestory Studio 160

Timber-frame Shed 174

Gothic Playhouse 186

Lean-to Tool Bin 200

Additional Barn & Shed Plans 210

Greenhouses & Garden Buildings 215

Types of Greenhouses 216

Hoophouse 220

A-frame Greenhouse 226

Hard-Sided Greenhouse Kit 232

Mini Garden Shed 240

Sunlight Garden Shed 248

Shelters, Arbors & Gazebos 263

Rustic Summerhouse 264

Contents (Cont.)

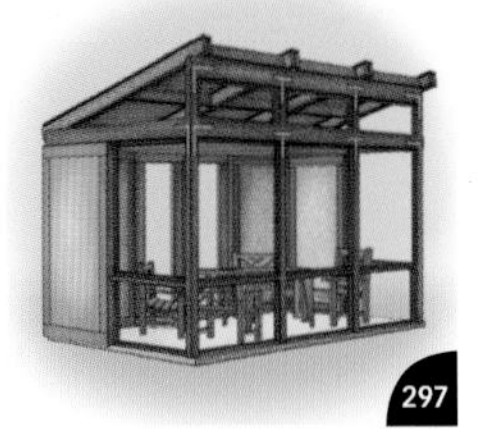

3-season Gazebo 282

Summer Pavilion 297

Lattice Gazebo 310

8-sided Gazebo 320

Pool Pavilion 336

Patio Shelter 346

Arbor Retreat 360

Classical Pergola 374

Corner Lounge 384

Garages 397

Single Detached Garage 398

Compact Garage 404

Carport 418

Additional Garage Plans 428

APPENDIX A: Foundations 432

APPENDIX B: Cedar Shingle Roofing 440

Resources & Credits 442

Conversion Charts 443

Index 445

Introduction

Building a new barn, shed, gazebo, or garage will definitely be one of the most rewarding do-it-yourself projects you will ever complete. Big or small, for work or play, decorative or functional (or both)—any type of outbuilding you build will help define and enhance your property for years to come. The chapters contained within The Complete Photo Guide to Sheds, Barns & Outbuildings provide you with the techniques, tips, plans, and know-how to help you take on a big project with the confidence that you can do it—and do it right.

The first chapter will guide you through the planning process, ensuring that you consider everything from building codes to daylight patterns and neighbors' opinions before starting construction on a major project. The next chapter, Construction Techniques, takes you through the process of building a typical outbuilding step-by-step. If this is your first large project, you'll likely reference this chapter throughout the construction of your building. If you're more experienced, these techniques and tips will be a great refresher course in those aspects of building construction you haven't practiced recently.

The remaining pages in this book are filled with complete plans and materials lists and detailed instructions for building more than 25 different outbuildings: barns, sheds, gazebos, garages, greenhouses, and more. Whatever specific criteria you have in mind for your new building—whether you need a new barn to house animals, a screened-in relaxation room to enjoy summer evenings, a storage shed for landscaping equipment, or a greenhouse for nurturing young plant life—the perfect building plan is contained in this volume. Utilizing timesaving tricks from the pros, proven techniques that have been refined over hundreds of years, and inspiring advice to help you make your building your own, these projects are designed to make sure that your new outbuilding will be built to last as a quality addition to your property.

There are few accomplishments that offer as much satisfaction as completing a big project and knowing that the fruits of your labor will improve the quality of life for you and your family for years to come. So the next time you find yourself staring out at the backyard, dreaming of just the right building to complete your landscape, pick up this book and start planning. Before you know it, you'll be staring out at your new building, satisfied that you made it happen and did it all yourself.

Note: Construction Diagrams ▸

If you find the construction diagrams accompanying the project plans difficult to read, you may find some benefit to making enlargements on a photocopy machine. Enlarged photocopies can be useful for jotting down notes, as well.

OUTBUILDINGS BASICS

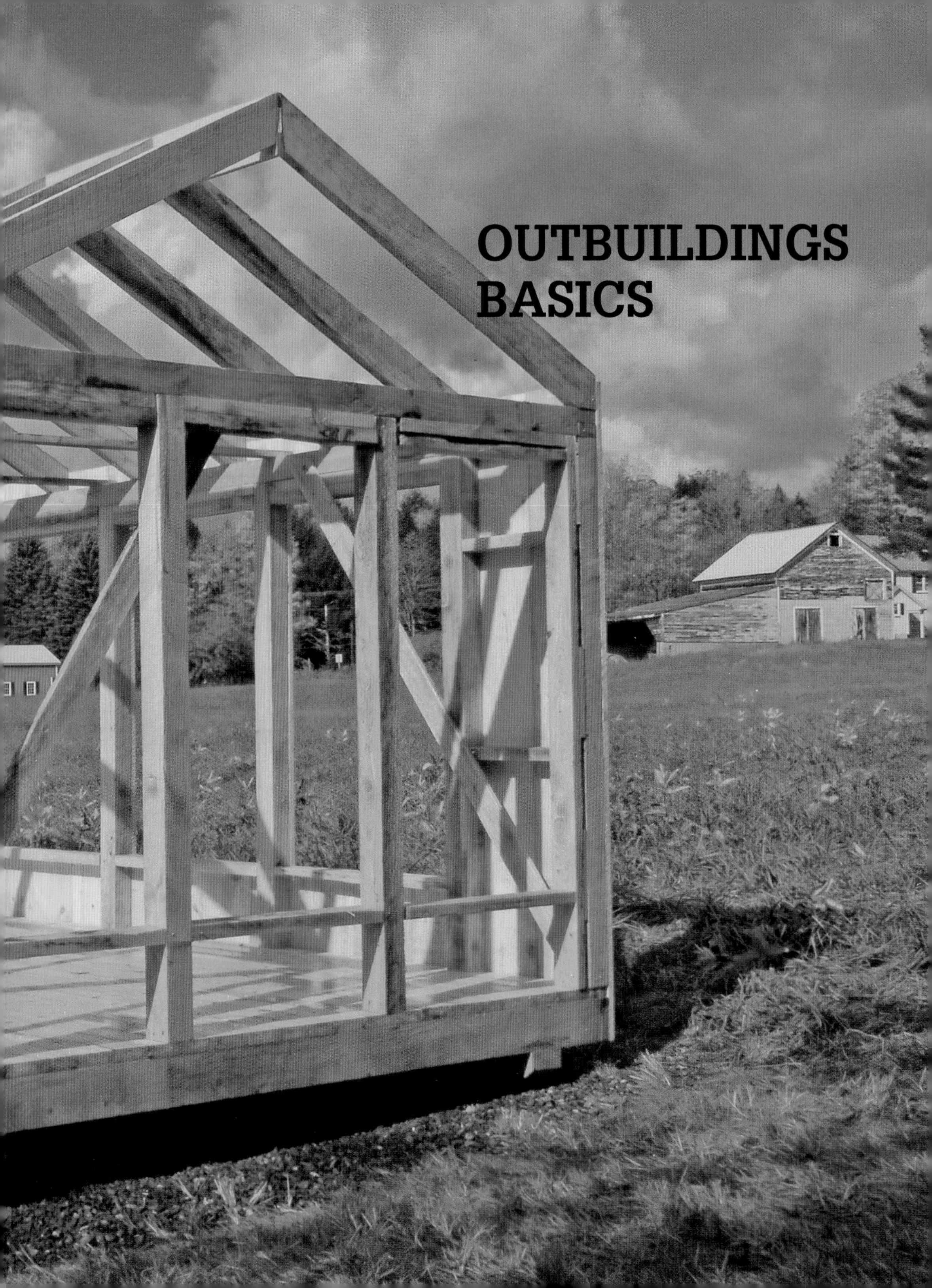

Getting Started

If you need a new building on your property—for storing equipment or tools, for growing seedlings for your garden, or just for enjoying the view—it may be difficult to know where to start. You aren't alone. Knowing the right questions to ask regarding what kind of building you need, where to erect it, how much of each material you should order, and how to get started isn't an intuitive process. There are multiple factors to consider that will affect your property and project, but also that affect your neighbors, the environment, and your municipality's rules and regulations.

This chapter will help you ask the right questions and get the answers you need to get started on your project. The tips from industry experts included here can help you with everything from choosing the right power tools to rent to figuring out where to place your new greenhouse so it receives the most sunlight year-round.

In this chapter:

- Choosing a Site for Your Building
- Building Codes & Zoning Laws
- Working with Plans
- Power & Rental Tools
- Fasteners & Hardware
- Selecting Lumber
- Working with Concrete
- Safety Considerations
- Environmentally Friendly Building Materials

Choosing a Site for Your Building

When selecting a site for a new outbuilding—whether it be an equipment storage shed, a backyard gazebo, a detached garage, or a new home for your animals—first consider the big picture of your property. Assess areas that could present problems, and document those trouble spots on your property map. Answer basic questions about your property, such as:

- What areas may be prone to snowdrifts, flooding, or direct sunlight?
- Where does your property slope, and where are the largest flat areas?
- Where are the natural drainage routes after heavy rains and snowmelt?

Then, address what is missing from your property, as well as your own needs:

- Will you need to add any roads or pathways to access your new building?
- Where is the most fertile soil?
- Are there areas of your home that you'll access frequently from this structure (e.g., the kitchen, pool, or garage)?
- How will adding a building improve or detract from the appearance of your property?
- How will adding a building affect your neighbors' view or their property's aesthetics?

To help you visualize some of these questions, make a few bubble diagrams like the one below. Make several versions of potential layouts, taking into account existing structures, property boundaries, and setback restrictions. When you have settled on a couple of good options for your new building's location, head to your municipality's building department (see page 14) to learn about restrictions that apply to your building and to finalize your site selection.

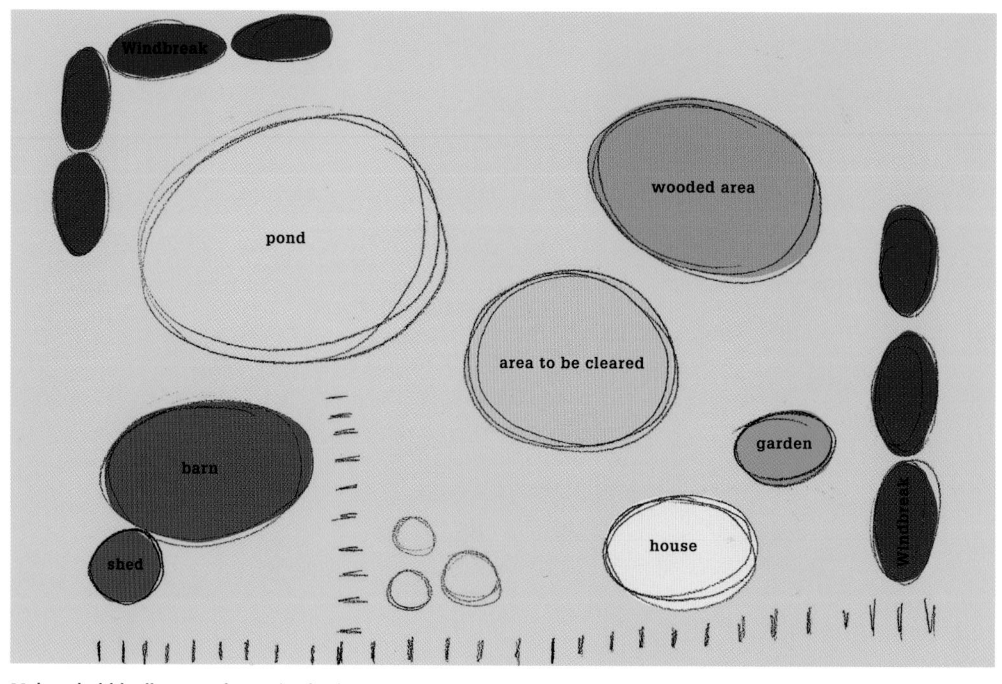

Make a bubble diagram of your plan for the land so you can visualize how your new project will fit into the big picture.

Siting for Sunlight

Barns, garden sheds, and recreational structures can all benefit enormously from natural light. To make the most of natural sunlight, the general rule is to orient the building so its long side (or the side with the most windows) faces south. However, be sure to consider the sun's position at all times of the year, as well as the shadows your building might cast on surrounding areas, such as a garden or outdoor sitting area.

SEASONAL CHANGES

Each day the sun crosses the sky at a slightly different angle, moving from its high point in summer to its low point in winter. Shadows change accordingly. In the summer, shadows follow the east-west axis and are very short at midday. Winter shadows point to the northeast and northwest and are relatively long at midday.

Generally, the south side of a building is exposed to sunlight throughout the year, while the north side may be shaded in fall, winter, and spring. Geographical location is also a factor: as you move north from the equator, the changes in the sun's path become more extreme.

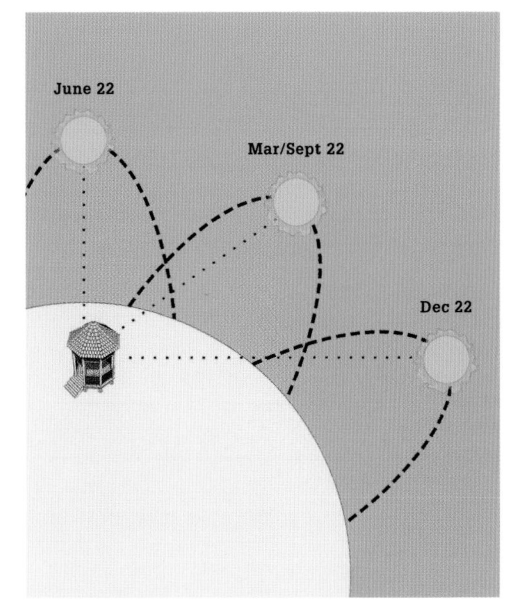

The sun moves from its high point in summer to its low point in winter. Shadows change accordingly.

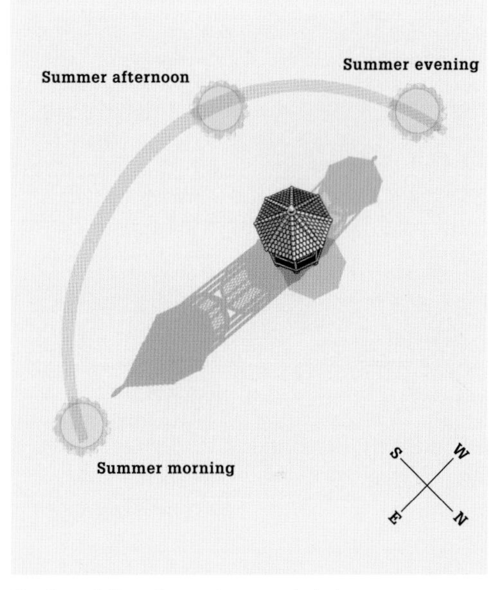

Shadows follow the east-west axis in the summer.

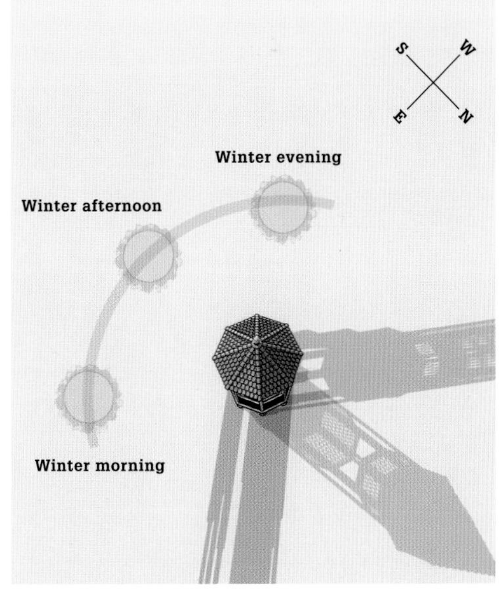

Winter shadows point to the northeast and northwest and are relatively long at midday.

Building Codes & Zoning Laws

Understanding your local zoning laws and building codes is an essential step when planning new construction on your property. Visit your municipality's building department to find out what kinds of structures are allowed on your property per local zoning regulations and what specific restrictions apply to your situation.

BUILDING CODES

Building Codes outline safety and structural standards for all types of buildings in a given municipality. Building Codes have specifications for the types of construction materials you may use, including minimum sizes for structural members and the size and spacing of fasteners. Building Codes will also define foundation or footing requirements for your project and the size of buildings you can build on your property.

ZONING LAWS

Zoning laws govern such matters as the building's location, its position relative to property lines and structures, the type of building permit(s) you'll need to secure before starting construction, and the total area or percentage of your lot that can be covered with buildings. It's also important to ask about easements, which are restricted zones on your property that must be left open for utility companies, emergency access, or other contingencies.

CALL BEFORE YOU DIG

Utility companies will come to mark all electrical, gas, and power lines, as well as telephone, cable, and sewage lines that are buried on your property—for free. Most states are part of the North American One Call Referral System (888-258-0808), which will contact all of the utilities in your area and notify them of your construction plans. Utility companies that have lines in your yard will automatically send out a representative to mark the lines after receiving this call.

Concrete frost footings are required for many outbuilding projects. The frost line is the first point below ground level where freezing will not occur. Frost lines may be 48" or deeper in colder climates. Always build frost footings 1 ft. deeper than the frost line or as specified by the local building code.

Once you have the architectural renderings, you must have the project approved by the city. Include as many details in your plan proposal as possible, including all materials you intend to use.

Working with Plans

The projects in this book include complete construction drawings in the style of architectural blueprints. If you're not familiar with reading plans, don't worry; they're easy to use once you know how to look at the different views. Flipping back and forth between the plan drawings and the project's step-by-step photos will help you visualize the actual structure.

Note: The drawings in this book are accurately proportioned, but they are not sized to a specific scale. Also, dimensions specified in the drawings are given in feet and inches (for example, 6'-8"), the standard format for architectural plans. For your convenience, the written instructions may give dimensions in inches so you don't have to make the conversion.

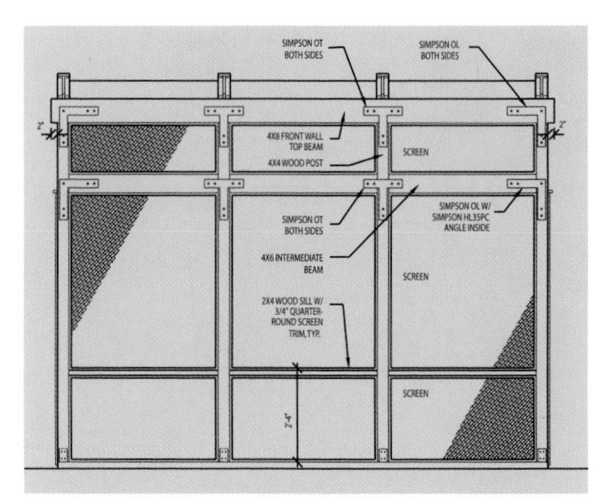

Elevations give you a direct, exterior view of the building from all sides. Drawings may include elevations for both the framing and the exterior finishes.

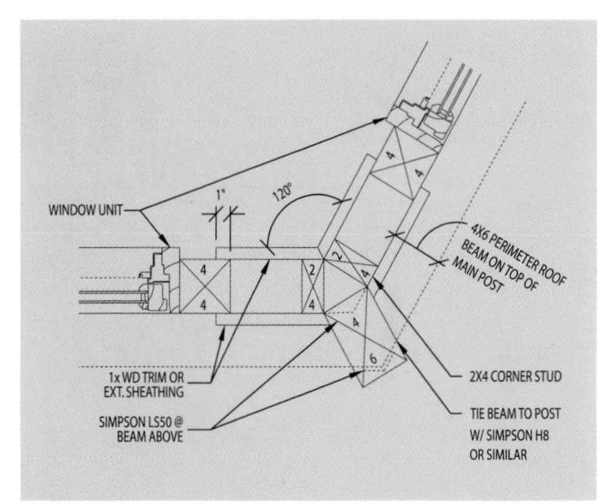

Detail drawings and templates show a close-up of a specific area or part of the structure. They typically show a side or overhead view.

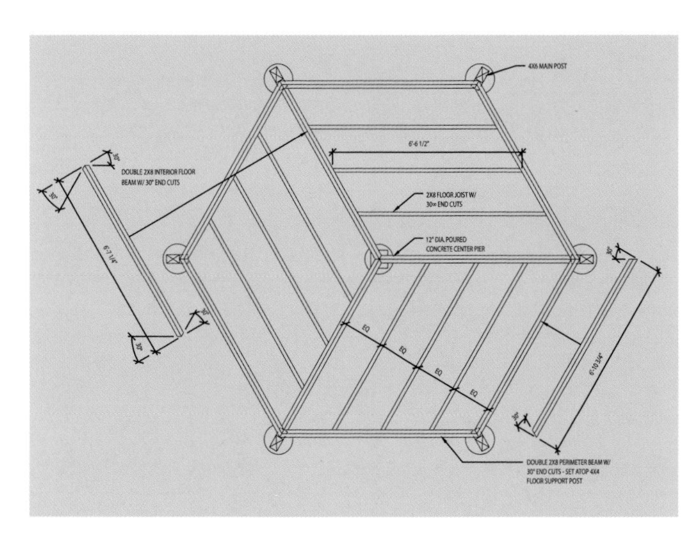

Plan views are overhead views looking straight down from above the structure. Floor plans show the layout of the walls or upright supports, with the top half of the structure sliced off. There are also foundation plans, roof framing plans, and other plan views.

Power & Rental Tools

When taking on a large project, there are few considerations more important than choosing the right tool for the job. Take a moment to familiarize yourself with many of the tools mentioned throughout this book.

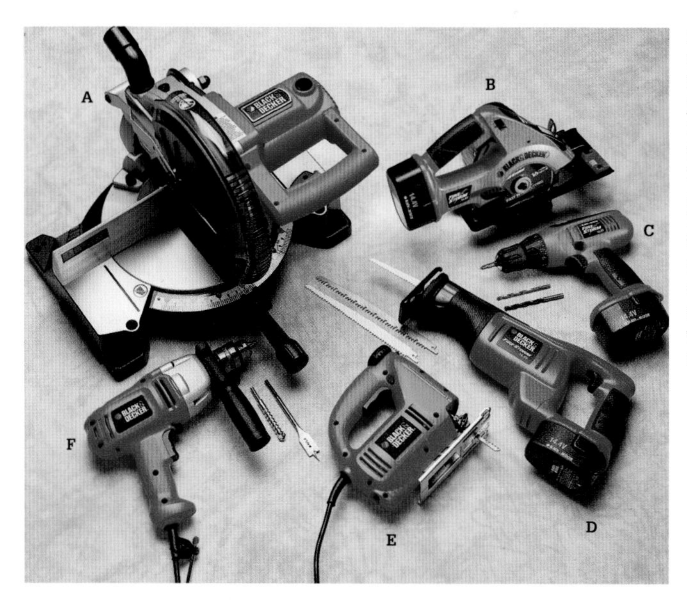

Power Tools: Some power tools are essential, including the circular saw (B) and drill with keyless chuck (C). Others just make jobs a lot easier: The power miter saw (A) makes quick, accurate cuts at any angle; the reciprocating saw (D), the ultimate multipurpose saw, is easily portable and makes straight or curved cuts in almost any material, including heavy timbers; a jigsaw (E) is best for clean, detailed curved cuts in various materials, especially thin sheets and fragile products; a ½" hammer drill (F) is a heavy-duty drill with a hammering motion for effective drilling into masonry, stone, and concrete.

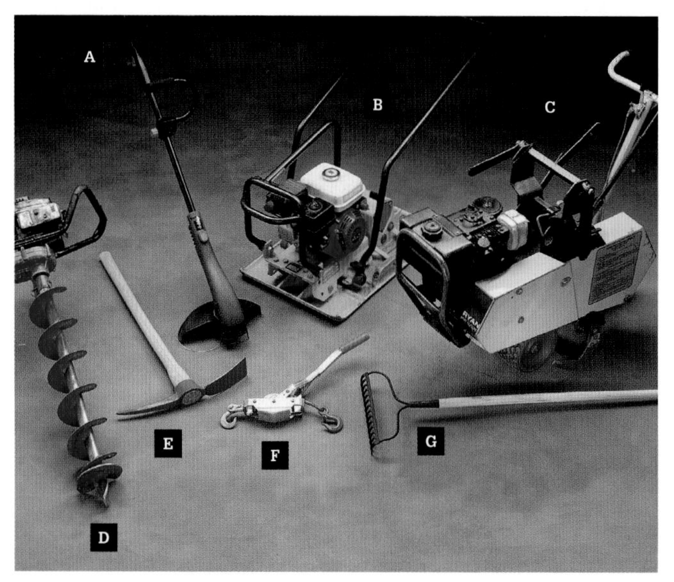

Rental & Landscape Tools: When you need the big guns, it usually makes more sense to rent than to buy. You might need any of these tools for clearing your project site or building the foundation: weed trimmer (A), power tamper or plate compactor (B), power sod cutter (C), power auger (D), pick (E), come-along (F), and garden rake (G).

Fasteners & Hardware

When selecting hardware for your project, remember one thing: all nails, screws, bolts, hinges, and anchors that will be exposed to weather or rest on concrete or that come in contact with treated lumber must be corrosion-resistant. The best all-around choice for nails and screws is hot-dipped galvanized steel, recognizable by its rough, dull-silver coating. Hot-dipped fasteners generally hold up better than the smoother, electroplated types, and they're the recommended choice for pressure-treated lumber. Aluminum and stainless steel are other materials suitable for outdoor exposure; however, aluminum fasteners corrode some types of treated lumber. While expensive, stainless steel is the best guarantee against staining from fasteners on cedar and redwood.

Another type of hardware you'll find throughout this book is the metal anchor, or framing connector, used to reinforce wood-framing connections. All of the anchors called for in the plans are Simpson Strong-Tie® brand (see Resources, page 442) that are available at most lumberyards and home centers. If you can't find what you need on the shelves, look through one of the manufacturer's catalogs or visit the manufacturer's website. You can also order custom-made hangers. Keep in mind that metal anchors are effective only if they are installed correctly—always follow the manufacturer's installation instructions, and use exactly the type and number of fasteners recommended.

Nailing Techniques

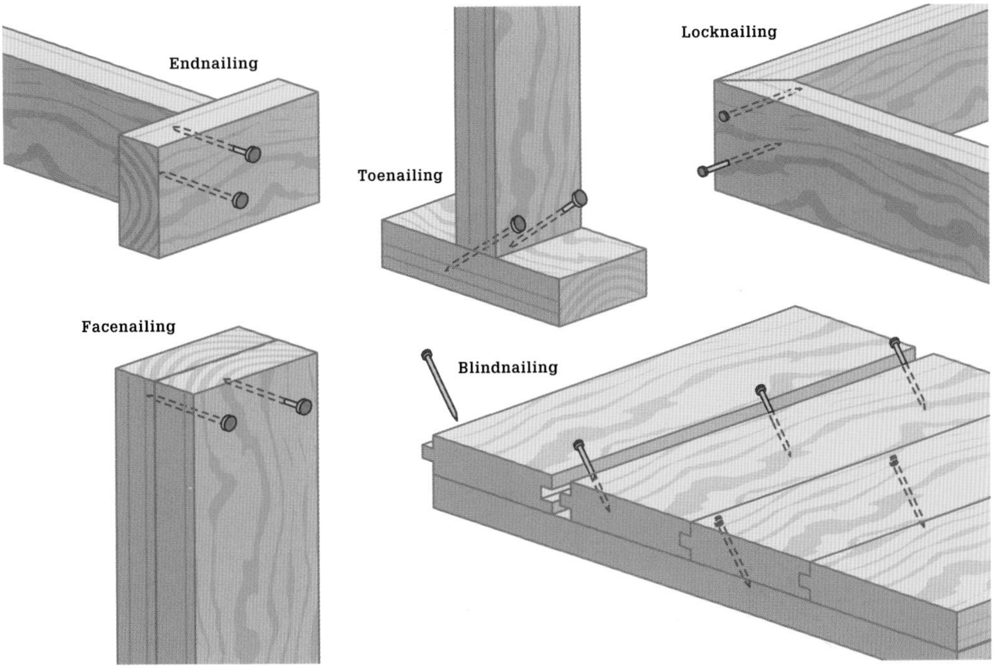

Endnailing

Toenailing

Locknailing

Facenailing

Blindnailing

Use the proper nailing technique for the task. Endnailing is used to attach perpendicular boards when moderate strength is required. Toenail at a 45° angle for extra strength when joining perpendicular framing members. Facenail to create strong headers for door and window openings. Blindnail tongue-and-groove boards to conceal nails, eliminating the need to set nails and cover them with putty before painting or staining. Locknail outside miter joints in trim projects to prevent gaps from developing as the trim pieces dry.

Selecting Lumber

Lumber types most commonly used in outbuildings are pine—or related softwoods—or cedar, which is naturally rot-resistant in specific grades and is less expensive than most other rot-resistant woods. For pine to be rot-resistant, it must be pressure-treated, typically with a chemical mixture called ACQ (ammoniacal copper quarternary).

Pressure-treated lumber is cheaper than cedar, but it's not as attractive, so you may want to use it only in areas where appearance is unimportant. Plywood designated as exterior-grade is made with layers of cedar or treated wood and a special glue that makes it weather-resistant. For the long run, though, it's a good idea to cover exposed plywood edges to prevent water intrusion.

Framing lumber—typically pine or pressure-treated pine—comes in a few different grades: Select Structural (SEL STR), Construction (CONST) or Standard (STAND), and Utility (UTIL). For most applications, Construction Grade No. 2 offers the best balance between quality and price. Utility grade is a lower-cost lumber suitable for blocking and similar uses but should not be used for structural members, such as studs and rafters. You can also buy "STUD" lumber: construction-grade 2 × 4s cut at the standard stud length of 92⅝". *Note: Treated lumber should be left exposed for approximately six months before applying finishes. Finishes will not adhere well to treated lumber that is still very green or wet. Lumber manufacturers likely have recommended times for their product.*

Board lumber, or finish lumber, is graded by quality and appearance, with the main criteria being the number and size of knots present. "Clear" pine, for example, has no knots.

All lumber has a nominal dimension (what it's called) and an actual dimension (what it actually measures). The chart on page 19 shows the differences for some common lumber sizes. Lumber that is greater than 4" thick (nominally) generally is referred to as timber. Depending on its surface texture and type, a timber may actually measure to its nominal dimensions, so check this out before buying. Cedar lumber also varies in size, depending on its surface texture. S4S (Surfaced-Four-Sides) lumber is milled smooth on all sides and follows the standard dimensioning, while boards with one or more rough surfaces can be over ⅛" thicker.

Materials for outbuildings include: Composite 2 × 4 (nonstructural) (A); Hardwood strips for screen frames (B); Fir floorboards (C); Construction lumber (D); Finish-grade lumber (E); T1-11 siding (F); Pre-primed plywood siding (G); Plywood (H); 6 × 6 treated post (I); 4 × 4 treated post (J); 2× treated construction lumber (K); 2× cedar construction lumber (L).

Lumber Types & Dimensions ▸

Type	Description	Common Nominal Sizes	Actual Sizes
Dimensional lumber	Used in framing of walls, ceilings, floors, and rafters, structural finishing, exterior decking, fencing, and stairs.	1 × 4	¾ × 3½"
		1 × 6	¾ × 5½"
		1 × 8	¾ × 7¼"
		2 × 2	1½ × 1½"
		2 × 4	1½ × 3½"
		2 × 6	1½ × 5½"
		2 × 8	1½ × 7¼"
Furring strips	Used as spacers and nonstructural blocking.	1 × 2	¾ × 1½"
		1 × 3	¾ × 2½"
Tongue-and-groove paneling	Used in wainscoting and full-length paneling of walls and ceilings.	5⁄16 × 4	Varies depending on milling process and application.
		1 × 4	
		1 × 6	
		1 × 8	
Finished boards	Used in trim, shelving, cabinetry, and other applications where a fine finish is required.	1 × 4	¾ × 3½"
		1 × 6	¾ × 5½"
		1 × 8	¾ × 7¼"
		1 × 10	¾ × 9¼"
		1 × 12	¾ × 11¼"
Micro-lam	Composed of thin layers glued together for use in joists and beams.	4 × 12	3½ × 11⅞"

Working with Concrete

Concrete is a versatile, indispensible building material that can be poured into footings or slabs to create a solid foundation that will last the life of your building. You may mix concrete yourself, but given the amount required for many of the projects in this book, you'll likely want to order ready-mix concrete delivered by truck to your site. When you order, explain how the concrete will be used to ensure you get the best mix for your project. For smaller projects, you could also rent a concrete trailer from a rental center or landscaping company; they fill the trailer with mixed concrete and you tow it home with your own vehicle.

If you're having your concrete delivered, be sure to have a few helpers on hand when the truck arrives; neither the concrete nor the driver will wait for you to get organized. Also, concrete trucks must be unloaded completely, so designate a dumping spot for any excess. Once the concrete form is filled, load a couple of wheelbarrows with concrete (in case you need it) and have the driver dump the rest. Be sure to spread out and hose down the excess concrete so you aren't left with an immovable boulder in your yard.

Timing is key to an attractive concrete finish. When concrete is poured, the heavy materials gradually sink, leaving a thin layer of water—known as bleed water—on the surface. To achieve an attractive finish, it's important to let the bleed water dry before proceeding with other steps. Follow these rules to avoid problems:

- Settle and screed the concrete immediately after pouring, and float the surface, if necessary.
- Let the bleed water dry before finishing or edging. Concrete should be hard enough that foot pressure leaves no more than a ¼"-deep impression.
- Do not overwork the concrete; it may cause the bleed water to reappear. Stop finishing if a sheen appears, and resume when it is gone.

If you've never worked with concrete, finishing a large slab can be a challenging introduction; you might want some experienced help with the pour.

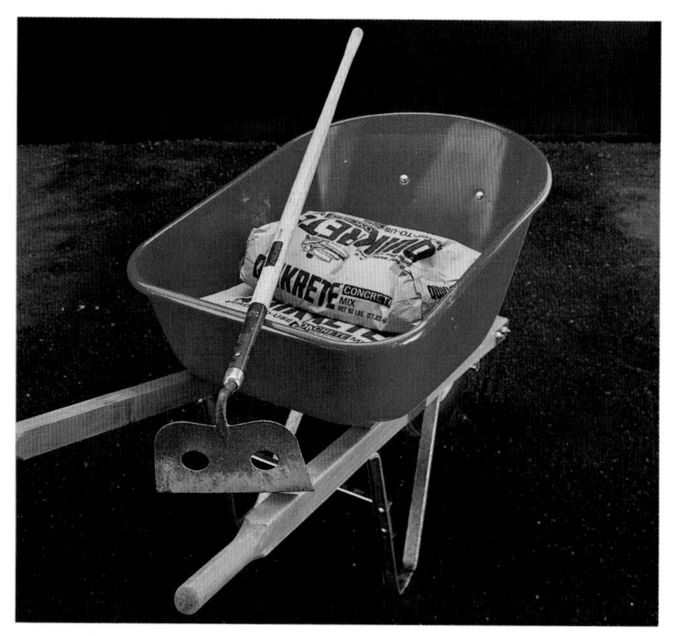

Buy premixed bags of dry concrete for small jobs; for bigger jobs, order premixed concrete to be delivered to your home when you're ready.

Safety Considerations

Building outbuildings means working outdoors. By taking commonsense precautions, you can work just as safely outdoors as indoors, even though doing construction work while exposed to the elements presents a few additional safety considerations. Bear the following in mind:

Use safety equipment when working up high. Building projects like the ones in this book frequently require that you use ladders or scaffolding or work on top of a roof. Learn and follow basic safety rules for working at heights and use personal fall-arresting gear if you will be spending considerable time on top of a roof.

Be mindful of the weather. Weather conditions should play a key role in just about every aspect of how you conduct your work—from the clothes you wear to the amount of work you undertake. Avoid extreme heat. During hot weather, take breaks as needed and drink plenty of fluids.

Get help. Work with a helper whenever you can. If you must work alone, inform a friend or family member so he or she can check up on you periodically. Always keep your cell phone with you.

Watch out for cords. Use cordless power tools when possible. When using corded tools, always plug them into a GFCI extension cord and unplug them when finished.

Protect yourself. Always wear eye and ear protection when working with power or pneumatic tools.

Keep it clean. Set up your work site for quick disposal of waste materials and clean up frequently.

Store tools safely. Set a sheet of plywood on top of a pair of sawhorses to make a surface for keeping tools off the ground, where they are a safety hazard and may be exposed to damage from moisture.

Wear sensible clothing and protective equipment when working outdoors, including a cap to protect against direct sunlight, eye protection when working with tools or chemicals, a particle mask when sanding, work gloves, full-length pants, and a long-sleeved shirt. A tool organizer turns a five-gallon bucket into a safe and convenient container for transporting tools.

Environmentally Friendly Building Materials

According to the U.S. Green Building Council, new construction uses over three billion tons of raw materials annually, roughly 40 percent of global raw materials. There are a wide variety of options available for DIYers who would like to make environmentally friendly decisions in their construction projects, including choosing reclaimed or recycled materials and making responsible energy decisions.

RECLAIMED MATERIALS

Reclaimed materials are perfectly usable structural members, siding, flooring, doors, windows, and other materials that have been salvaged from old construction or donated by companies that never used them. Many companies are now beginning to collect and sell reclaimed materials to DIY builders. Some of the reclaimed materials available on the market today offer aesthetic appeal unattainable with new materials. For example, siding reclaimed from an existing barn will add rustic charm and weathered appeal to your new barn or shed, enhancing the nostalgic charm of your property.

RECYCLED MATERIALS

Many new raw materials, including flooring, roofing, and even concrete, have recycled alternatives that match the strength, aesthetic appeal, and durability of their all-new counterparts. Recycled rubber flooring mats are a comfortable and attractive floor surface choice for your garden shed, barn, or gazebo. Many types of shingles and roofing surfaces are also made from recycled materials using composite technologies. Some companies are even using fly ash, a recovered resource from coal-fired power plants, to make specialized concrete blocks that have excellent load-bearing and thermal properties and are resistant to mold and fungi. See Resources for Environmentally Friendly Materials, on page 23, for details.

Siding your shed or barn with reclaimed barn siding adds rustic charm that will stand the test of time.

Recycled roofing materials, like the tiles shown here, are made of 80% postindustrial rubber and plastic. They come in many colors and shapes and are strong, durable, and long-lasting.

ENERGY CONSERVATION

Outbuildings are great places to utilize alternative energy sources, such as solar or wind power. Although initial installation costs may be higher, solar or wind power often can provide all the energy needed for the electricity and, in some cases, heat in a barn or shed and often pay for themselves within a reasonable time frame. Check with your local utility companies to learn about options available in your area. If your outbuilding will be heated or cooled, make sure you select EnergyStar certified doors and windows and explore energy-efficient heating and cooling systems as well.

Solar panels can reduce energy expenses and are ideal for south-facing roof surfaces.

If the structure of your building's roof can support extra weight (and local ordinances approve it), consider planting a living roof. This innovative solution reduces water runoff, adds insulation to save on energy costs, and improves air quality. Be sure to consult with a qualified architect and your insurance provider first.

Resources for Environmentally Friendly Materials ▸

Reclaimed timbers, siding, and building materials: trestlewood.com, heritagesalvage.com
Recycled "cedar shake" roofing tiles: ecostar.com
Rubber flooring: dinoflex.com
Green building materials showroom: ecospaces.net
Living Roof Resources: greenroofs.org

To minimize shipping distances from specialty suppliers, search online for a manufacturer near your home.

Construction Techniques

Most outbuilding construction follows the same basic steps: foundation, floors, wall framing, roof framing, sheathing, roofing, windows and doors, and then finishing steps such as siding and trim. Although every project in this book has a slight variation on these steps, the basic techniques you'll need to complete them is very similar.

This chapter demonstrates each of these steps through construction of a typical outbuilding. If you haven't built a shed or outbuilding before, read through this chapter thoroughly for a crash course in outbuilding construction. If you're a veteran builder, you will find this chapter useful to help refine your technique in those aspects of outbuilding construction you haven't practiced for awhile, or it may introduce you to new, innovative approaches to time-tested techniques that you may want to try out on your next project. Regardless of your skill level, spend some time with this chapter before starting your next outbuilding project; you'll find these pages to be rich with step-by-step instructions, tips, and advice from the experts.

In this chapter:

- Constructing an Outbuilding Overview
- Building the Foundation
- Framing & Raising Walls
- Framing the Roof
- Sheathing Walls
- Installing Fascia & Soffits
- Completing the Roof
- Installing Windows & Service Doors
- Installing Overhead Doors
- Installing Siding & Trim

Constructing an Outbuilding Overview

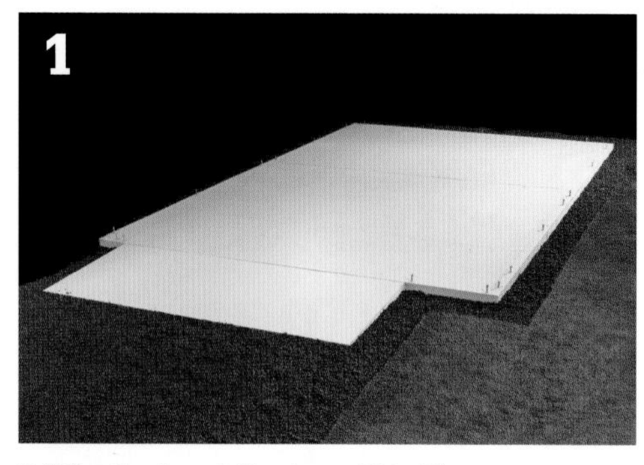

Building the Foundation (pages 28 to 33)

Framing & Raising Walls (pages 34 to 43)

Framing the Roof (pages 44 to 51)

Sheathing Walls (pages 52 to 55)

Installing Fascia & Soffits (pages 56 to 59)

Completing the Roof (pages 60 to 67)

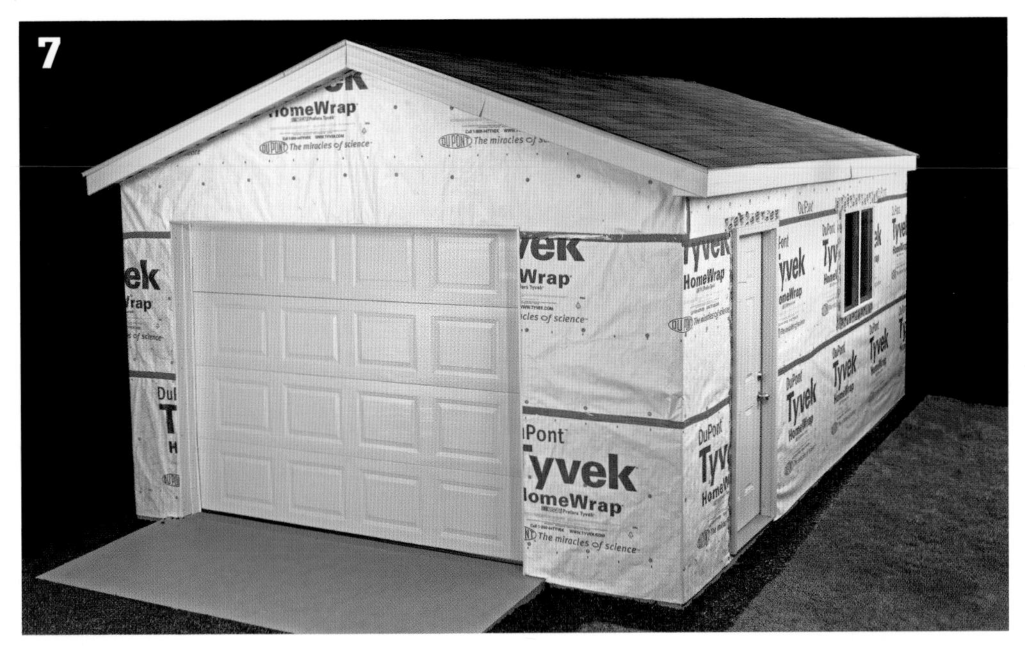

Installing Windows & Service Doors (pages 68 to 75).
Installing Overhead Garage Doors (pages 76 to 81)

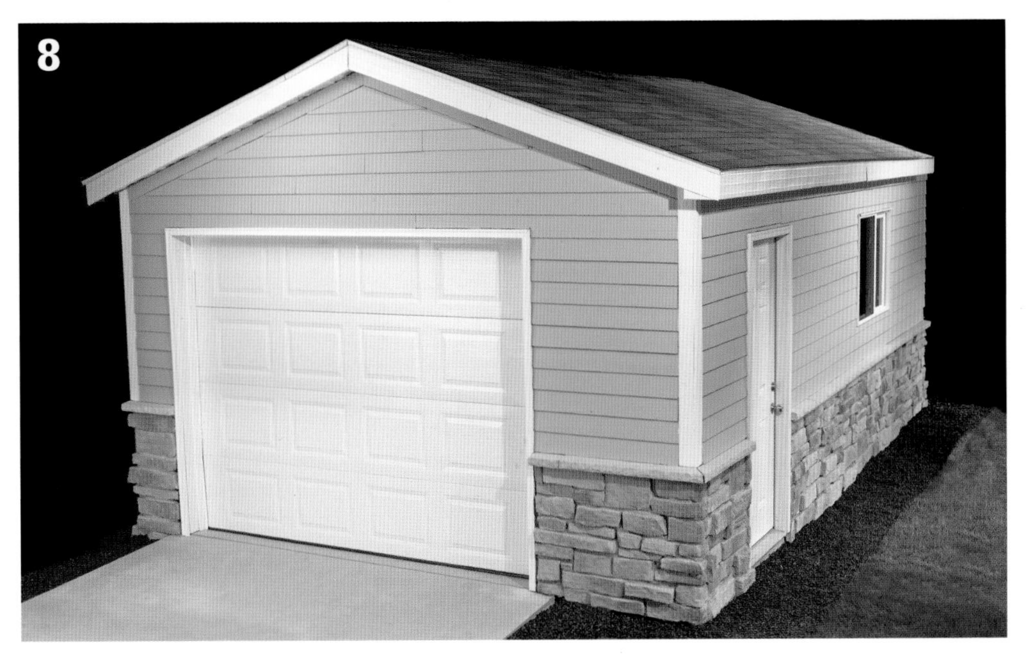

Installing Siding & Trim (pages 82 to 89)

Building the Foundation

The slab foundation commonly used for outbuildings is called a slab-on-grade foundation. This combines a 3½- to 4"-thick floor slab with an 8- to 12"-thick perimeter footing that provides extra support for the walls of the building. The whole foundation can be poured at one time using a simple wood form.

Because they sit above ground, slab-on-grade foundations are susceptible to frost heave; in cold-weather climates they are suitable only for detached buildings. Specific design requirements also vary by locality, so check with the local building department regarding the depth of the slab, the metal reinforcement required, the type and amount of gravel required for the subbase, and whether a plastic or other type of moisture barrier is needed under the slab.

The slab shown in this project has a 4"-thick interior with an 8"-wide × 8"-deep footing along the perimeter. The top of the slab sits 4" above ground level (grade). There is a 4"-thick layer of compacted gravel underneath the slab, and the concrete is reinforced internally with a layer of 6 × 6" 10/10 welded wire mesh (WWM). In some areas, you may be required to add rebar in the foundation perimeter (check the local code). After the concrete is poured and finished, 8"-long J-bolts are set into the slab along the edges. These are used later to anchor the wall framing to the slab.

A slab for an outbuilding requires a lot of concrete. Considering the amount involved, you'll probably want to order ready-mix concrete delivered by truck to the site (most companies have a one-yard minimum). Order air-entrained concrete, which will hold up best, and tell the mixing company that you're using it for an exterior slab.

Tools & Materials ▸

Work gloves & eye protection	Drill	J-bolts	Wood or magnesium concrete float
Stakes & boards	Wheelbarrow	Concrete groover tool	Rewire mesh
Mason's lines	Bull float	Concrete edging tool	Concrete cure & seal
Plumb bob	Paint roller	Compactable gravel	Brick chunks or metal chairs
Shovel	Hand maul	2× lumber for forms	Plate compactor
Line level	Tie wire	3" deck screws	Excavation tools
Tape measure	Concrete	Metal mending plates	

A concrete slab with an optional concrete apron is the most common garage foundation setup and is usable for most other outbuildings.

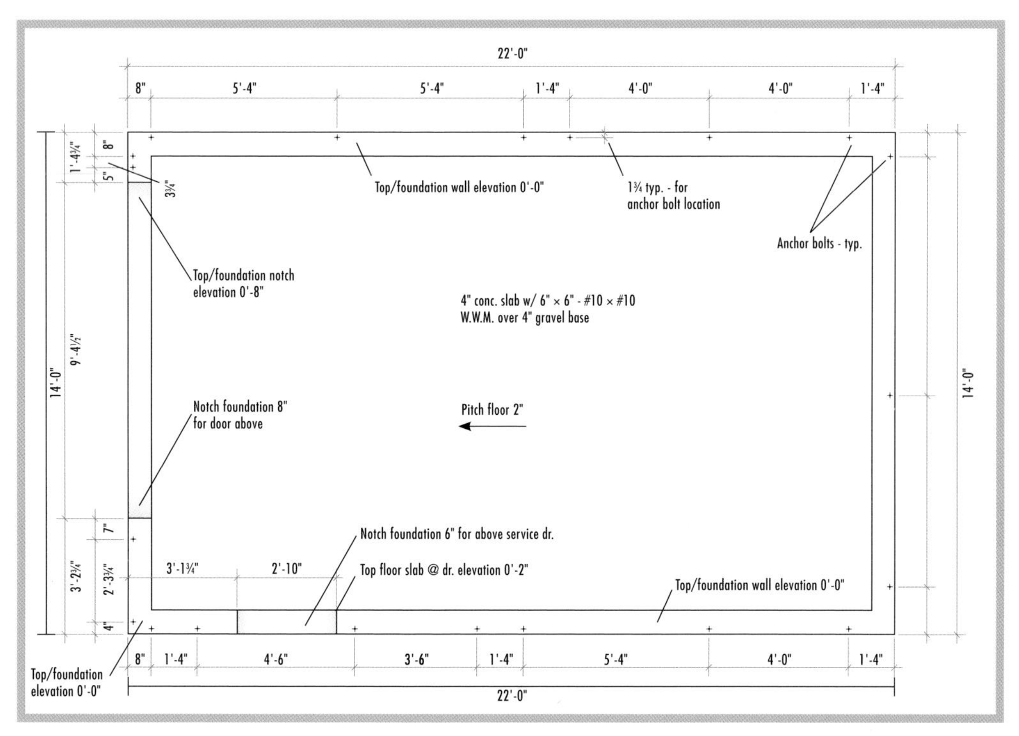

A plan view of the slab should include J-bolt locations, door locations, and footing sizes (as applicable). Also indicate the overall dimensions and the direction and height of the floor pitch.

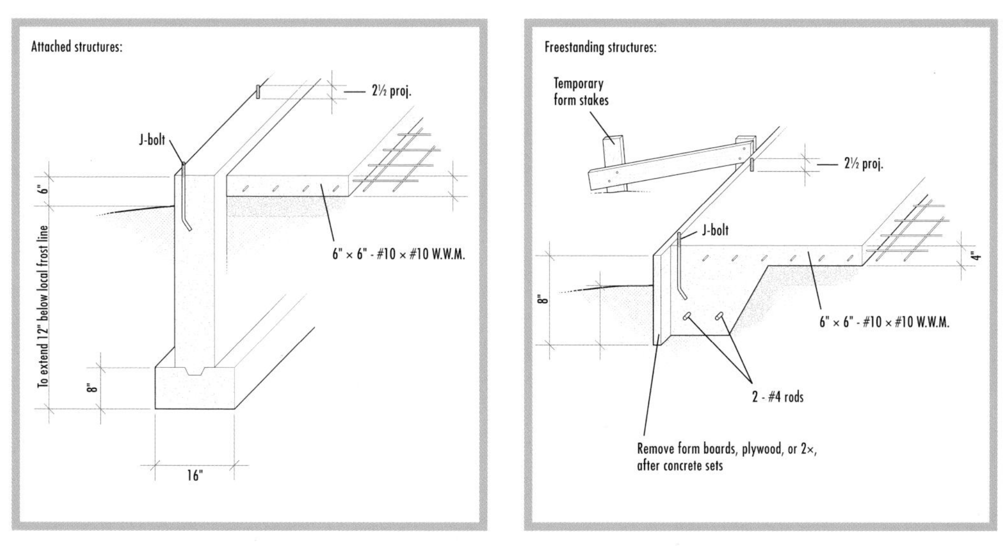

The outbuilding slab cannot simply float on the ground. It requires footings around the perimeter. For detached outbuildings, an 8 × 8" footing will comply with some local codes. For outbuildings attached to other buildings, the footings must extend past the frost line. In both cases, an ample layer of drainage gravel is required to help minimize movement from freezing and thawing.

How to Pour a Slab-On-Grade Foundation

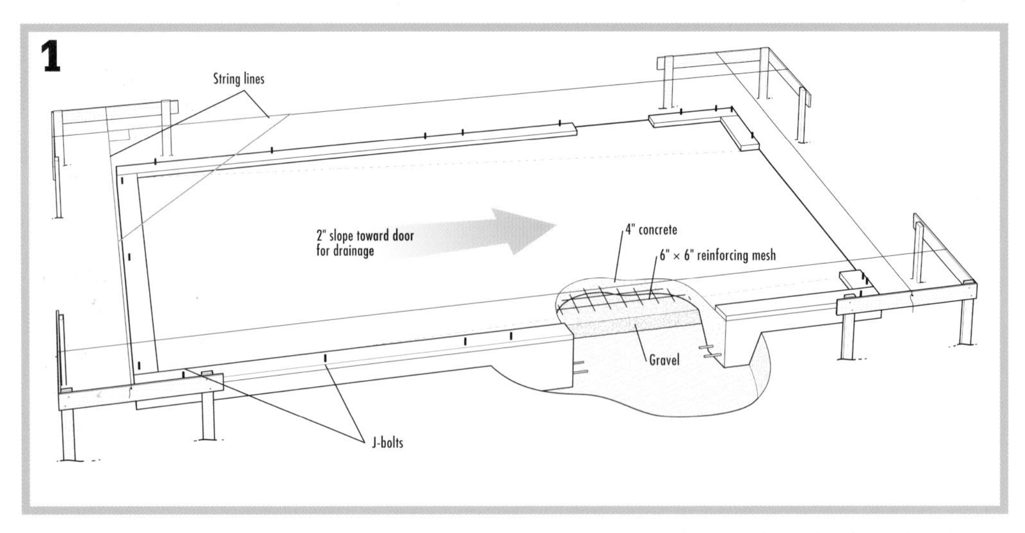

Begin to lay out the excavation with pairs of batterboards installed at each corner of the outbuilding slab site. Position them about 2 ft. outside the perimeter of the slab area so you'll have plenty of room to work. Run level mason's lines between the batterboards to establish the final size of the slab. Drop a plumb bob down from the intersections of the strings, and drive a stake at each corner.

Excavate the area about 2 ft. wider and longer than the staked size of the slab. The poured slab should slope 2" total from the back wall to the overhead door wall to facilitate drainage. Remove 3 to 4" of soil from the excavation area, and dig a deeper trench around the perimeter for the footing. The outside of the footing should line up with the mason's lines. Slope the soil to create a transition between the excavated interior and the footing. Check your local building codes to determine the correct footing size and depth for your climate and soil conditions. Fill the excavation area with 4" of compactable gravel, letting it spill down into the deeper footings that frame the perimeter. Tamp the gravel level and smooth with a rented plate compactor. The gravel surface should maintain the 2" total back-to-front slope. Depending on your soil conditions, some concrete contractors recommend laying 6-mil polyethylene sheeting over the compacted base to form a moisture barrier. *Tip: Install electrical conduit underneath the slab if you will be providing underground electrical service.*

Build a form for pouring the slab using 2× lumber or strips of exterior-rated plywood. The inside dimensions of the form should match the final slab size. If necessary on long runs, join the lumber end-to-end, reinforcing the butt joints with metal mending plates screwed to the outside surfaces. Fasten the form pieces together at the corners with 3" deck screws. Position the form so it aligns with the mason's lines. The form should also follow the 2" total back-to-front slope.

Drive wood stakes along the outsides of the form at 4-ft. intervals. Place two stakes at each corner. Set the tops of the stakes flush with the top edges of the form (or slightly below the tops). As you drive the stakes, periodically measure from corner to corner to ensure that it's square. Measure down from the mason's lines to position the form 4" above grade. Attach the stakes to the form with deck screws to hold the form in place.

Add rewire reinforcement according to the requirements in your area. Here, rows of 6 × 6" 10/10 wire mesh are set onto spacers (chunks of brick or metal "chairs") in the pour area. Overlap the sheets of mesh by 6", and stop the rows about 2" in from the insides of the form. Fasten the mesh together with tie wire. *Option: Reinforce the footings by laying out two rows of #4 rebar 2" above the bottom of the trench by wire-tying it to shorter pieces of rebar driven into the gravel. Space the rows about 4" apart.*

(continued)

Pour the concrete. Coat the inside surfaces of the forms with a release agent. Have ready-mix concrete delivered to your job site and place it into the forms with wheelbarrows and shovels (make sure to have plenty of help for this job). Fill the form with concrete, starting at one end. Use a shovel to settle the concrete around the reinforcement and to remove air pockets. Fill the form to the top. *Note: In most municipalities you must have the forms and subbase inspected before the concrete is poured.*

Strike off the concrete once a section of the form is filled. The best way to do this is to have two helpers strike off (screed) the wet concrete with a long 2 × 6 or 2 × 8 that spans the width of the form. Drag the screed board back and forth along the top of the form in a sawing motion to level and smooth the concrete. Use a shovel to fill any voids ahead of the screed board with shovelfuls of concrete.

Smooth the surface further with a bull float as soon as you're finished screeding, working across the width of the slab. Floating forces aggregate down and draws sand and water to the surface to begin the smoothing process.

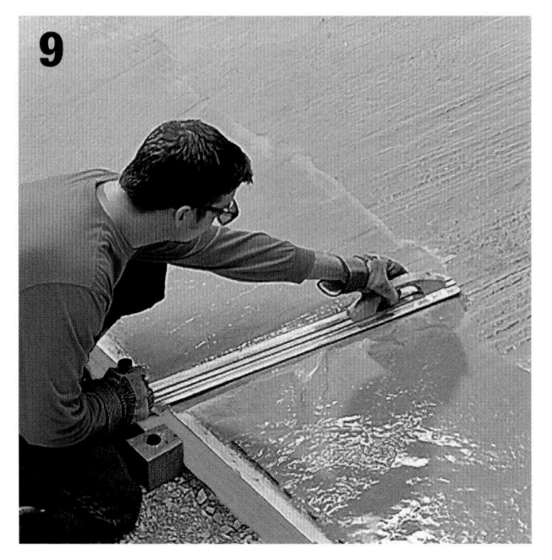

Use a magnesium or wood handheld float (darby) to refine the slab's finished surface as soon as the bleed water evaporates. Work the float back and forth, starting from the middle of the slab and moving outward to the edges. If desired, finish the surface further with a finishing trowel.

Push J-bolts down into the concrete, wiggling them slightly to eliminate air pockets. Twist the bottom hooked ends so they face into the slab. Position the J-bolts $1\frac{3}{4}$" from the edges of the slab. Leave $2\frac{1}{2}$" of bolt exposed, and make sure the bolts are plumb. Smooth the surrounding concrete with a wood or magnesium concrete float.

Cut control joints using a groover if your local codes require them (dividing slabs into 10 × 10-ft. sections is standard). Lay a long 2× to span the slab and line up one edge so it's centered on the slab's length. Use the 2× as a guide for cutting across the slab with a groover tool.

Round the edges of the slab next to the forms using an edging tool.

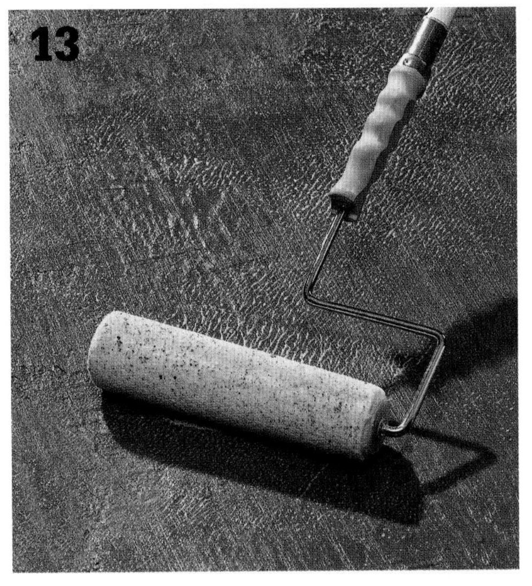

Apply a coat of cure and seal product (See Resources, page 442) to the surface once it dries so you do not have to moisten the concrete surface during the curing stage. After a couple of days, strip off the forms. Wait at least one more day before you begin building on the slab.

Framing & Raising Walls

Framing and raising walls should prove to be one of the more enjoyable aspects of your new outbuilding project. You'll be able to assemble the skeletal sides of the building fairly rapidly, especially if you work with a helper or two and use a pneumatic nail gun for fastening and a power miter saw for cutting. Assembling walls isn't a complicated process. In fact, if you set aside a full day for the job, you'll probably have all the walls assembled and standing on the slab before sundown—maybe even sooner.

We'll use fundamental stick-framing techniques and 2 × 4s to assemble the walls of this outbuilding. In terms of the tools you need, be sure to have a circular saw or power miter saw on hand with a quality (carbide-tipped) crosscutting or combination blade installed. You also need a framing square, speed square, or combination square; a 4-ft. level, a 25- or 50-foot tape measure, string line, and a framing hammer or pneumatic framing nailer.

As you lay out each wall section, carefully inspect the studs and the top and bottom plates to make sure they're straight and free of large splits, knots, or other defects. Separate your lesser-quality lumber for use as wall braces or shorter pieces of blocking. If you end up with a lot of bad studs, call your supplier and request a better supply.

Tools & Materials ▸

Work gloves & eye protection
Combination square
Drill & spade bit
Miter saw
Speed square
Tape measure
Hammer (or pneumatic nailer)
Caulk gun
Mason's line
Reciprocating saw
Stakes
Level
Pressure-treated 2× lumber for sole plates
2× lumber (2 × 4, 2 × 8, 2 × 12)
Galvanized common nails (8d, 10d, 16d)
1 × 4 bracing
Deck screws
Galvanized washers & nuts for J-bolts
½" plywood
Construction adhesive
Clamps

Raising the outbuilding walls is an exciting time in your project, as the structure begins to take shape rapidly with relatively little effort.

Framing Tips ▸

Pole framing is a completely different framing system used for large structures, such as barns. In pole framing, the weight of the structure hangs entirely on a matrix of poles or posts embedded deep in the ground, rather than setting the weight of the building on the walls, which rest on the foundation. Since the walls are not load-bearing in this framing style, they do not require as intensive an infrastructure as in the typical 2 × 4 framing shown here.

A pneumatic framing nailer makes fast work of frame carpentry. Typical collated strips have nails varying in length between 2⅜ and 2½". Framing nailers can be relatively expensive but are also available for rent at larger rental centers.

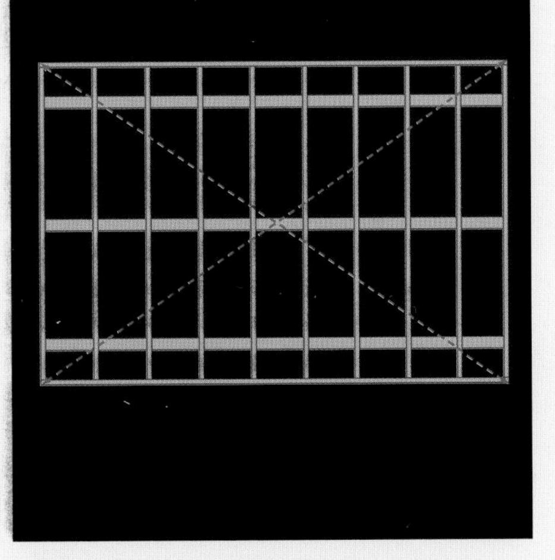

Measure the diagonal distances once you have assembled each wall. The distances between opposite corners will be equal when the walls are square.

How to Frame Outbuilding Walls

Prepare the sole plates. Select straight lumber for the wall sole plates and cut them to length (use pressure-treated lumber for concrete floors). Position the bottom plates on the slab up against the J-bolts. Follow your plans to determine which walls run to the edges of the slab or floor (through walls) and which butt into the other walls (butt walls). Use a combination square and pencil to extend a line across the bottom plate at each J-bolt location.

Drill holes for J-bolts. Make a tick mark on the J-bolt layout marks 1¾" in from the outside edge of the bottom wall plates to determine where to drill the J-bolt through-holes. Drill through the bottom plate at each hole location with a ⅝ or ¾" spade bit to allow some room for adjusting the plate on the slab. Slip a backer board beneath the workpiece before finishing the hole.

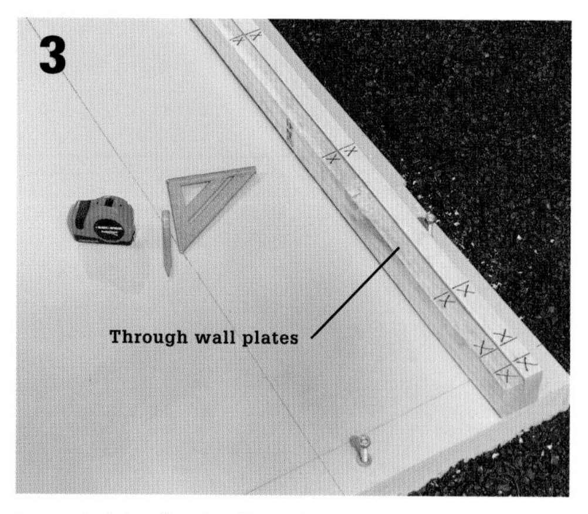

Lay out plates for the through walls: Cut a cap plate for each wall so its length matches the sole plate. Stand both plates on edge and line up the ends. If the first wall is a through wall, make marks at 1½ and 3" to indicate the end stud and extra corner stud. Mark the next stud at 15¼" according to your stud layout. Step off the remaining studs at 16" on center. Mark double studs at the opposite end of the wall. Draw Xs to the side of each of these marks to designate on which side of the marks the studs should go. Extend these stud layout marks across both edges of the cap and sole plates.

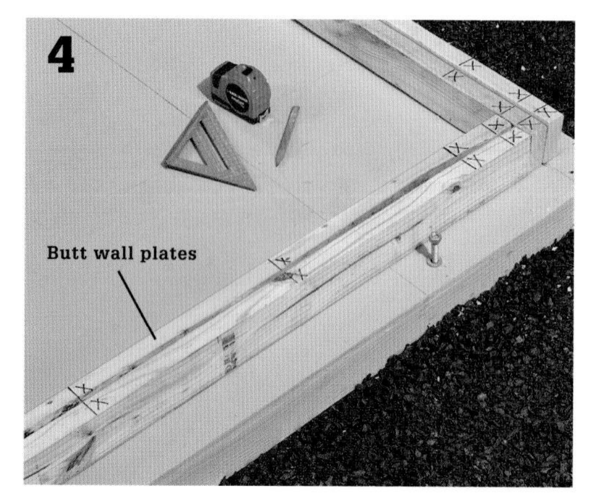

Lay out plates for butt walls: Cut the cap plates for the butt walls. Mark each end stud so it is aligned with the ends of the top and bottom plates. Mark each second stud 15¼" from the plate ends, and step off the rest of the studs at 16" on center. Extend the lines across both wall plates and draw Xs to the right of your stud marks.

Cut wall studs to length. Select the studs you'll need to build the first wall, and sight down their edges to make sure they're straight. Inspect for deep end checks or loose knots (a check is a lengthwise separation of the wood; an end check is one occurring on an end of a piece). Set defective studs aside for use as blocking.

Assemble the front wall. Position the marked wall plates about 8 ft. apart with the stud markings facing up. Lay out the studs between the plates, and start by nailing the bottom plate to the wall studs with pairs of 16d galvanized common nails or pneumatic framing nails. *(Note: Use hot-dipped galvanized nails for fastening treated bottom plates.)* Make sure the edges of the studs and plate are flush and the studs line up with their layout marks on the plate. Drive two nails through the plate into the stud ends to secure them. Nail the top plate to the studs the same way.

Add end blocking for a through wall. Cut three 12" lengths of 2 × 4s to serve as blocking between the end and second studs on through walls. Space the blocking evenly top to bottom along the inside face of the end studs. Nail the blocking in place.

Nail the second stud in place. Butt the second stud against the blocking, and nail the top and bottom plates to it. Drive more nails through the second stud and into the blocking.

(continued)

Square up the wall. Check the wall for squareness by measuring from corner to corner and comparing the diagonals. If the measurements are not equal, push the longer-dimension corners inward as needed until the diagonals are the same.

Install temporary bracing. Once the wall is square, install a temporary 1 × 4 brace across the wall plates and studs to stabilize the wall and keep it square. Use deck screws or 8d nails to tack the brace diagonally across the wall, driving two fasteners into the top and bottom plates and one nail into every other stud. Leave these braces in place until the walls are ready to be sheathed.

Set up the first wall. Before standing the wall up, nail a temporary brace to each end stud to hold the wall in position after it is raised. Drive one 16d nail through the brace and into the end stud about 7 ft. up from the bottom plate to act as a pivot. Tip the wall up and onto the J-bolts with the aid of a helper. Swing the end braces out into the yard, and attach them to stakes in the ground. Check the wall for plumb with a level held against the studs before fixing the braces to the stakes.

Anchor the wall plates. Use a hammer to tap the bottom plate into final position on the slab, and attach it to the J-bolts with galvanized washers and nuts.

Mark window and door openings. For walls with windows or a service door, mark the positions of king and jack studs when you are laying out the top and bottom plates. Identify these studs with a K or J instead of an X to keep them clear. Mark the cripple studs with a C as well. Install the king studs.

(continued)

Frame window and door openings. Measure and cut the jack studs to length following your outbuilding plans. For either window or door jack studs, make the jack stud length equal to the height of the rough opening minus 1½" for the bottom plate (door framing) or 3" for a double rough sill (window framing). Facenail the jack studs to the king studs with 10d common nails spaced every 12".

Make the headers. The header seen here is assembled from doubled-up 2 × 8 lumber sandwiched around a piece of ½" plywood sized to match. Fasten the header pieces together with wavy beads of construction adhesive and 16d nails spaced every 12". Make sure the ends and edges are aligned. Drive the nails at a slight angle to keep them from protruding, and nail from both sides of the header.

Install the headers. Set the headers in position on top of the jack studs and drive 16d nails through the king studs and into the ends of the header to fasten it in place. Use six nails (three per end) for 2 × 8 headers.

Install cripple studs above. First, cut the cripple studs to fit between the header and the wall's top plate, and then toenail them in place with three 8d nails on each end. Fasten the bottom end of the header by toenailing.

Install cripple studs below. Cut the rough sill to length from 2 × 4 and toenail it to the jack studs with 16d nails.

Join wall sections. For long walls, your outbuilding plans may require you to build the wall in two sections and nail these together before erecting the wall. Facenail the wall sections with pairs of 8d nails spaced every 12" along the adjacent end-wall studs.

Raise the window/door wall. You'll need three or four helpers to tilt the heavy wall up and into position on the slab. Adjust the wall as needed so it butts against the adjoining wall and lines up properly on the slab. Check the wall for plumb along several studs, and attach a temporary staked brace to the unsupported end. Install washers and nuts on the J-bolts to fasten the wall to the slab.

(continued)

Nail walls together. Drive 16d nails through the end stud of the butted wall into the end studs and blocking of the through wall. Space these nails every 12" along the length of the studs. Prior to nailing the second long wall, you can remove the temporary brace and stake that to hold the back wall in position.

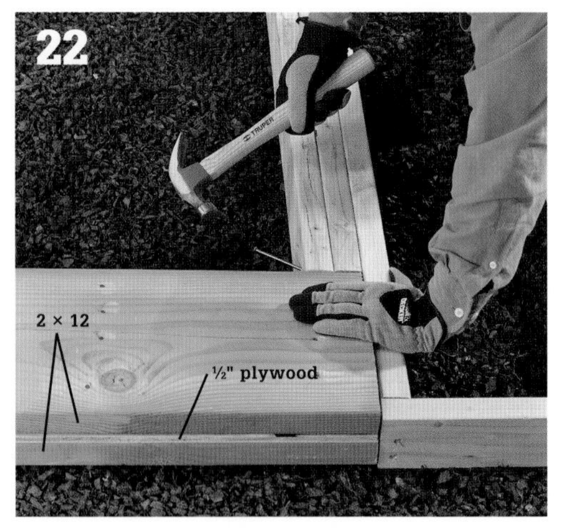

Assemble the outbuilding door wall. Follow your outbuilding plans to frame the front wall and rough framing for the sectional door (if applicable). Sectional doors typically have a doubled-up 2 × 12 header sandwiching a piece of ½"-thick plywood. Build the header just as you would a window or service door header. The header will be supported by double jack studs. This wall may or may not have a continuous top wall plate and cripple studs above the header, depending on the height of your outbuilding walls.

Position the front wall. Remove temporary braces and stakes supporting the side walls, then tip the front wall up and into position against the side walls. Line up the ends of the side walls with the front wall, and nail the walls together through the end studs with 16d nails. Install washers and nuts on the front wall J-bolts.

Test the walls for straightness. Check the long walls for bowing by tacking a scrap block of 2 × 4 at the top outside corner of each wall. Drive another nail partially into these blocks, and then string a mason's line between the nails. Pull the line taut, and measure the distance between the string and the wall's top plate. The distance should be 1½" all along the wall.

Lock the walls together. Cut top plates to length. Make the through-wall tie plates 7" shorter than the through-wall top plate (3½" on each side). Cut the tie plates for butt walls 7" longer than the butt-wall top plates. This way, the double top plates on butted walls will overlap the through-wall top plates, locking the walls together. Facenail the tie plates to the wall top plates with 10d nails.

Cut out the threshold. Cut away the bottom plate from the rough opening of the service door with a reciprocating saw, with the blade installed upside down. Make these cuts flush with the edge of the jack studs so the door jamb will fit properly in the opening.

Frame the overhead door opening. *Note: If you have already purchased your sectional outbuilding door, check the door opening requirements in the installation manual and compare them to these instructions before proceeding with this step.* Facenail a 2 × 6 around each side and the top to frame the rough opening on the inside face of the front wall. These boards form blocking for installing the door and automatic opener later. Position the blocking flush with the faces of the jack studs and the bottom edge of the door header. Fasten the blocking with 10d nails.

Framing the Roof

This outbuilding has a simple gable-style roof consisting of only two roof planes with flat gable-end walls. For that reason, we'll frame the roof using rafters as the principal structural members. Rafters extend from the wall top plates and meet at a ridgeboard at the roof's peak. They're a traditional form of roof construction on both simple and complex roof designs, and rafters are also a more economical option than custom-built trusses. If you're unfamiliar with roof framing, constructing this rafter roof will be an excellent opportunity to learn some important basic skills.

Building the roof frame is a departure from wall framing because you can't nail whole sections of the roof together at once and set them in place. Instead, you'll cut all the rafters to size and shape to match the slope of the roof, and then install them in pairs "stick built" style. For an outbuilding this small, 2 × 6s spaced 24 inches on center provide sufficient strength for normal roof loads. You may need closer spacing or deeper rafters if you live in a heavy-snow climate. Since the outbuilding's ridge runs from front to back, rafters are installed perpendicular to the length of the building. A third important component of rafter framing—horizontal rafter ties—span the width of the structure and can function as ceiling joists. Rafter ties help keep the walls from spreading apart by locking pairs of rafters together into triangulated frames, similar to a roof truss.

Tools & Materials ▸

Work gloves & eye protection	Circular saw
Carpenter's pencil	Hammer
Speed square	Level
Tape measure	Lumber (2 × 4, 2 × 6, 2 × 8)
Miter saw	Metal framing connectors
Framing square	Galvanized common nails (10d, 16d)
Ladders	
Jigsaw	

A system of rafters, ridgeboard, and rafter ties creates the framework for this outbuilding's simple gable-style roof. Rafters are a traditional, sturdy, and economical option for this project, but custom-built trusses are another viable option.

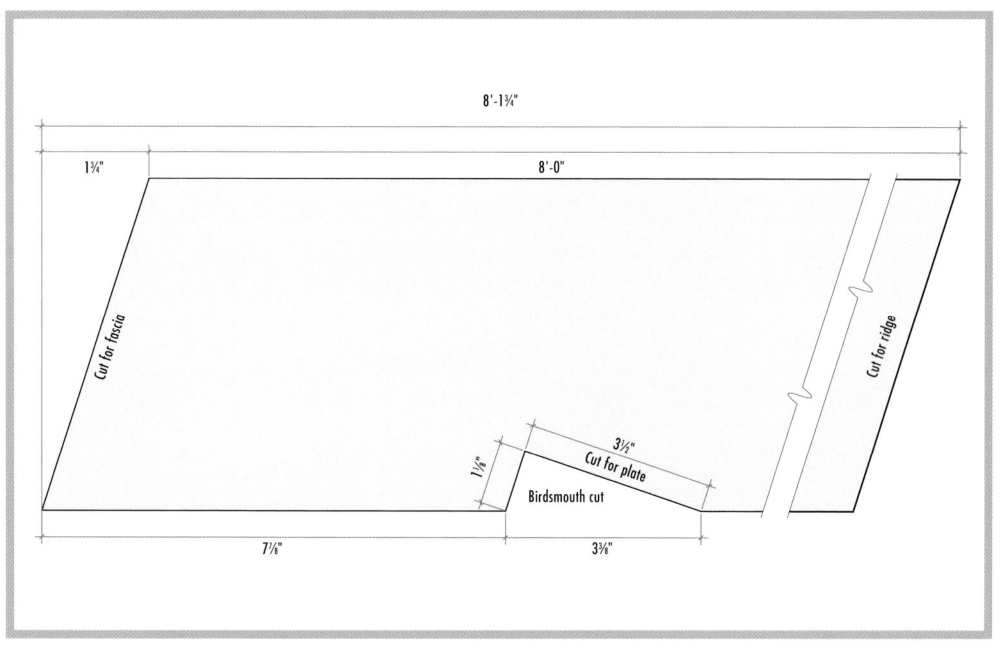

This template may be used as a guide for laying out the birdsmouth cuts on the rafter ends for the outbuilding project seen here.

Using a Speed Square ▸

A speed square is a handy tool for marking angled cuts (like end cuts on rafters) using the degree of the cut or the roof slope. Set the square flange against the board edge and align the pivot point with the top of the cut. Pivot the square until the board edge is aligned with the desired degree marking or the rise of the roof slope, indicated in the row of common numbers. Mark along the right-angle edge of the square.

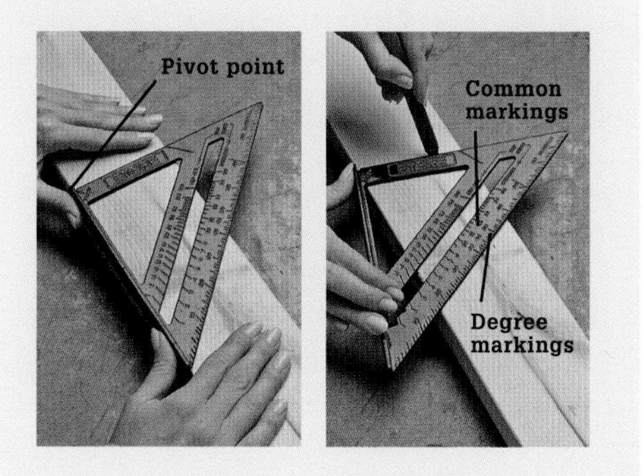

Metal framing connectors add strength to the connection between the rafter and the top plate of your outbuilding walls. They also help with alignment and minimize any splintering of the rafter caused by toenailing. In some areas of the country where hurricanes and tornadoes are common, metal connectors are required by local codes.

How to Frame a Gable Roof

Make a pair of pattern rafters. Choose two straight 2 × 6s to create full-size pattern rafters for each rafter pair. Mark a cutting line for the top on one end of each pattern, with the correct angle to meet the ridgeboard (left). Refer to your outbuilding plans to determine the correct roof pitch, which determines the cutting angle. Then, measure from the top of the ridge angle along the rafter to determine its overall length, and draw a second line for the plumb cut at the eave end. Make the plumb cuts with a power miter saw or a circular saw. Lay out and cut the birdsmouths on the pattern rafters using a speed square. Use a square to create the level and plumb lines that form the birdsmouth cuts. The birdsmouth enables the rafters to rest squarely on the wall cap plates. Make the birdsmouth cuts with a jigsaw (right).

Check the fit. Set your pattern rafters in position on top of the side walls with a 2 × 8 spacer block tacked between them to represent the ridgeboard. You'll know you have a good fit if the top plumb cuts meet the ridgeboard flush and the birdsmouth cuts sit flush on the wall plates. Have a helper position and check the fit of these parts. Adjust the angles, if necessary, to improve the fit of the parts.

Cut all the rafters. Use the pattern rafters to trace the plumb cuts and birdsmouths onto the workpieces for all of the rafters. Set the cutting angle on your power miter saw to match the plumb cut, and cut each rafter at the cutting lines. Then, finish the rafters by cutting the birdsmouths with a jigsaw.

Plot the rafter locations. Mark the location of each rafter on the doubled top plates. The rafters begin at the ends of the walls, and the intermediate rafters should line up over the wall studs. Use a speed square to extend a rafter layout line up from each wall stud layout line to the top plate. Mark an X next to the line to indicate which side of the line the rafter should go. Mark the position of all the rafters.

Option: Install framing connectors. If building codes in your area require it or if you simply want a stronger structure, nail metal connectors to the wall top plates before installing the rafters.

Mark the ridgeboard. Select a straight, flat 2 × 8 for the ridgeboard. It should be several feet longer than the roof length. The ridge length should be specified in your plans. Cut it to this length to ensure an accurate layout. Adjust the ridgeboard so it overhangs the end walls evenly. Use a square to transfer the rafter layout lines and X marks from the wall double top plate to the ridgeboard. Then, flip the ridgeboard over and mark the rafter locations on the opposite face.

(continued)

Install the ridgeboard. To make it easier to begin the rafter installation, nail the first end rafter to the ridgeboard before lifting it into place on the walls. Nail the ridgeboard to one end rafter through the top plumb cut with three 16d nails. Make sure the rafter is properly lined up with the ridgeboard layout line. Toenail the opposite rafter to the ridgeboard. Then, with several helpers, lift the end rafters and ridgeboard into position on the wall plates. Have helpers hold up the opposite end of the ridgeboard while you toenail the end rafters to the wall plates.

Install a temporary brace. Toenail a temporary 2 × 4 brace vertically to the opposite end wall. Choose a brace longer than the roof will be high. Rest the ridgeboard against the brace and adjust it until it is in position parallel to the slope of the foundation. Use 10d nails to nail the ridgeboard temporarily to the brace to hold it in position.

Install the remaining rafters. With the ridgeboard braced, fit and install the rest of the rafters, fastening them with 16d nails. Toenail the rafters to the metal connectors at the birdsmouths (if applicable), and either endnail or toenail the rafters to the ridgeboard, depending on which rafter you are installing for each pair. When you reach the opposite end of the roof, remove the temporary ridge brace and install the end rafters.

Install the rafter ties. Follow your outbuilding plans to lay out and cut rafter ties to size. Angle-cut the top ends of each rafter tie, if necessary, to match the roof slope. Install the rafter ties by facenailing them to the rafters with three 10d nails at each end.

Install the gable top plates. On the gable ends of the roof, you'll need to install additional studs under the rafters to provide nailing surfaces for wall sheathing. Start by cutting a pair of 2 × 4 gable wall top plates that will extend from the sides of the ridgeboard down to the wall double top plates.

Lay out and install the gable studs. Locate these by holding a level against the wall studs and transferring layout lines to the edges of the gable top plates. Plan for a gable stud to line up over each wall stud. Cut the gable studs to fit and toenail them to the gable and wall top plates.

Install lookouts. Follow your plans to lay out the locations of the lookout blocking that will form gable overhangs on the roof. Cut the blocking to size, and endnail it through the end rafters. Make sure the top edges of the blocking and rafters are flush.

Complete the overhang. Lay out and cut the gable overhang rafters to size and shape using your pattern rafter as a template. *Note: Gable end rafters do not have birdsmouths. Nail these rafters to the lookout blocking and the ridgeboard to complete the roof framing.*

Option: Roof Trusses ▸

Custom-made roof trusses save time and practically guarantee that your roof will be square and strong. They add considerably to the project cost, however, and must be ordered well in advance.

Trusses are engineered roof support members that can be used instead of hand-cut rafters to support your roof. You can build them yourself or you can order them premade to match your building size and preferred roof pitch. A truss has a triangular shape with two matching top chords that meet a horizontal bottom chord. Diagonal crossbracing, called webs, are fitted between the top chords and the bottom chord. Typically, the joints between chords and web members are reinforced with metal or plywood gusset plates.

Trusses are designed so the ends of the bottom chord rest on the top plates of the side walls. Consequently, you don't have to cut tricky birdsmouths or rafter angles—you simply fasten the bottom chord by toenailing or using metal connectors. The relative ease of the installation may make up for the higher costs compared to rafters. But unless your outbuilding is very small, you will likely need to rent a crane, forklift, or other mechanical assistant to raise the trusses into position.

Most professional outbuilding contractors employ trusses because they go up quickly and don't require complicated cutting. There are limitations, however. If you are purchasing the truss premade, chances are the quality of the lumber won't be as high as the dimensional lumber you'd use to make rafters. The presence of the bottom chord reduces the open space in an outbuilding, potentially limiting the storage options. But if you are planning to install a ceiling in your outbuilding, the chords can be put to work as ceiling joists.

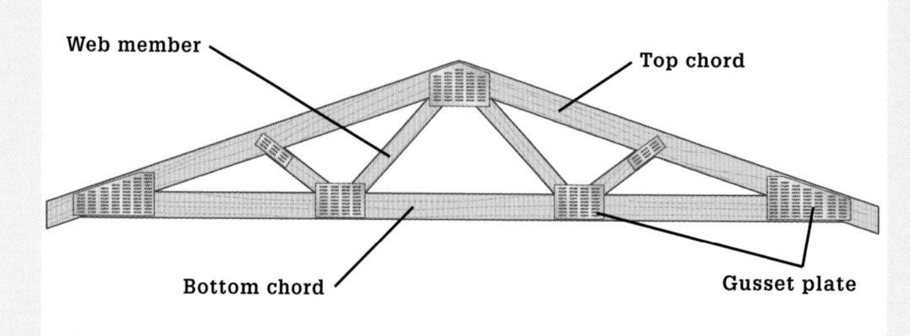

A manufactured truss consists of two top chords and a bottom chord, with web members installed between chords for strength. The joints are usually reinforced with metal or plywood gussets. Unlike rafter roofs, a truss roof does not have a ridgeboard.

Working with Trusses

Use long 2 × 4 braces clamped to the end wall to temporarily clamp or tack the end truss in position. If the truss is sized correctly, there should be no need to adjust it side to side, but you'll need to make sure it is flush with the end wall and plumb before you nail it into place.

Secure the trusses to the walls with metal framing connectors. These are required in high-wind areas but are a good idea anywhere because they strengthen the roof and help with alignment.

Toenail trusses to wall plates with 16d nails. Typically, the two end trusses are installed first and then a mason's line is stretched between the tails of the top chords to use as an alignment reference. A temporary brace marked with the truss layout guides the installation and helps stabilize the trusses as you go. Remove the brace before installing the roof decking.

Do's & Dont's for Working with Trusses ▸

- DO set trusses on blocking for their protection when storing.
- DO have plenty of help when it's time to raise the trusses.
- DO NOT cut trusses for any reason.
- DO NOT exceed the span for which the truss is rated.
- DO provide your truss dealer with an accurate plan drawing of your outbuilding.
- DO NOT walk on trusses if they are being stored lying flat.
- DO NOT install trusses in high winds.
- DO use temporary braces to ensure that trusses stay plumb during installation.

Sheathing Walls

Once the outbuilding walls are framed and erected, all exterior wall surfaces should be covered with a layer of oriented strand board (OSB) or CDX plywood sheathing. Wall sheathing serves two basic purposes: it strengthens the wall framing by locking the studs to a stiff outer "skin," and it provides a uniform backing for nailing the siding and trim in place. The minimum sheathing thickness for 16" O.C. stud walls is ⅜", and ½" material is even better.

Provided you've framed your outbuilding walls correctly, you should be able to install sheathing in full 4 × 8 sheets because the stud spacing will enable the sheets to be nailed along the edges and ends evenly. You can hang sheathing horizontally or vertically, but generally the horizontal approach makes large sheets easier to manage. Install a bottom course of sheathing first all around the building so you can use the top edge as a handy ledger for resting and nailing off the top course. To speed the process along, you can sheath right over the service door and window openings, and then cut out these openings once all the sheathing is in place for each wall.

Even exterior-rated sheathing isn't immune to the effects of wind-driven rain, especially around nail holes. It's good practice to cover sheathing with 15-pound building paper or housewrap. Install it horizontally, working from the bottom of the walls up and overlapping the seams by at least 2". If you use housewrap, be sure to tape all seams with housewrap tape recommended for the brand of wrap you are using. Housewrap will begin to degrade from sunlight in just a few weeks, so be sure to get your permanent siding on promptly.

Tools & Materials ▸

- Work gloves & eye protection
- Chalk line
- Tape measure
- Hammer
- Drill
- Reciprocating saw
- Utility knife
- OSB sheathing
- 6d common nails
- Housewrap
- Housewrap tape
- Circular saw
- Straightedge
- Housewrap nails

Wall sheathing stiffens wall framing and creates a uniform backing nailing surface for siding and trim. A layer of building paper or housewrap seals the sheathing from moisture infiltration.

How to Install Wall Sheathing

Snap a layout line. Use a chalk line to create a level line 47" up the walls, measured from the bottom of the bottom plate. Snap a line the full length of each wall. At this height, the bottom course of sheathing will cover the bottom wall plate and overlap the foundation by 1", minimizing water infiltration. A few inches of slab should still be visible after the sheathing is installed. Sheathing should not contact the soil.

Install the first sheet. Position the first full sheet of sheathing in one corner so the top edge is on the chalk line. One end of the sheet should align with the edge of the framed wall and the other should fall midway across a stud. Attach the sheathing with 6d common nails. Space the nails every 6" around the perimeter and every 12" at the intermediate studs. Before nailing, snap chalk lines across the sheet to show the centerlines of every wall stud. Install all first-course panels.

Install the second course. Begin this course with a half sheet of OSB to establish a staggered pattern between courses. Trim the second-course panels so their top edges touch the bottoms of the rafters. Maintain a gap of ⅛" between the first- and second-course panels to allow for expansion and contraction (6d nails can be used as spacers between panels).

Mark the door and window openings. Drill through the sheathing at all corners of the door and window openings (you can drive nails if you prefer), and then connect the holes (or nails) with straight cutting lines.

(continued)

Cut out the door and window openings using a reciprocating saw. Cut carefully so the sheathing does not extend into the opening.

Sheath the next wall. The panels for the adjoining wall should overlap the ends of the panels on the first wall without extending beyond them. Complete installing full panels on all four walls.

Install sheathing in gable areas. After the first courses are installed on the end walls, lay out and cut second-course panels that follow the eave line. Mark stud locations and attach the panels with 6d nails, maintaining ⅛" gaps between panels.

Begin installing housewrap. Begin at the bottom courses if the product you're using is not wide enough to cover a wall in one piece. *Note: Housewrap is a one-way permeable fabric that helps keep moisture from entering the structure from the exterior. Installing it makes sense only if you are planning to finish the interior walls in the outbuilding.*

Attach the housewrap with cap nails. Drive at least three housewrap nails spaced evenly along each wall stud.

Finish installing the housewrap. All seams should overlap by at least 6 to 12", with horizontal seams overlapping from above.

Cut out windows and doors. Make a long X cut in the housewrap, connecting corners diagonally at window and door openings. Use a utility knife to make the cut. Staple down the housewrap in the rough openings so it wraps around the jack studs, header, and rough sill.

Tape the seams. To seal the housewrap, apply housewrap tape along all horizontal and vertical seams. *Note: Housewrap is not rated for long-term exposure to the sun, so do not wait more than a few weeks after installing it before siding the outbuilding.*

Installing Fascia & Soffits

Fascia and soffits form transitions from the roof to the wall siding. Fascia usually consists of 1× wood boards or metal strips that cover the ends of the rafters at the roof eaves to keep moisture and pests out. It also serves as an attachment surface for gutters. The faces of the gable end rafters are also covered with fascia boards to continue the roof trim pattern all around the building. Generally, fascia boards are installed before the roof sheathing to ensure that the roof sheathing will overlap them once it's in place. You must paint fascia to protect it from the elements, or you can cover it with manufactured aluminum or vinyl fascia trim that matches the soffit color.

A soffit extends from the fascia to the wall. It encloses the bays between the rafters or trusses and provides an important means of ventilation beneath the roof deck. Sometimes a soffit is made of exterior plywood with vents cut into it, but the soffit we show here is made with ventilated aluminum strips, available in a range of colors to match aluminum or vinyl siding. Install your outbuilding soffit material before hanging the siding so you can nail it directly to the wall sheathing.

Tools & Materials ▸

- Work gloves & eye protection
- Miter saw
- Hammer
- Speed square
- Chalk line
- Aviation snips
- Caulk gun
- 16d common nails
- Siding nails
- 8d galvanized casing nails
- Vented aluminum soffit panels with mounting strips
- Rolled aluminum flashing with color-matched nails
- Fascia covers
- Color-matched caulk
- Lumber (1 × 8, 2 × 2, 2 × 6)

Fascia and soffits enclose roof rafters to keep weather and pests out while providing a means of roof ventilation and a graceful transition from the roof to the walls.

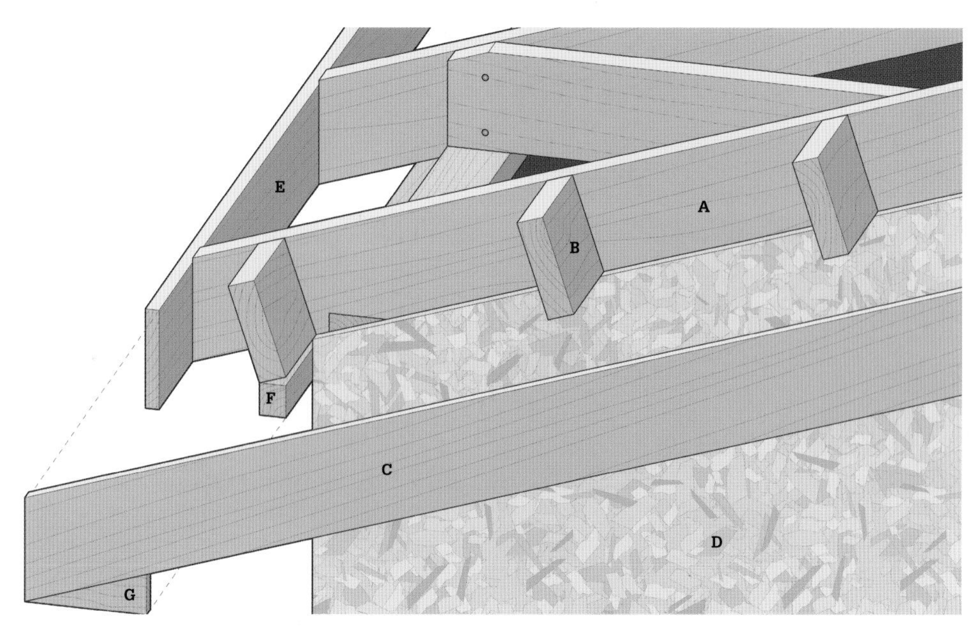

Components of the cornice system built here include: (A) End rafters, (B) 2× lookout blocking, (C) Gable overhang rafters, (D) Wall sheathing, (E) 1 × 8 fascia (eaves), (F) 2× soffit blocking-eaves (continuous strip along wall), (G) 2× gable rafter trim. Metal fascia cover not shown.

How to Install Fascia and Soffits

Install the fascia. Cut pieces of 1 × 8 to make fascia strips and attach them to the rafter tails with 8d galvanized casing nails. The ends of the fascia should be flush with the faces of the gable overhang rafters. Use a speed square held against the top edges of the rafters to adjust the fascia up or down until the square meets it halfway through the fascia's thickness. This will allow the roofing to overhang the fascia for proper drainage. Once the fascia is properly adjusted, drive three nails per rafter tail to secure it.

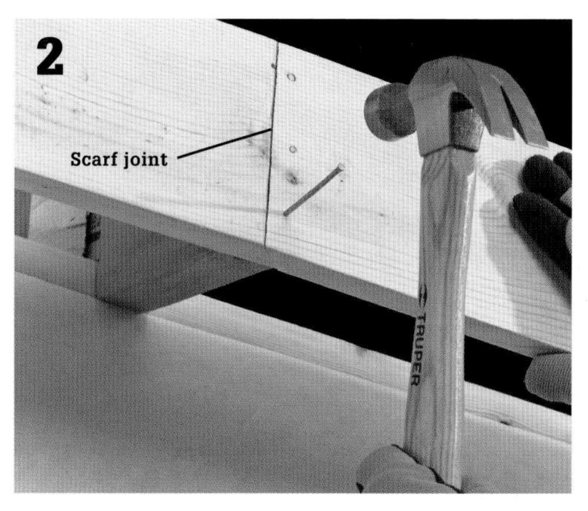

Make vertical joints. If your fascia boards are not long enough to make full runs in one piece, use overlapping scarf joints to join the ends. Miter cut the ends of the scarf joint parts so they overlap and fall over a rafter tail. Drive three 8d nails through both joint parts to secure them to the rafter.

(continued)

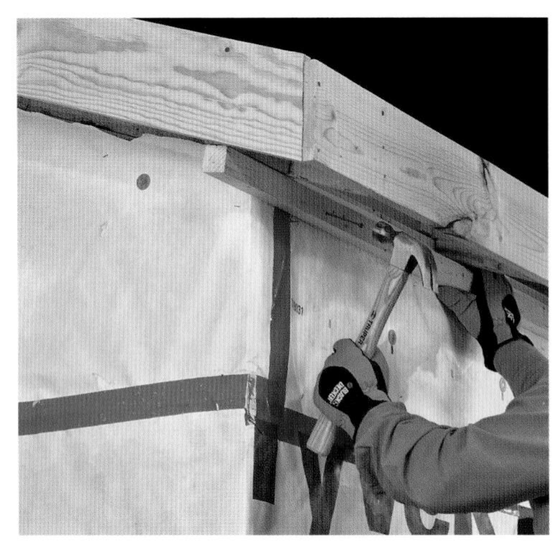

Option: If you will be installing wood soffit panels, install 2 × 2 soffit blocking. (The outbuilding seen here will be equipped with metal soffits that do not need backer blocking.) The blocking should be positioned so the bottom edge is flush with the soffit groove or backer in the fascia. Cut the soffit blocking so it extends beyond the ends of the walls to create a nailing surface for any filler pieces that will be installed with the cornice. Nail the soffit blocking to the wall studs with 10d nails, one nail per stud.

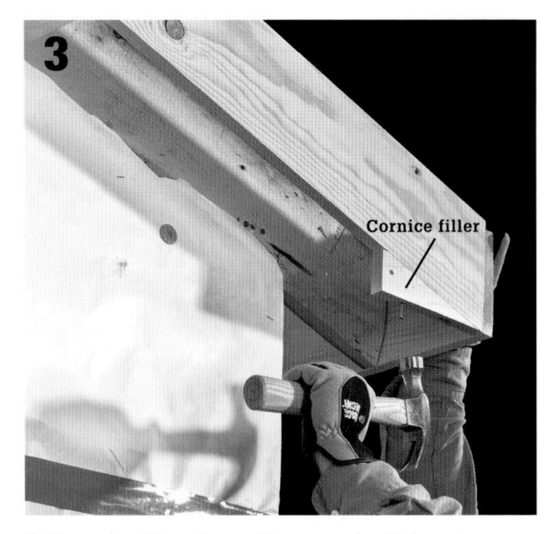

Add cornice filler pieces. Measure and cut triangular blocking to fit underneath the gable end rafter tails. Lay out the blocking so it forms a plumb bottom to the rafter tails. Toenail this blocking to the rafters. If soffit blocking is present, screw or nail the cornice blocking to the end of the soffit blocking. Lay out, cut, and nail 1× subfascia boards to cover the gable rafters and the ends of the ridgeboard. Miter cut the ends of the subfascia where they meet at the roof ridge.

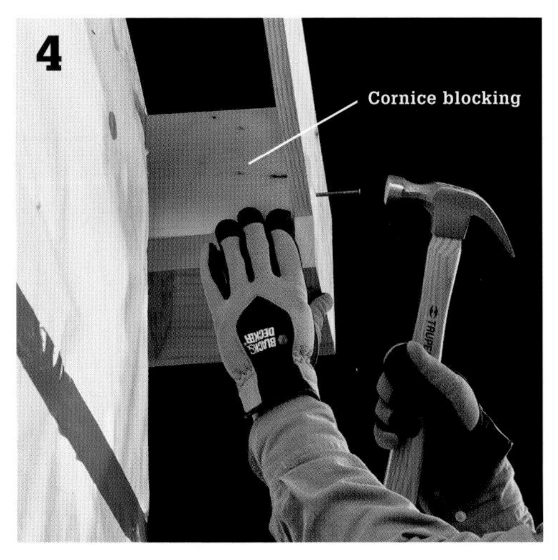

Install cornice blocking. Cut and fit short lengths of 2 × 6 scrap between the gable-end rafters and the wall to box in the cornice. Drive 16d nails through the fascia and end rafters to attach the blocking.

Enclose the eaves. Cut strips of vented aluminum soffit to enclose the eaves of the roof. Hang mounting strips for the soffit panels on the outbuilding walls (if you did not install backer boards, see Option, above). Insert the inside end of each soffit panel above the mounting strip. Attach the outside end of the panel to the bottom of the fascia with siding nails. The soffit panels should stop flush with the fascia.

Install soffit in the gables. Lay out and snap chalk lines on the gable walls for installing soffit mounting strips, and then mount the strips. Cut, fit, and nail the soffit panel strips to the fascia and soffit blocking to enclose the gable overhang.

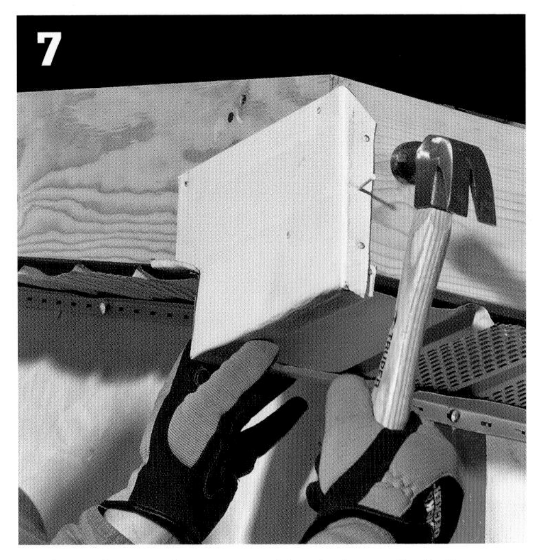

Enclose the cornices. Cut and bend pieces of rolled aluminum flashing to fit over the roof cornices and cover the blocking. Nail this flashing to the cornices with color-matched siding nails. Wrap this flashing around the eave fascia boards by 1 to 2" so you can install metal fascia to overlap it.

Install fascia covers. Measure the width of the fascia boards, and cut fascia covers to fit. Fit the fascia in place over the fascia boards so the bottom lip overlaps the soffits. Nail through the lip every 16" into the fascia with color-matched siding nails. Fasten the top of the fascia cover within ½" of the cut edge so the nail heads will be covered by drip edge molding later. At the cornice, bend the last piece of fascia cover at a right angle to turn the corner (make relief cuts with aviation snips first).

Finish installing fascia covers. Install the fascia covers on the gable ends, stopping just short of the cornices. At the cornices, bend a piece of fascia cover to turn the corner, and trim the end so it will make a straight vertical seam. Caulk the seam with color-matched caulk.

Completing the Roof

Now that your fascia and soffits are installed, it's time to sheath the roof deck, install roofing, and add a ridge vent (optional). The purpose of roof sheathing is obvious: it completes the structural deck of the roof and provides a flat, continuous surface for installing the roofing. As with wall sheathing, you can use either oriented strand board (OSB) or CDX plywood for roof sheathing. Be sure to use sheathing that's rated for your rafter/truss spacing and snow loads in your area (if applicable). If you accurately placed your rafters at the roof framing stage, the sheathing should install quickly, with minimal waste, and all seams should fall at the rafter locations. Stagger the joints from one row of sheathing to the next.

After constructing the roof deck, install a layer of 15# or 30# roofing felt (also called building paper). Roofing felt protects the sheathing and serves as an important second line of defense against leaks beneath the roofing. Roll out and nail the felt horizontally, starting at the eaves and overlapping the felt as you work your way up to the peak. Once the felt is in place, you can install a metal drip edge around the roof perimeter and then proceed with the roof covering. We used asphalt shingles for this project, but feel free to use roofing material to match your home's roof—cedar, fiber-cement, or metal roofing are other good options.

Finally, you can provide excellent ventilation by topping off your outbuilding roof with a continuous ridge vent. A ridge vent combined with vented soffits allows convection to draw cool air in through the eave or gable vents and exhaust hot air out at the roof peak.

Tools & Materials ▸

- Work gloves & eye protection
- Tape measure
- Hammer (or pneumatic nailer)
- Circular saw
- Aviation snips
- Stapler
- Utility knife
- Chalk line
- Roofing hammer
- Framing square
- ½" CDX or OSB sheathing panels
- 8d box nails
- Metal drip edge
- Roofing nails
- Building paper (15# or 30#)
- Shingles
- Continuous ridge vent (optional)
- Metal flashing

A top-notch roof includes roof deck sheathing, drip edge, roofing felt, shingles, and a continuous ridge vent. When properly installed, your outbuilding roof should last as long as your house roof.

Asphalt Shingles ▸

Asphalt shingles are usually rated by life span, with 20-, 25-, and 40-year ratings the most common (although some now claim to be 50-year shingles). Functionally, these ratings should be used for comparison purposes only. In fact, the average life span of an asphalt shingle roof in the United States is 8 to 10 years.

The term multitab shingle refers to any asphalt shingle manufactured with stamped cutouts to mimic the shapes of slate tile or wood shakes. Multitab cutouts are made and installed in single thickness 3-ft. strips, so these tabbed reveals show up. The ubiquitous term for them is three-tab, but two- and four-tab styles are also available. Generally, the tabs are spaced evenly along each sheet of shingle to provide a uniform appearance and a stepped, brick-laid pattern on the roof. However, some manufacturers also offer styles with shaped corners or randomly spaced tabs trimmed to different heights for a more unique look.

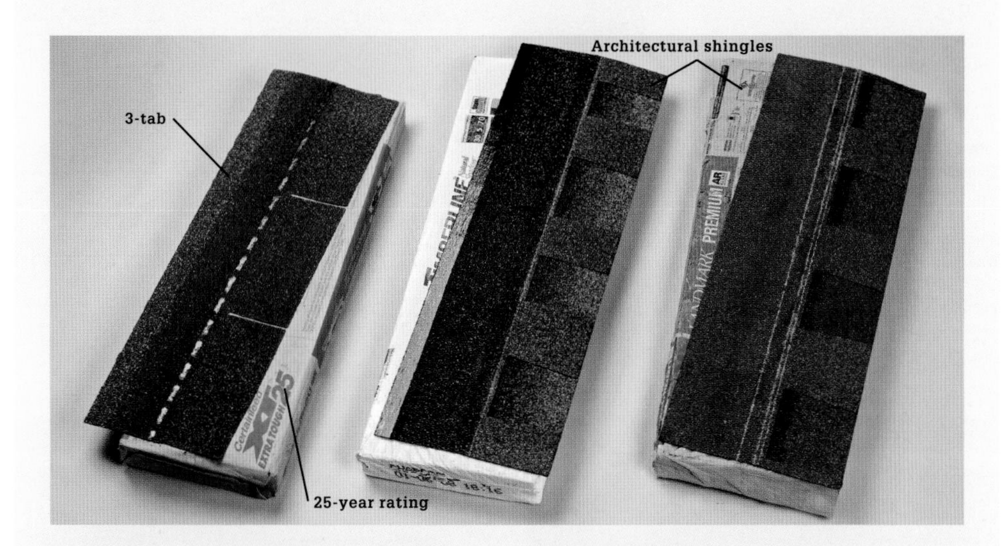

How to Install & Prepare the Roof Deck

Install the first course of roof decking. Start sheathing the roof at one of the lower corners with ½" CDX plywood or oriented strand board (OSB) sheathing. Where possible, use a full 8-ft.-long sheet or a half sheet with the seam still falling midway across a rafter or truss. Align the sheet so it overlaps the gable fascia and touches the eave fascia. Fasten the sheet to the rafters with 8d box nails spaced every 6" along the edges and 12" along the intermediate rafters. Lay out and install the rest of the sheathing to complete the first row, spacing the sheets ⅛" apart to allow for expansion.

Install the second row of decking. Start with a half sheet (approximately) to stagger the vertical joints between rows. Make sure the end of the half sheet falls midway along a rafter. Continue to sheath the roof up to the ridge, but stop nailing within 6" of the ridge. This area will be cut away to install a continuous ridge vent later. Add decking to the other side of the roof up to the ridge.

Install drip edge on eaves. Cut a 45° miter at the end of a piece of drip edge flashing and position it along one eave edge of the roof. The mitered end should be positioned to form a miter joint with the drip edge that will be installed on the gable end after the building paper is laid. Attach the drip edge with roofing nails driven every 12" and overlap any butt joints by 2". Flash both eave edges.

Begin installing building paper. Snap a chalk line across the roof sheathing 35⅝" up from the roof edge. At this location, the first row of building paper will overhang the drip edge by ⅜". Roll out 15# or 30# building paper along the eaves with the top edge aligned with the chalk line. Staple it to the sheathing every 12" along the edges and one staple per sq. ft. in the field area. Trim the gable ends of the paper flush with the edges of the sheathing. If you live in a cold climate and plan to heat your outbuilding, install self-adhesive ice-guard membrane for the first two courses.

Install the second course of paper. Snap another chalk line across the first row of paper, 32" up from the eaves. Roll out the second row of building paper with the bottom edge following the chalk line to create a 4" overlap. Staple it in place. Cover the entire roof up to the ridge with underlayment, overlapping each row by 4".

Install drip edge on the gable ends. Cut a 45° miter at the end of the first piece of drip edge, and install it along the gable edge of the roof, covering the building paper. Fit the mitered end over the eave's drip edge, overlapping the pieces by 2". The gable drip edge should be on top. Nail the drip edge all the way to the peak, and then repeat for the other three gable edges.

How to Install Asphalt Shingles

Mark starting lines. Snap a chalk line for the starter course, 11½" (½" less than the height of the shingle) up from the eave edge to mark the top edge of the starter course of shingles for each roof deck. This will result in a ½" shingle overhang at the eave edge for standard 12" three-tab shingles.

Install the starter course. Trim off one half of an end tab on a shingle. Position the shingle upside down so the tabs are aligned with the chalk line and the half tab is flush against the gable edge of the roof. Drive roofing nails near each end, 1" down from each slot between the tabs. Continue the row with full shingles nailed upside down to complete the starter course. Trim the last shingle flush with the opposite rake edge.

(continued)

Install the first full course. Apply the first full course of shingles over the starter course with the tabs pointing down. Start from the same corner you began the starter course. Place the first shingle so it overhangs the gable edge by ⅜" and the eaves by ½". The top edge of the first course should align with the top of the starter course.

Create a vertical reference line. Snap a chalk line from the eave's edge to the ridge to create a vertical line to align the shingles. Choose a spot close to the center of the roof, located so the chalk line passes through a slot or a shingle edge on the first full shingle course. Use a framing square to make sure the line is perpendicular to the eave's edge.

Working on Roofs ▸

When working on the roof and staging heavy bundles of shingles, it's a good idea to share the job with a helper. Set up ladders carefully, stay well clear of overhead power lines, and work cautiously near the eaves and gable ends of the roof to prevent accidents. Get off the roof if you are tired, overheated, or if impending bad weather threatens your safety.

Set the shingle pattern. If you are installing standard three-tab shingles, use the vertical reference line to establish a shingle pattern with slots that are offset by 6" in succeeding courses. Tack down a shingle 6" to one side of the vertical line and 5" above the bottom edge of the first-course shingles to start the second row. Tack down a shingle for the third course 12" from the vertical line. Begin at the vertical line for the fifth course. Repeat.

Fill in the shingles. Add shingles in the second through the fifth courses, working upward from the second course and maintaining consistent exposure. Insert lower-course shingles under any upper-course shingles left partially nailed, and then nail them down.

Test shingle alignment regularly. After each three-course cycle, measure from the bottom edge of the top row of shingles to the closest layout line on the building paper, and take several of these measurements along the course. If the row is slightly out of alignment, make incremental adjustments over the next few courses to correct it—don't try to get it back all in one course.

Cutting Ridge Caps ▸

Cut three 12"-sq. cap shingles from each three-tab shingle. With the back surface facing up, cut the shingles at the tab lines. Trim the top corners of each square with an angled cut, starting just below the seal strip to avoid overlaps in the reveal area.

Shingle up to the ridge. At the ridge, shingle up the first side of the roof until the top of the uppermost exposure line is within 5" of the ridge (for standard three-tabs). Trim the shingles along the peak. Shingle the other side of the roof up to the peak. If you plan to install a continuous ridge vent, skip to step 1 on page 66.

(continued)

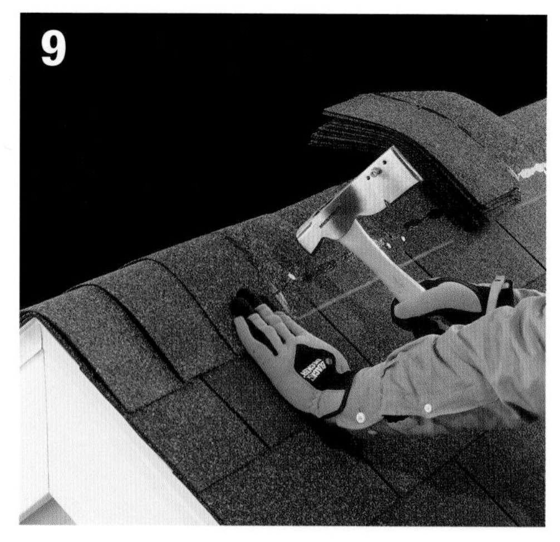

Install ridge cap shingles. Start by installing one cap shingle at one end so equal amounts hang down on each side of the ridge. Measure this distance and snap straight chalk lines to the other end of the roof, extending the lines formed by the edges of the shingles. Nail in the tapered area of each shingle so the next shingle will cover the nail heads. Complete the installation of the ridge cap shingles.

Trim shingles. Mark and trim the shingles at the gable edges of the roof. Snap a chalk line down the roof ⅜" from the drip edge. Use old aviation snips to cut the shingles. You may use a utility knife with backer board instead.

How to Install a Continuous Ridge Vent

Mark cutting lines. Measure from the ridge down each roof plane the distance recommended by the ridge vent manufacturer. Mark straight cutting lines at this distance by snapping a chalk line on each side of the ridge.

Cut out the ridge. Using a circular saw equipped with an old blade, cut through the shingles and sheathing along the cutting lines. Be careful not to cut into the rafters. Stop both cuts 6 to 12" from the gable ends. Make two crosscuts up and over the ridge to join the long cuts on the ends. Remove the shingles and sheathing from the cutout area. Drive additional nails through the shingles and sheathing near the cut edges to secure them to the rafters.

Mark installation reference lines. Test-fit the vent at one end, measuring down from the ridge half the width of the ridge vent, and marking that distance on both ends of the roof. Join the marks with two more chalk lines to establish the position for the edges of the continuous ridge vent.

Attach the ridge vent. Center the ridge vent over the opening, aligning the end with the gable edge of the roof. The side edges of the vent should be even with the chalk lines. Drive 1½" roofing nails through the vent and into the roof where indicated by the manufacturer.

Add vent sections. Butt subsequent pieces of continuous ridge vent against the pieces you have installed and fasten as before. Install the vent along the full length of the roof, including the end areas with shingles still intact.

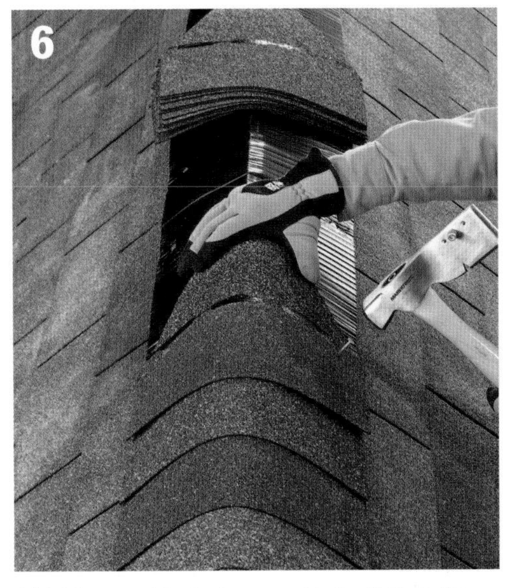

Add ridge cap shingles (see page 66, step 9). Cover the ridge vent with ridge cap shingles, nailing them with two 1½" roofing nails per cap. Overlap the shingles as you would on a normal ridge. Trim the end ridge cap shingle flush with the other gable edge shingles.

Installing Windows & Service Doors

Many outbuildings, like the one featured here, have a service door for added safety and accessibility. A window improves ventilation and is a pleasant source of natural light. This section will show you how to install both features. If you already have experience hanging doors and windows, you'll find that the process for installing them in an outbuilding is no different from installing them in a home. However, it's a good idea to review these pages to refamiliarize yourself with the proper techniques.

Installing doors and windows are similar operations. First, you'll need to seal the rough openings in the walls with self-adhesive flashing tape to prevent moisture infiltration. Tape should be applied from the bottom of the doorway or windowsill first, working up to the header and overlapping the tape to shed water. Once you've inserted the window or door in its opening, you'll need to shim it, adjusting for level and plumb, before nailing the jamb framework (or brickmold) in place.

When you have the option, hang the service door and window before proceeding with the siding (which we'll cover in the next section). That way, you'll be able to fit the siding up tight against the brickmold for a professional-looking finish.

Tools & Materials ▸

- Utility knife
- Caulk gun
- Level
- Hammer
- Screwdriver
- Silicone caulk
- Window
- 2" roofing nails
- Shims
- 6d casing nails
- Drip cap
- Low-expansion foam insulation
- Prehung service door
- Lockset
- 8d galvanized casing nails
- 9" self-adhesive flashing tape
- Work gloves & eye protection

A sturdy service door and deadbolt lock will give your new outbuilding added accessibility without compromising security. Installing one is a fairly simple project. A vinyl- or aluminum-clad window will bring a breath of fresh air and improve your ambient lighting throughout the day.

Tips for Sizing & Framing ▸

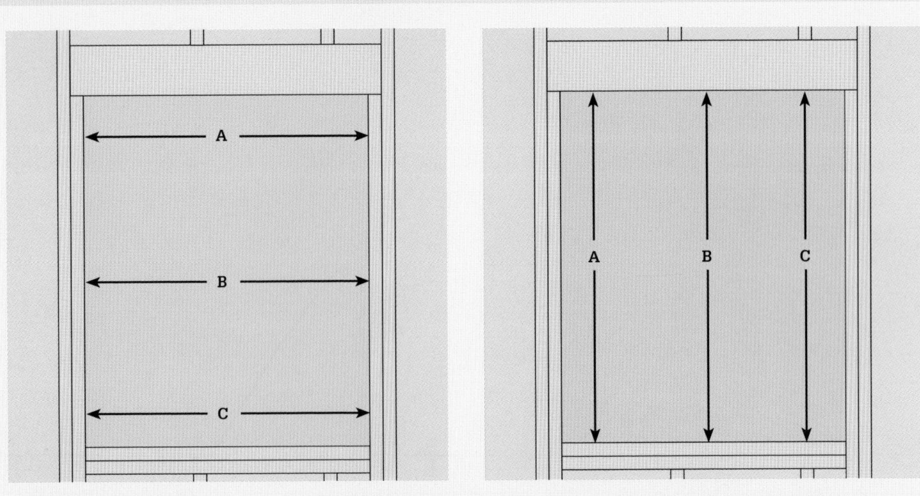

Determine the exact size of your new window or door by measuring the opening carefully. For the width (left illustration), measure between the jack studs in three places: near the top, at the middle, and near the bottom of the opening. Use the same procedure for the height (right illustration), measuring from the header to the sill near the left edge, at the middle, and near the right edge of the opening. Use the smallest measurement of each dimension for ordering the unit.

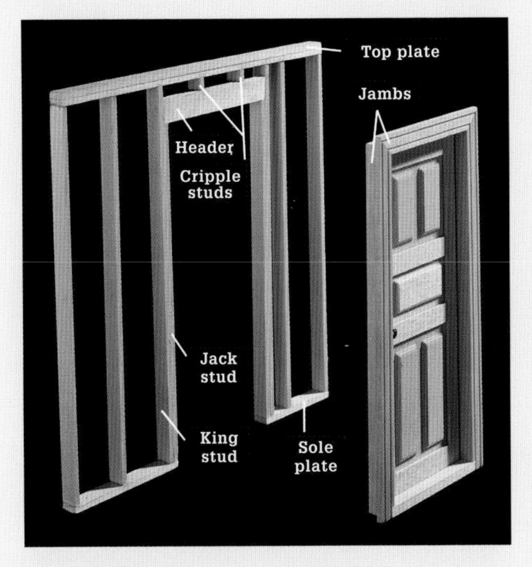

Door opening: The structural load above the door is carried by cripple studs that rest on a header. The ends of the header are supported by jack studs (also known as trimmer studs) and king studs that transfer the load to the sole plate and the foundation of the building. The rough opening for a door should be 1" wider and ½" taller than the dimensions of the door unit, including the jambs. This extra space lets you adjust the door unit during installation.

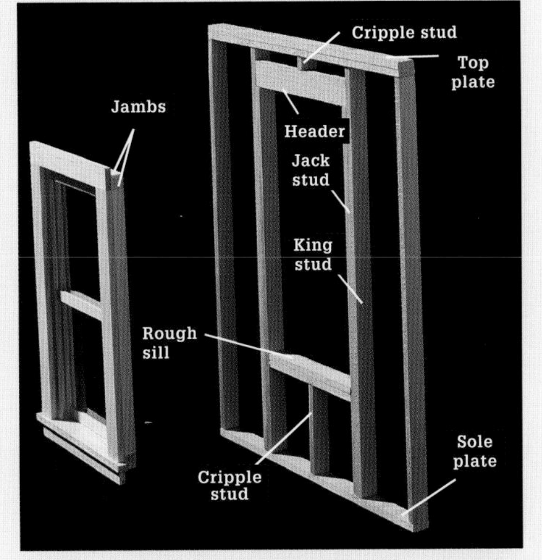

Window opening: The structural load above the window is carried by a cripple stud resting on a header. The ends of the header are supported by jack studs and king studs that transfer the load to the sole plate and the foundation of the building. The rough sill, which helps anchor the window unit but carries no structural weight, is supported by cripple studs. To provide room for adjustments during installation, the rough opening for a window should be 1" wider and ½" taller than the window unit, including the jambs.

How to Install an Outbuilding Window

Flash the rough sill. Apply 9"-wide self-adhesive flashing tape to the rough sill to prevent moisture infiltration below the window. Install the flashing tape so it wraps completely over the sill and extends 10 to 12" up the jack studs. Fold the rest of the tape over the housewrap to create a 3" overlap. Peel off the backing and press the tape firmly in place. Install tape on the side jambs butting up to the header, and then flash the header.

Option: You can save a step (and some material) by installing the flashing on the sides and top after the window is installed, as seen in this skylight installation. The disadvantage to doing it this way is that the inside faces of the rough frame will not be sealed against moisture.

Caulk the opening. Apply a ½"-wide bead of caulk around the outside edges of the jack studs and header to seal the window flange in the opening. Leave the rough sill uncaulked to allow any trapped moisture to escape.

Position the window. Set the window unit into the rough opening, and center it side to side. Check the sill for level.

Tack the top corners. Drive a roofing nail through each top corner hole of the top window flange to tack it in place. Do not drive the rest of the nails into the top flange yet.

Plumb the window. Have a helper hold the window in place from outside while you work inside the outbuilding. Check the window jamb for square by measuring from corner to corner. If the measurements are the same, the jamb is square. Insert shims between the side jambs and rough opening near the top corners to hold the jambs in position. Use additional shims as needed to bring the jamb into square. Recheck the diagonals after shimming.

Nail the flange. Drive 2" roofing nails through the flange nailing holes and into the rough frame to secure it. Handnail this flange, being careful not to damage the flange or the window cladding.

Nail the jambs (wood-frame windows only). Drive 6d (2") casing nails through the jambs and top corner shims to lock them in place. Add more shims to the centers and bottom corners of the jamb, and test the window action by opening and closing it. If it operates without binding, nail through the rest of the shims.

(continued)

Flash the side flanges. Seal the side flanges with flashing tape, starting 4 to 6" above the top flange flashing and ending 4 to 6" below the bottom flange/sill flashing. Press the tape firmly in place.

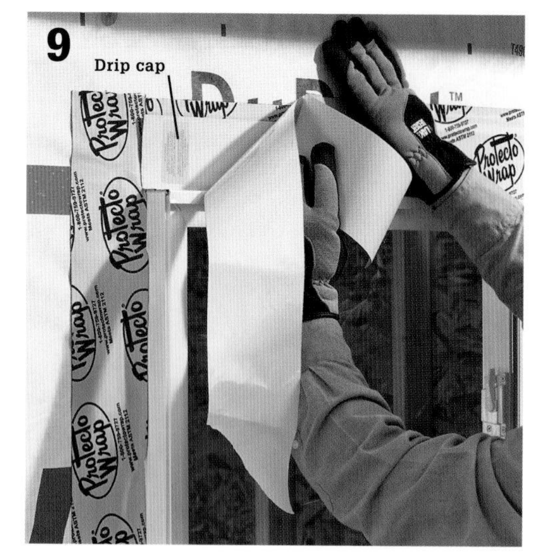

Install the drip cap. Cut a piece of metal drip cap to fit over the top window jamb. This is particularly important if your new window has an unclad wooden jamb with preinstalled brickmold. Set the drip cap in place on the top jamb, and secure the flange with a strip of wide flashing tape. Do not nail the drip cap. Have the tape overlap the side flashing tape by 6". *Note: If you plan to trim the window with wood brickmold or other moldings, install the drip edge above that trim instead.*

Finish the installation. Cut the shim ends so they are flush with the jambs using a utility knife or handsaw.

Spray expanding foam insulation around the perimeter of the window on the interior side if you will be insulating and heating or cooling your outbuilding.

Service Door Buyer's Tips ▸

- If you plan to use your new outbuilding as a workshop, buy the widest service door that will suit your building. That way, you won't have to open your sectional door every time you want to pull out the lawnmower or trash cans.
- Although primed-wood service doors are less expensive than aluminum- or vinyl-clad doors, they're generally not a better value in the long run. Normal wear and tear and the effects of the elements will mean you'll need to keep up with regular scraping and painting in order to keep your wooden door in good condition. A clad door, on the other hand, requires little or no maintenance over the life of the door.
- Another option for many of today's quality service doors is a jamb made of composite materials instead of wood. A composite jamb will not absorb water when it rains, and it's impervious to rot and insects.

How to Install a Service Door

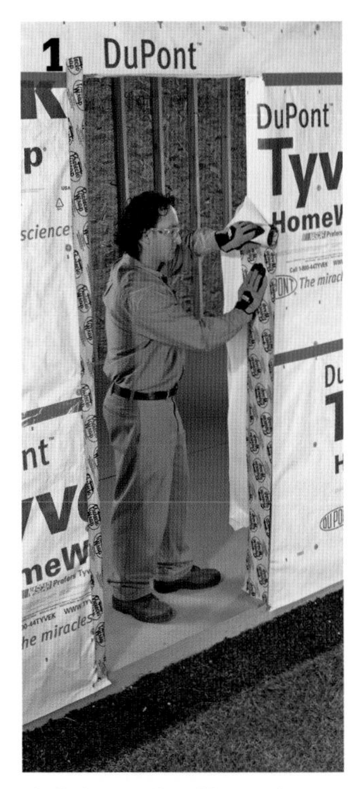

Flash the opening sides. Apply two strips of 9"-wide self-adhesive flashing tape to cover the jack studs in the door's rough opening. Cut a slit in the tape and extend the outer ear 4 to 6" past the bottom edge of the header. Fold the tape over the housewrap to create a 3" overlap. Peel off the backing and press the tape firmly in place.

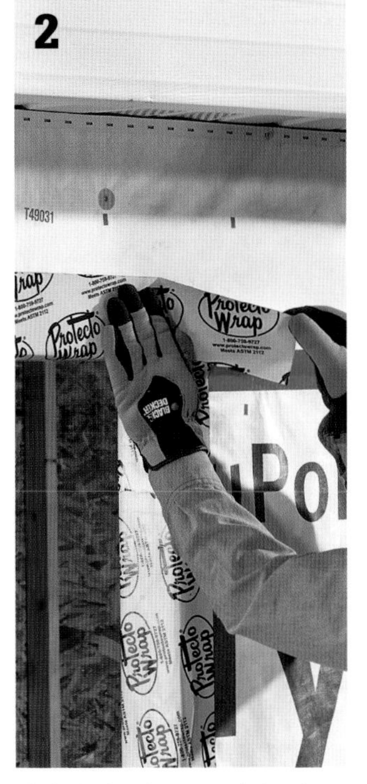

Flash the header. Cover the header with a third piece of self-adhesive flashing tape, extending the ends of the tape 6" beyond the side flashing.

Seal the opening. Apply a ½"-wide bead of caulk up the outside edges of the jack stud area and around the header to seal the brickmold casing.

(continued)

Position the door in the opening. Set the bottoms of the side jambs inside the rough opening, and tip the door into place. Adjust the door so it's centered side-to-side in the opening.

Adjust the door. Orient pairs of shims so the thick and thin ends are reversed, forming a rectangular block. Insert the shims into the gap between the rough framing and the hinge-side jamb. Spread the shims closer together or farther apart to adjust the total thickness until they are pressure-fitted into the gap. Space the shims every 12" along the jamb, and locate two pairs near the hinges. Check the hinge jamb for plumb and to make sure the shims do not cause it to bow. Drive pairs of 6d casing nails through the jambs at the shim locations.

Shim the latch side. Insert pairs of shims every 12" in the gap between the latch-side jamb and the rough framing. With the door closed, adjust the shims in or out until there's a consistent gap from top to bottom between the door and the jamb. Then drive pairs of 6d casing nails through the jamb and shims to secure them.

Attach the brickmold. Drive 8d galvanized casing nails through the brickmold to fasten it to the jack studs and header. Space the nails every 12". Trim off the shims so they are flush with the jambs using a utility knife or handsaw.

How to Install a Lockset

Insert the lock bolt for the lockset (and deadbolt, if installing one) into their respective holes in the door. These days, new exterior doors are almost always predrilled for locksets and deadbolts. Screw the bolt plates into the premortised openings.

Assemble the handles and lock mechanisms by tightening the screws that draw the two halves together. Do not overtighten.

Door Security

Add metal door reinforcers to strengthen the area around the lockset or deadbolt. These strengthen the door and make it more resistant to kick-ins.

Add a heavy-duty latch guard to reinforce the door jamb around the strike plate. For added protection, choose a guard with a flange to resist pry-bar insertion. Attach the guard with 3" screws that will penetrate through the jamb and into the wall studs.

Installing Overhead Doors

Your sectional overhead door will bear the brunt of everything Mother Nature and an active household throws at it—seasonal temperature swings, moisture, blistering sunlight, and the occasional misfired half-court jump shot. If that isn't enough, the average sectional door typically cycles up and down at least four times per day, which totals up to around 1,300 or more uses every year. For all of these reasons, it pays to install a high-quality door on your new outbuilding so you can enjoy a long service life from it.

These days, you don't have to settle for a drab, flat-panel door. Door manufacturers provide many options for cladding colors, panel texture and layout, exterior hardware, and window styles. Today's state-of-the-art doors also benefit from improved material construction, more sophisticated safety features, and enhanced energy efficiency. When you order your new door, double-check your building's rough opening and minimum ceiling height to be sure the new door will fit the space properly.

Installing a sectional overhead door is easier than you might think, and manufacturers make the process quite accessible for average do-it-yourselfers. With a helper or two, you should have little difficulty installing a new overhead door in a single day. The job is really no more complex than other window and door replacements if you work carefully and exercise good judgment. Overhead door kits come with all the necessary hardware and detailed step-by-step instructions. Since overhead door styles vary, the installation process for your new door may differ from the photo sequence you see here, so always defer to the manufacturer's instructions. This will ensure the door is installed correctly and the manufacturer will honor the product warranty.

Tools & Materials ▸

- Work gloves & eye protection
- Tape measure
- 4-ft. level
- Drill with nut drivers
- Stepladder
- Ratchet wrench with sockets
- 16d nails
- Adjustable wrench
- Hammer
- Sectional overhead door with tracks & mounting brackets
- Doorstop molding & galvanized nails or siding nails

The sectional overhead door you choose for your shed, barn, or garage will go a long way toward defining the building's appearance and giving you trouble-free performance day in and day out.

How to Install an Overhead Door

Measure for the door. Measure the width of the header, the headroom clearance to the rafter ties (or bottom truss chords), and the rough opening of the doorway. Check these measurements against the minimum requirements outlined in the instruction manual that comes with your sectional garage door.

Assemble the door tracks. Working on the floor, lay out and assemble the vertical tracks, jamb brackets, and flag angle hardware. Install the door bottom seal.

Install the first section. Set the bottom door section into position against the side jambs, and adjust it left or right until it overlaps the side jambs evenly. Check the top of the door section for level. Place shims beneath the door to level it, if necessary. Have a helper hold the door section in place against the jambs until it is secured in the tracks.

Attach the tracks. Slip a vertical track over the door section rollers and against the side jamb. Adjust it for plumb, then fasten the jamb brackets to the side jamb blocking with lag screws. Carefully measure, mark, and install the other vertical track, following the manufacturer's specifications.

(continued)

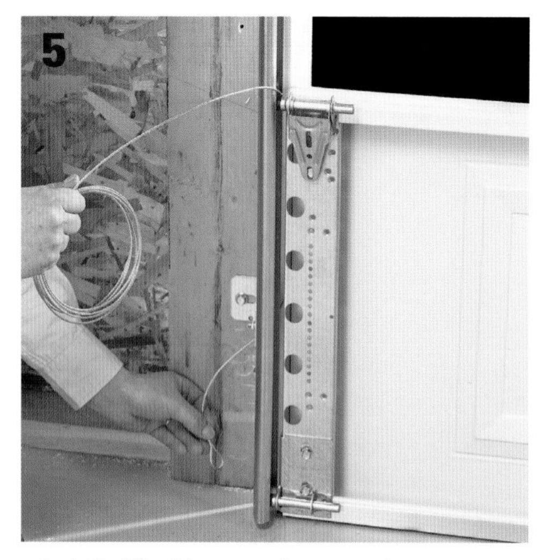

Attach the lift cables. Depending on your door design, you may need to attach lift cables to the bottom door section at this time. Follow the instructions that come with your door to connect these cables correctly.

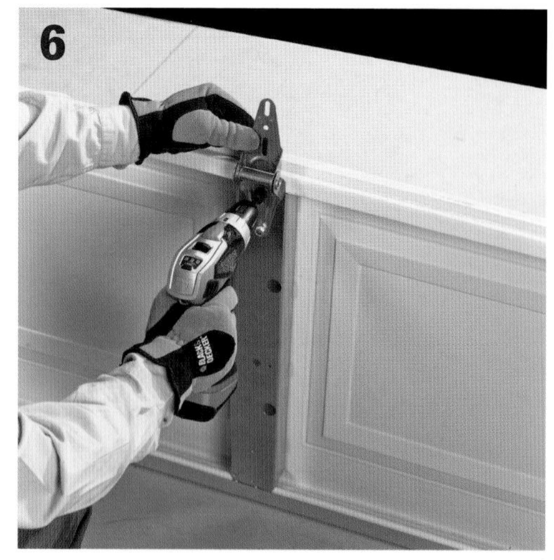

Install the door hinges. Fasten the end and intermediate hinges to the bottom door section, and then install roller brackets and hinges on the other door sections. Attach hinges to the top edges of each door section only. This way you'll be able to stack one section on top of the next during assembly.

Add the next sections. Slip the next door section into place in the door tracks and on top of the first section. Connect the bottom hinges (already attached to the first section) to the second door section. Repeat the process until you have stacked and installed all but the top door section.

Option: The top door section may require additional bracing, special top roller brackets, and a bracket for securing a garage door opener. Install these parts now following the door manufacturer's instructions.

Install the top section. Set the top door section in place and fasten it to the hinges on the section below it. Support the door section temporarily with a few 16d nails driven into the door header blocking and bent down at an angle.

Complete the track installation. Fasten the horizontal door tracks to the flag angle brackets on top of the vertical track sections. Temporarily suspend the back ends of the tracks with rope so they are level.

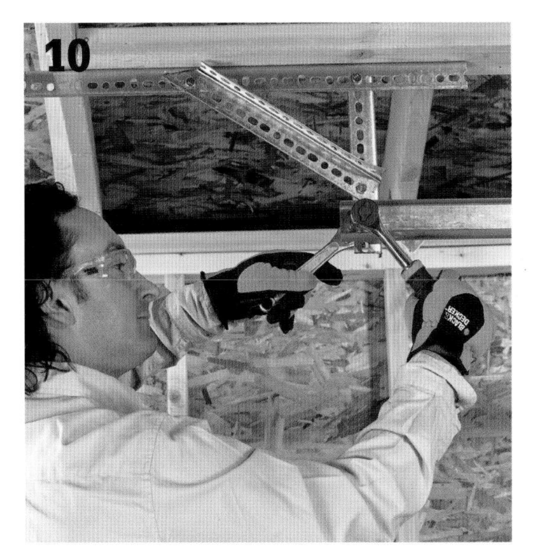

Install the rear hanger brackets. This step will vary among door opener brands. Check your door instruction manual for the correct location of rear hanger brackets that will hold the horizontal door tracks in position. Measure, cut, and fasten sections of perforated angle iron together with bolts, washers, and nuts to form two Y-shaped door track brackets. Fasten the brackets to the rafter tie or bottom truss chord with lag screws and washers following the door manufacturer's recommendations.

Attach the extension springs. The door opener here features a pair of smaller springs that run parallel to the horizontal door tracks (not parallel to the door header, as larger torsion springs are installed). The springs are attached to cables that attach to the door and the rear hanger brackets.

(continued)

Test the door operation. Raise it about halfway first. You'll need at least one helper here. Slide a sturdy support underneath the door bottom to hold the door and then inspect to make sure the rollers are tracking and the tracks are parallel.

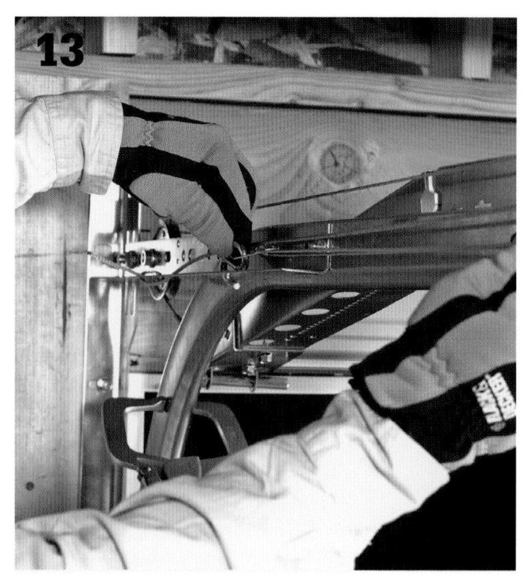

Attach the spring cables. Raise the door fully and hold it in place with C-clamps tightened onto the tracks to prevent it from slipping down. The tension in the springs should be relieved. The cables in this case are tied off onto a 3-hole clip that is then hooked onto the horizontal angle bracket near the front of the tracks.

Attach the doorstop molding. Measure, cut, and nail sections of doorstop molding to the door jambs on the outside of the door to seal out weather (inset). A rolled vinyl doorstop may come with your door kit. If not, use strips of 1 × 2 treated wood or cedar for this purpose.

Option: Install a garage door opener, following the manufacturer's instructions.

Garage Door Opener Safety Tips ▸

Whether you're adding an opener to a new or an old garage door, these tips will help make it a safe part of your building.

- Before beginning the installation, be sure the garage door manually opens and closes properly.
- If you have a one-piece door, with or without a track, read all additional manufacturer's installation information.
- The gap between the bottom of the garage door and the floor must not exceed ¼". If it does, the safety reversal system may not work properly.
- If the building has a finished ceiling, attach a sturdy metal bracket to the structural supports before installing the opener. This bracket and hardware are not usually provided with the garage door opener kit.
- Install the wall-mounted garage door control within sight of the garage door, out of reach of children (at a minimum height of 5 ft.), and away from all moving parts of the door.
- Never use an extension cord or two-wire adapter to power the opener. Do not change the opener plug in any way to make it fit an outlet. Be sure the opener is grounded.
- When an obstruction breaks the light beam while the door is closing, most door models stop and reverse to full open position, and the opener lights flash 10 times. If no bulbs are installed, you will hear 10 clicks.
- To avoid any damage to vehicles or equipment entering or leaving the barn or garage, be sure the door provides adequate clearance when fully open.
- Garage doors may include tempered glass, laminate glass, or clear-plastic panels—all safe window options.

Make sure your garage door opener is securely supported to trusses or ceiling framing with sturdy metal hanging brackets.

Use the emergency release handle to disengage the trolley only when the garage door is closed. Never use the handle to pull the door open or closed.

Installing Siding & Trim

Siding will protect your new outbuilding from the elements, of course, but it also serves as a way to visually tie the outbuilding to your home. Ideally, you should choose the same siding for the outbuilding as you have on your house, but if you decide to go with a different material it should mimic the same pattern, such as horizontal laps, overlapping shingles, or vertical boards and battens. These days, material options for siding are more varied than ever. You might choose wood, vinyl, aluminum, fiber-cement lap siding; cedar or vinyl shingles; faux brick and stone; or stucco. Or, depending on your home's siding scheme, it might be a combination of two different siding materials that complement each other.

Each type of siding will typically have its own unique installation process, and each application requires the correct underlayment, fasteners, and nailing or bonding method. The installation process can even vary among manufacturers for the same product type.

For the project shown here, we install a combination of fiber-cement lap siding and cast veneer stone.

Tools & Materials ▸

Work gloves & eye protection
For stone veneer:
Aviation snips
Hammer
Mixing trough
Stiff-bristle brush
Angle grinder with diamond blade
Mason's hammer
Trowel
Grout bag
Jointing tool
Expanded metal lath
Building paper
Type N mortar
Masonry sand
Veneer stone
Sill blocks
Metal flashing
2 × 2" zinc-coated L-brackets
For fiber-cement siding:
Tape measure
Bevel gauge
Circular saw
Paintbrush
Chalk line
Cementboard shears
Jigsaw with masonry blades
Fiber cement corner boards
Drill with bits
Caulk gun
6d casing nails
Fiber cement frieze boards
Primer
Fiber cement siding
Dust respirator
Paintable exterior caulk
Paint (optional)
Screws

A combination of faux stone and lap siding, with accenting corner trim, transforms what could otherwise be an ordinary outbuilding into a structure that adds real curb appeal to your home.

Siding Types

Vinyl lap siding is inexpensive, relatively easy to install, and low maintenance. Some styles can be paired with custom profiled foam insulation boards. Matching corner trim boards are available, but you can also make your own wood trim boards and paint them.

Wood lap siding comes in wide or narrow strips and is normally beveled. Exterior-rated wood that can be clear coated is common (usually cedar or redwood). Other wood types are used, too, but these are usually sold preprimed and are suitable for painting only.

Fiber-cement lap siding is a relative newcomer but its use is spreading quickly. It is very durable but requires some special tools for cutting. This siding takes paint well but is also available with a baked-on painted finish in a range of colors.

Specialty siding products like these cast veneer stones are often used as accents on partial-wall applications (see the photo on previous page). They can also be used to side one wall of a structure. For the most part, their effectiveness (and your budget) would be diminished if they were used to cover the entire structure.

Cast Veneer Stones

Cast veneer stones are thin masonry units that are applied to building walls to imitate the appearance of natural stone veneer. They come in random shapes, sizes, and colors, but they are scaled to fit together neatly without looking unnaturally uniform. Outside corner stones and a sill block (used for capping half-wall installations) are also shown here.

How to Apply Cast Veneer Stone

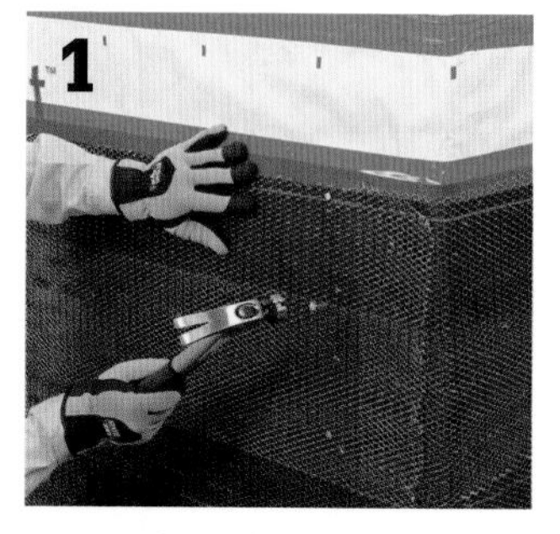

Prepare the wall. Veneer stones can be applied to a full wall or as an accent on the lower portion of a wall. A top height of 36 to 42" looks good. A layer of expanded metal lath (stucco lath) is attached over a substrate of building paper.

Apply a scratch coat. The wall in the installation area should be covered with a ½- to ¾"-thick layer of mortar. Mix one part Type N mortar to two parts masonry sand and enough water to make the consistency workable. Apply with a trowel, and then let the mortar dry for 30 minutes. Brush the surface with a stiff-bristle brush.

Test layouts. Uncrate large groups of stones and dry-lay them on the ground to find units that blend well together in shape as well as in color. This will save an enormous amount of time as you install the stones.

Cut veneer stones, if necessary, by scoring with an angle grinder and diamond blade along a cutting line. Rap the waste side of the cut near the scored line with a mason's hammer or a maul. The stone should fracture along the line. Try to keep the cut edge out of view as much as you can.

Apply the stones. Mix mortar in the same ratios as in step 2, but instead of applying it to the wall, apply it to the backs of the stones with a trowel. A ½"-thick layer is about right. Press the mortared stones against the wall in their position. Hold them for a few seconds so they adhere.

Fill the gaps between stones with mortar once all of the stones are installed and the mortar has had time to dry. Fill a grout bag (sold at concrete supply stores) with mortar mixture and squeeze it into the gaps. Once the mortar sets up, strike it smooth with a jointing tool.

Install sill blocks. These are heavier and wider than the veneer block so they require some reinforcement. Attach three 2 × 2" zinc-coated L-brackets to the wall for each piece of sill block. Butter the backs of the sill blocks with mortar and press them in place, resting on the L-brackets. Install metal flashing first for extra protection against water penetration.

How to Install Fiber-cement Lap Siding

Install corner boards. Nail one board flush with the wall corner and even with the bottom of the wall sheathing using 6d galvanized casing nails. Keep nails 1" from each end and ¾" from the edges. Drive two nails every 16". Overlap a second trim board on the adjacent side, aligning the edge with the face of the first board, and nail in place.

Trim windows and doors. Measure and cut brickmold or other trim to fit around the windows and doors. The trim joints can either be butted or mitered, depending on your preference. For miter joints, cut corners at 45°. Nail trim with galvanized casing nails driven every 16".

Install frieze boards. Cut the frieze boards to match the width of the corner boards. Butt them against the corner trim, and nail them to the wall studs directly under the soffits on the eaves with 6d galvanized casing nails.

Install gable frieze boards. Use a bevel gauge to transfer the gable angle to the frieze boards, and miter cut the ends to match. Install the gable frieze boards so they meet neatly in a miter joint at the roof peak. Nail them to the gable wall plates and studs with pairs of 6d galvanized casing nails every 16".

Install starter strips. Install strips cut from the thick edge of the siding boards along the bottoms of the walls, flush with the bottom edges of the wall sheathing. The strips will tip the first row of siding out to match the overlap projection of the other rows. Attach the strips to the wall studs with 6d galvanized casing nails. Snap vertical chalk lines to mark wall stud locations.

Install the first course. Cut the first siding board so it ends halfway over a stud when the other end is placed ⅛" from a corner trim board (see Tip, below). Prime the cut end before installing it. Align the siding with the bottom edge of the starter strip, keeping a ⅛" gap between the siding and the corner board. Nail the siding at each stud location 1" from the top edge with siding nails. Keep nails at least ½" in from the panel ends to prevent splitting.

Cutting Fiber Cement Siding ▸

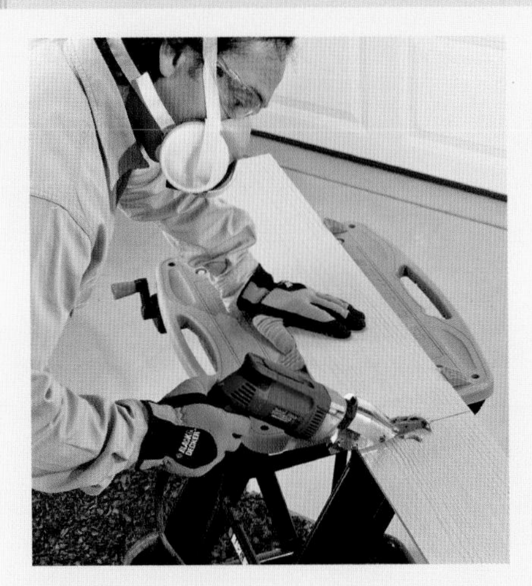

Wear a dust respirator when cutting the siding, especially if you use a circular saw instead of electric cementboard shears.

Install the next course. Set the second siding boards in place over the first course. Nail the siding to the wall at stud locations. Snap level chalk lines across the wall to mark layout lines for the remaining courses of siding. Set this pattern so each course of siding will overlap the course below it by 1¼".

(continued)

Install the remaining courses. As you install each course of siding, stagger the joints between the butt joints to offset the joints by at least one wall stud.

Work around windows and doors. Slide a piece of siding against the horizontal trim, and mark the board ⅛" from the outside edges of the trim. Use these marks to draw perpendicular lines on the board, and make a mark on the lines to represent the correct overlap. Connect these marks with a long line, and make the cutout with a jigsaw equipped with a masonry blade. Fit and nail the notched panel around the opening.

Install the top course. Unless you get lucky or have planned very carefully, the top course of siding boards will likely require rip-cutting to make sure that your reveals and setbacks are maintained. With a circular saw and a straightedge guide, trim off the top of the boards so the cut tops butt up against the frieze. Nail the cut boards in place.

Transfer gable angles. Use a bevel gauge to determine the roof angle on the gable-end walls. Transfer the angle to the siding panels that butt against the gable frieze boards, and cut them to fit.

Fill in under gables. Drill pilot holes though the angled ends of the gable siding pieces to keep them from splitting. Drive the nails through the holes to install the boards.

Caulk gaps. Fill all gaps between boards and between boards and trim with flexible, paintable caulk. Paint the siding and trim as desired.

Wood Shakes & Shingles ▸

Wood shingles and shakes are another good siding option that add a warm, organic look to your structure. Shingles and shakes are typically made from various species of cedar and cypress, white oak, and pressure-treated yellow southern pine. The principal difference between wood shakes and shingles is how they are cut and processed. Shakes are sawn or hand split in varying thicknesses and lengths; these irregularities give shakes an organic appeal, but can increase the chances for water infiltration. Shingles are manufactured to specific sizes and have a tapered edge.

Shingles and shakes can be installed in a double-course (which utilizes an undercourse) or staggered butt course, which features a random, three-dimensional look (shown). See page 440 for cedar shingle installation instructions.

PLANS & PROJECTS

HONDA
14
LP
APA

Barns & Sheds

Whether you need a space to store large farm equipment, board animals, or just tuck away out-of-season holiday decorations, this chapter has the project for you. The first building in this chapter, a 36- by 56-foot pole barn with a partial lean-to, is the perfect solution for a variety of needs, whether yours be agricultural, equestrian, or any of multiple others. Learn the process of pole framing, the versatile technique that barn builders have been using in America for centuries, through the clearly laid-out steps and complete project plans on the following pages.

If your needs are less expansive or your ambitions less grand, perhaps the Mini Gambrel Barn or Simple Storage Shed is a better project for you. Though smaller-scale, these sturdy structures are durable, attractive, and easily adaptable to many purposes. If you're seeking a hobby area or would like to create a fun play space for children, look into the Clerestory Studio or the Gothic Playhouse. The projects included in this chapter fit a wide variety of budgets, skill levels, and purposes—one of them is undoubtedly the perfect building for you.

Always consult your local building department for information on building permits, codes, and other laws as they apply to your projects.

In this chapter:

- Pole Barn
- Mini Gambrel Barn
- Simple Storage Shed
- Metal & Wood Kit Sheds
- Clerestory Studio
- Timber-frame Shed
- Gothic Playhouse
- Lean-to Tool Bin
- Additional Barn & Shed Plans

Pole Barn: CASE STUDY

Pole structures have been a mainstay of American barn construction since the country's beginnings. Farmers originally built pole-framed buildings because of their economical use of lumber and their strength and durability. And, the materials were readily available: the poles used in early barns were usually tree trunks from elsewhere on the property that were felled to clear the land.

The main difference between pole barns and traditionally framed barns lies in the way the building's weight is distributed. In traditional frame construction, the building's weight rests on top of its foundation and a series of load-bearing walls or other structural elements. In pole construction, the building's weight hangs on the poles instead of resting on the foundation. This means that pole buildings can easily be quite large without a lot of foundation materials or interior supports. And, the builder can choose a variety of different flooring options, from bare earth to compacted gravel to poured concrete. Also, because the poles or posts are embedded deep in the ground (typically four feet or more), the ground can shift around them with seasonal freeze/thaw cycles without affecting the building at all. The builder also benefits from flexibility with regard to the interior of the barn—stalls for animals, walls, or storage space can easily be added, or the interior can be left completely open to make space to store large pieces of machinery or other large projects.

Today, pole structures are always built on poles or posts that have been chemically treated to protect the wood from rot and to ensure the long-term life of the building. Often, any framing elements that are attached near the ground—such as the skirt board in this project—are chemically treated as well. The barn featured here utilizes 29-gauge steel siding and roofing for a sturdy, industrial-quality building, but these materials can easily be substituted for cedar board-and-batten or tongue-and-groove siding and traditional roof sheathing and asphalt shingles to match other structures on your property.

Pole barns allow for great design flexibility, from the choice of finishing materials to the building's actual purpose and design. This version features a partial lean-to overhang on one side that could be used to temporarily shelter vehicles or to stack firewood and building materials.

Tools, Materials & Cutting List

Excavation tools
Stakes
Mason's string
Hammer
Tape measure
Marking paint
Power skid steer with 22"-dia. auger dig-in attachment
Hand tamper
Maul
Shovel
Chainsaw
Reciprocating saw
Circular saw
4-foot level
Chalk line
Laser level on a tripod
Combination square
Framing square
Cordless drill
Tin snips or nibbler
Eye and ear protection

Description	Qty/Size	Material
Foundation		
Precast concrete footing pad	27@20"-dia × 6"-thick	
Eave post	16@16'	3-ply 2 × 6, treated on one end
Lean-to post	4@14'	3-ply 2 × 6, treated
Gable post 1	1@18'	3-ply 2 × 6, treated on one end
Gable post 2	3@20'	3-ply 2 × 6, treated on one end
Gable post 3	3@22'	3-ply 2 × 6, treated on one end
Uplift blocks	54@12"	treated 2 × 4
Floor		
Foundation base (optional)	31 cu. yds.	Compactable stone gravel
Wall Framing		
Skirt board	8@16'	Treated 2 × 6
Ribbon board	7@16'	2 × 4
Wall framing members	2@10', 6@12', 48@16', 12@18'	2 × 4
Truss spacer	14@8'	2 × 2
Lean-to eave nailers	24'	2 × 4
Corner bracing	16@16'	2 × 4
Roof Framing		
Trusses	8	Preassembled, engineered for your building's design
Purlins	84@18'	2 × 4
Lateral truss bracing	28@18'	2 × 4
Lean-to ledger	24'	2 × 10
Lean-to rafter	4@10'	2 × 10
Lean-to 90° return board	2@102"	2 × 6
Knee braces	2@12'	2 × 6
Exterior Finishes		
Tails	8@12'	2 × 4
Subfascia	13@18'	2 × 8
Eave molding trim	12@10'6"	29-gauge steel
Soffit starter	19@10'6"	1¾" 29-gauge steel
Base trim	18@10'6"	29-gauge steel
Eave wall siding	38@136"-long	29-gauge steel, 36"-wide panels
Gable wall siding	4@218¼", 4@206¼", 4@194¼", 4@182¼", 4@170¼", 4@158¼"	29-gauge steel, 36"-wide panels 36"-wide panels
Wall corner trim	4@14'6"	29-gauge steel
Outside roof corner trim	6@14'6"	29-guage steel
Vented soffit	140 ft.	12 × 16" panel
Fascia	20@10'6"	5½" angle trim
Door jamb trim	2@10'6", 1@12'6"	7½" 29-gauge steel
J-channel trim	6@10'6"	29-gauge steel
Roofing		
Roofing	40@239"-long	29-gauge steel, 36"-wide panels
Ridge cap	6@10'6"	20" metal ridge
Windows & Doors		
Windows	3	3' × 4'
Service Doors	2	
Garage Door		12' × 10'
Hardware		
4" steel pole barn nails	30 lbs	
6" steel pole barn nails	20 lbs	
16d common nails	20 lbs	
PVC Nails	2 lbs	
2" stainless steel screws	5 lbs	
Joist hangers	4	
Joist hanger nails		
1½" galvanized trim nails	5 lbs	
1½" gasket washer bell-cap screws	3000	

Floor Plan

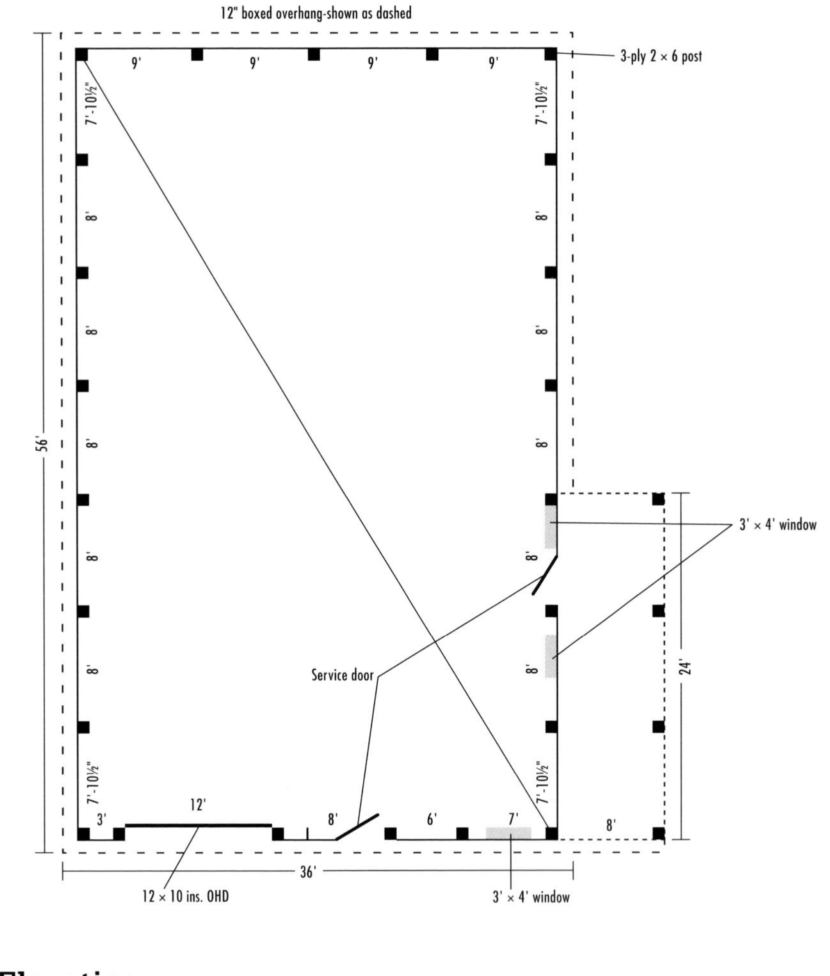

Front Elevation

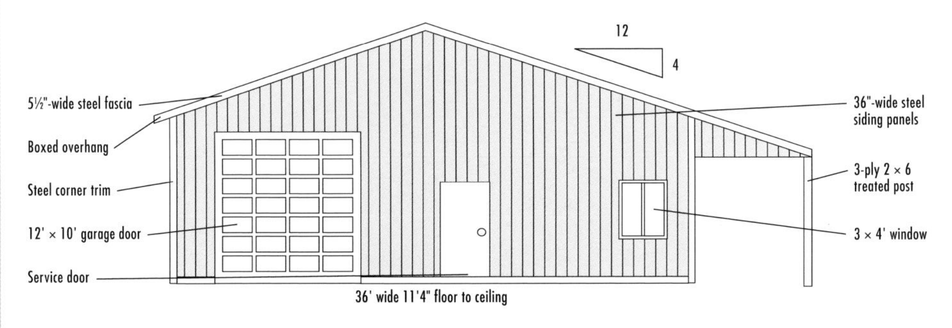

Rear Elevation

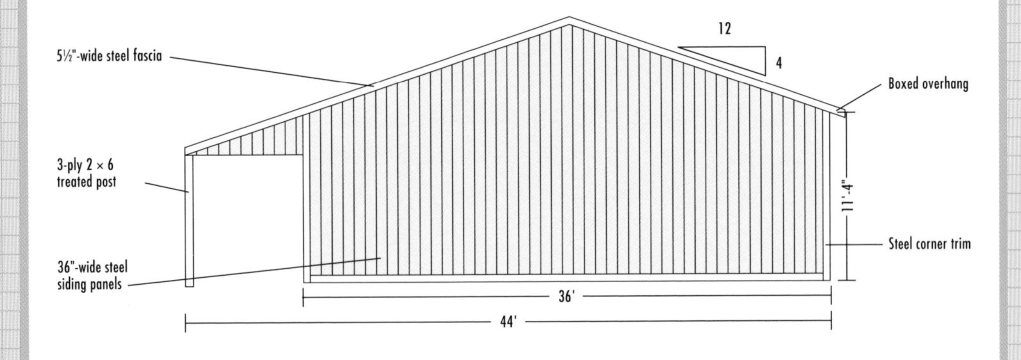

Right Side Elevation

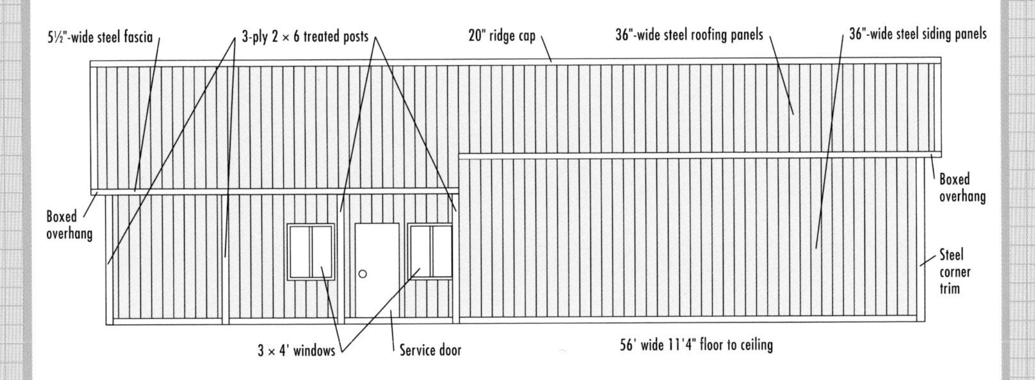

Left Side Elevation

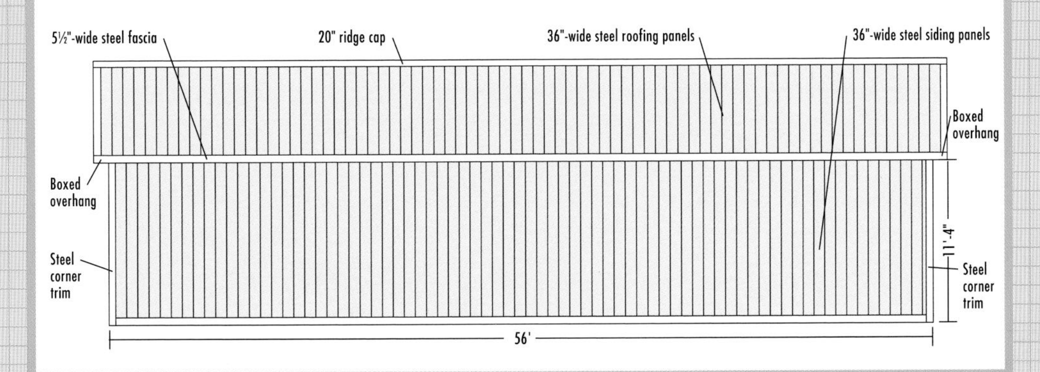

Left Side Framing Plan

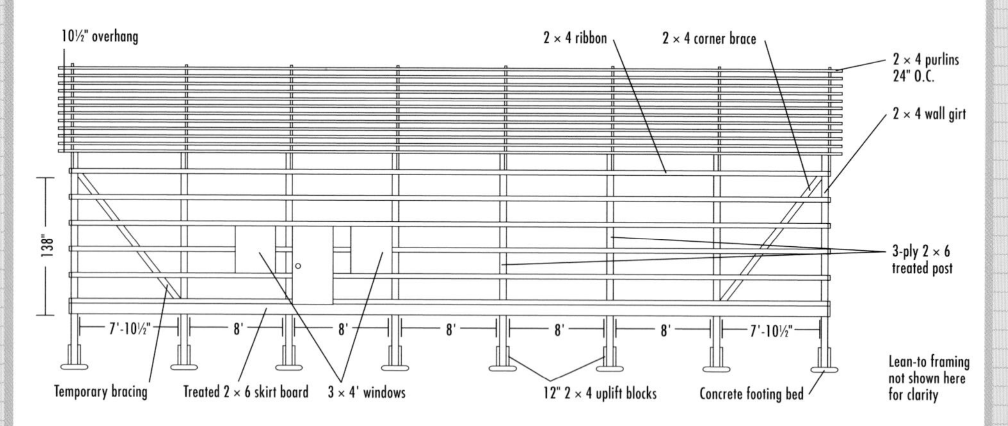

Right Side Framing Plan

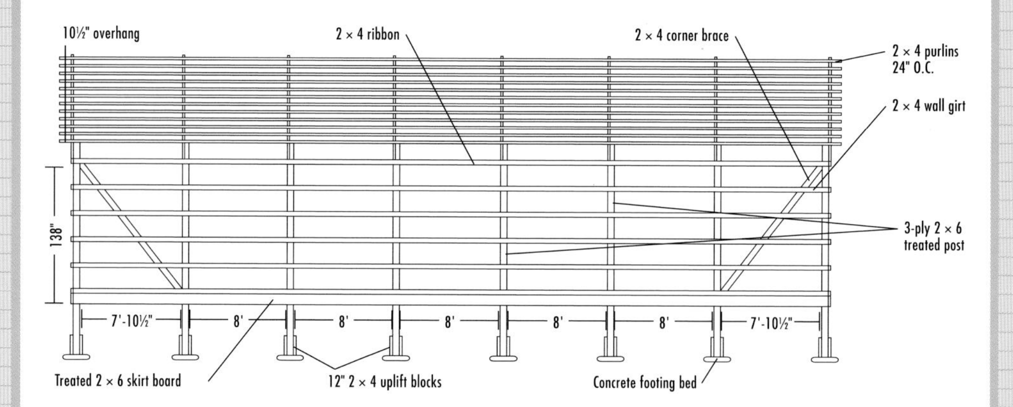

Lean-to Framing Plan

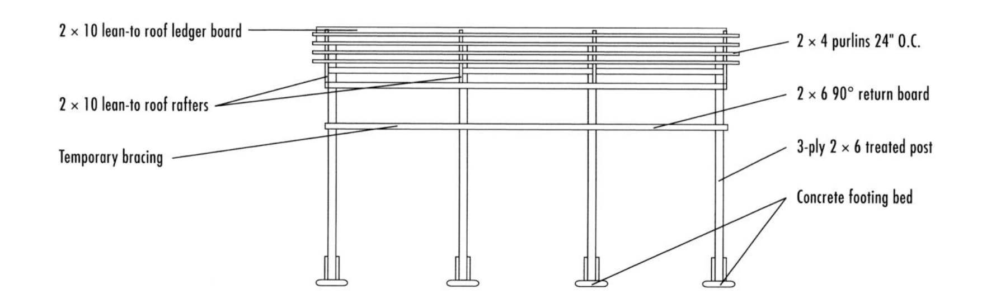

Front Framing Plan

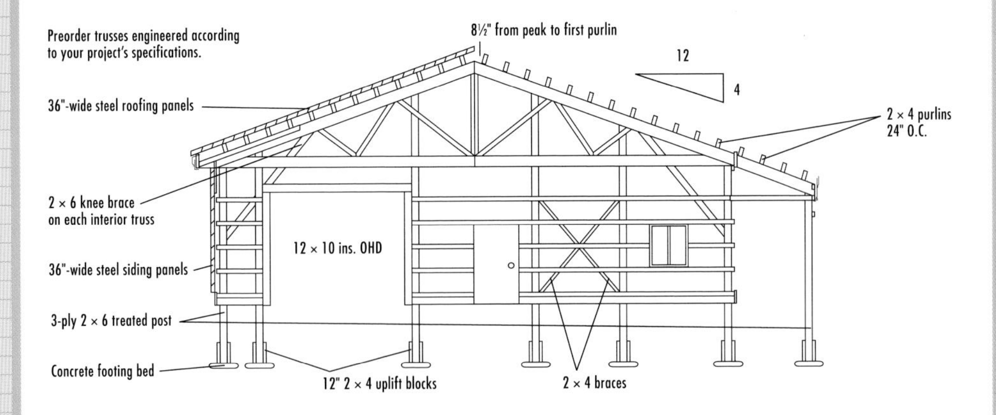

Rear Framing Plan

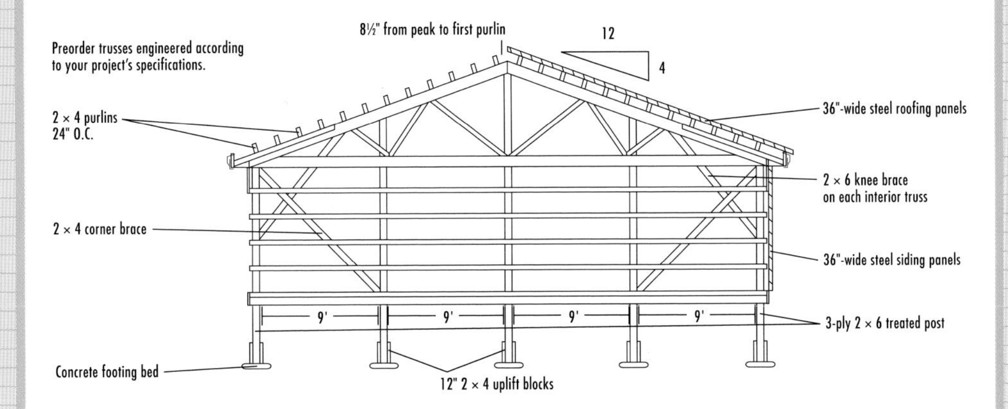

Lateral Truss Bracing Detail

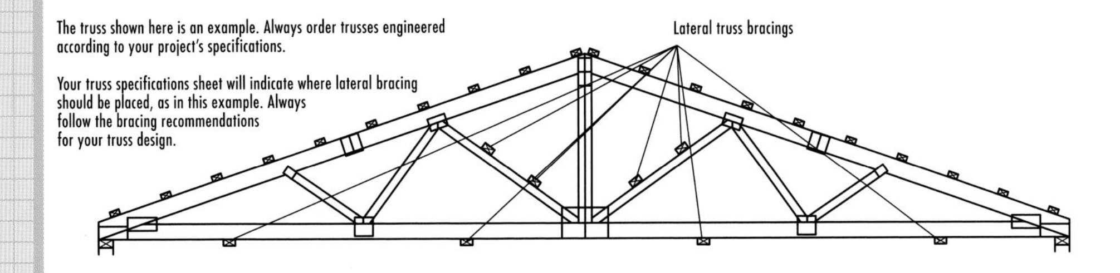

Lean-to Roof Detail

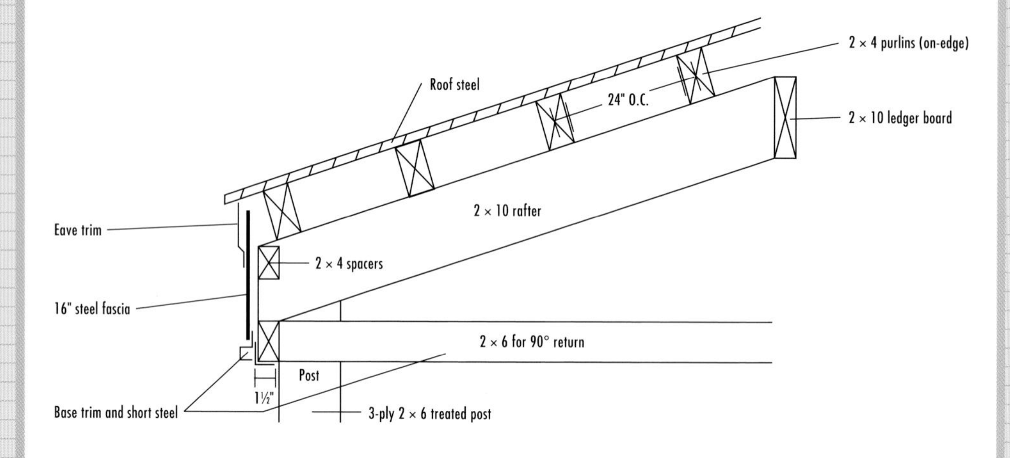

Boxed Overhang Detail

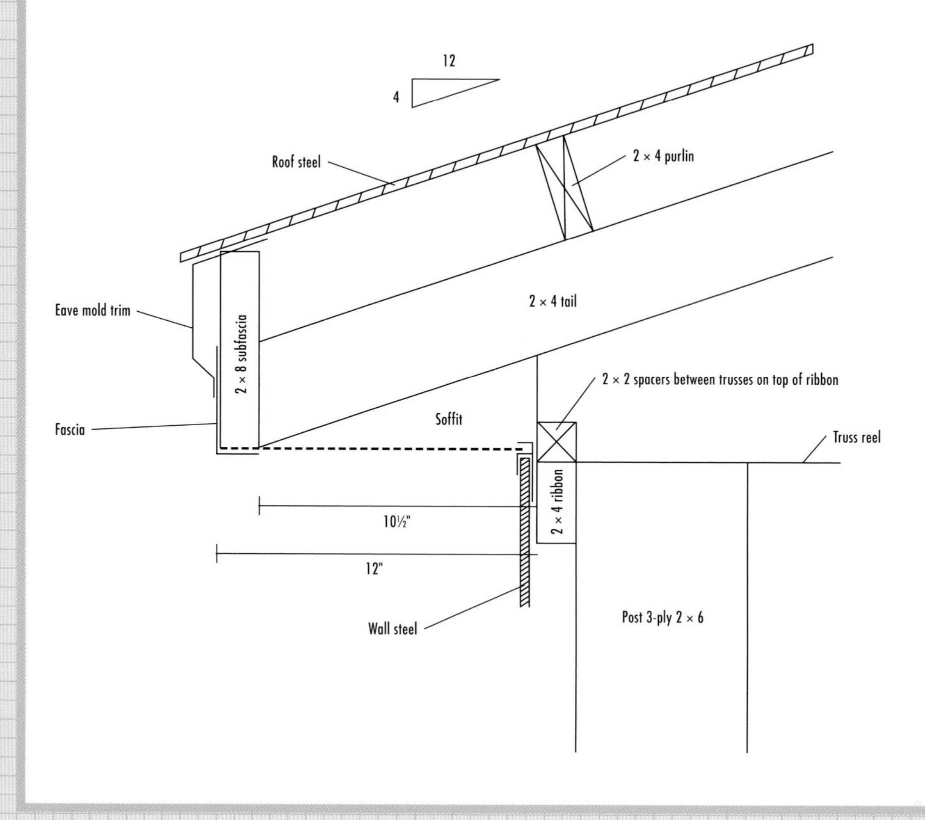

CASE STUDY: Pole Barn with Partial Lean-to Overhang

Warning ▸

The following photos feature construction professionals who are very experienced working at dangerous heights. Do not attempt these techniques yourself. For your safety, always use ladders or scaffolding to access the upper work areas.

Excavate the barn site approximately 3 ft. longer and wider than the building's footprint (44 × 56-ft. as shown in this project). Make sure the area is flat and all surrounding growth is removed. If you prefer to work on a gravel base, add a 4"-layer of compactable gravel now, or you can leave the excavated area bare. The foundation for this barn will be a perimeter of posts embedded in concrete beds within 4-ft. footings.

Stake out the four corner post locations to form a rectangle. Posts should be positioned 1½" inside the building's actual footprint to leave room for the wall girts, which will be installed on the outside of the posts.

Set up the batter boards at least 12" outside of each corner. Pound a nail partially into the top of each batter board so that the strings intersect at the stake's location; tie a mason's string to one of these nails and run to the location of the next corner post. Wrap the string around this nail to form a line for the first wall. Then, wrap the string around the perpendicular batter board nail and run to the next corner.

Adjust batter boards, stakes, and string for square, as necessary. Finish by tying the end of the string to the second batter board outside the first corner post. Lay out post locations for lean-to corners according to the FLOOR PLAN.

(continued)

Check the mason's string layout for level. Mount a laser level on a tripod in one location where you can easily shoot all four corner post locations. Hold a tape measure running up from the mason's strings along with the laser level receiver; find the laser level beam on the receiver and check that it shoots each corner at the same measurement on the tape measure. Adjust strings until the measurements in all four corners are equal.

Measure diagonally from corner to corner of the building outline to confirm square. After adjusting all measurements, confirm the measurements one last time by repeating steps 5 and 6. Setting up a properly leveled and square foundation is key to the success of the building. Unload the 2 × 6 skirt boards and place them around the perimeter of the building.

Mark wall post locations. Start by marking all corner posts on the ground with marking paint at the intersection of the mason's strings. Measure out with a tape measure from the first corner and mark post location. Continue marking post locations around the perimeter according to FLOOR PLAN. After post locations are marked, double-check corners for square and check that post locations on the eave sides are marked directly across from one another for proper alignment. Adjust and remark as needed.

Dig post footings. In each corner, measure the distance from the mason's strings to the ground. Add the required post hole depth: here, 54" (48" for the post footing, plus 6" for the concrete footing bed) to the smallest of these measurements. Use a rented power skid steer with a 22"-dia. auger dig-in attachment to dig holes for the post footings to this depth.

9

Tamp post holes with a hand tamper or make your own tamper by attaching scrap lumber to the end of a long 2 × 4, as shown. Tamp hole bottoms, measure for depth, and check footing base for level multiple times. When the footing base is solid and level and the hole measures the correct depth, insert a precast, circular concrete footing bed into the bottom of each hole (inset).

Tip: Working Around Rocks ▸

As you dig with an auger, you may encounter large stones, roots, compacted clay, or other hard ground material. When you confront this, it may be necessary to use a maul or jack hammer to break up the material and then shovel out the loosened soil and/or rocks.

10

Center 12"-long 2 × 4 uplift blocks on two opposing sides of the treated ends of the posts, flush with the bottom ends. Attach with two to three 16d galvanized common nails. After backfilling footings, these blocks will help to anchor the post in the ground and protect it from shifting due to weather or other environmental changes. Cut the posts to the correct length.

11

Drop all posts into holes. Drop the chemically treated end of each post into the footing. Rest the post in the footing so the end leans in toward the building's interior. Double-check post location measurements and make sure mason's strings are level before proceeding to step 12. This is also a good time to unload purlins, wall girts, and temporary bracing lumber around your work site.

(continued)

Raise the first corner post. Set the wall posts so the 4½"-side faces the exterior of the building (the corner posts should be positioned so the 4½"-side faces the eave wall). Use a 4-ft. level to check for plumb.

Backfill around the plumb post with the soil you removed to dig the footings. As you hold the post, maintaining plumb, have one or two helpers backfill and tamp soil firmly. When set, measure to the next post to ensure accurate placement. Continue around the structure and lean-to until all posts have been plumbed and backfilled.

Mark level reference lines. Set a laser level on a tripod inside the building footprint. Set the level's height and shoot one post at a time; use a combination square to make a mark on each post. Be careful not to jostle the level or change its height or placement at all during this process. Then, measure down on each post from this mark to the ground and record the measurement. The smallest measurement will determine the placement of the skirt board around the structure.

Attach the skirt board. Mark the measurement from step 14 onto the exterior face of the two adjacent corner posts and pound a nail about halfway in at the mark. Connect the nails on the outside of the posts with mason's strings (as shown). Pound a nail halfway into the exterior face of each wall post between the corners at the string's location. Align the first 2 × 6 skirt board with its bottom edge flush to the placement nails and attach to posts with two 4" steel pole barn nails. Continue installing skirt boards to finish the wall. *Note: On eave walls, position the board so its end extends beyond each corner post 1½". On gable walls, cut the skirt board to fit flush to the edge of the post, creating a 1½" corner.*

Attach the bottom 16-ft. 2 × 4 wall girt nailer directly above the skirt board. Attach to each post with two 4" steel pole barn nails.

Attach the four remaining wall girt nailers every 24" on-center up the posts. Each wall girt will create a 1½" corner, like the skirt board. Repeat steps 15 to 17 to install skirt boards and wall girts on the opposite eave wall. Attach a temporary wall girt to the outside of the lean-to posts to keep them square during construction.

Attach the ribbon board to the eave sides. Hook your measuring tape to the bottom of the skirt board and measure up to 138". Mark this location on all eave-side posts. Align a 2 × 4 ribbon board with its bottom edge flush to this mark and attach to each post with two 4" steel pole barn nails. *Note: Always use ladders and safety equipment. The worker pictured here is a professional.*

Install 2 × 4 corner bracing between the corner post and the adjacent wall post, from the bottom of one post to the ribbon of the next. Nail bracing to wall girts.

(continued)

Notch the top of each eave wall post to support the trusses. Cut notches to the top of the ribbon. Then, measure up from the ribbon board to the height of the truss heel and make a mark. Starting at this mark, cut the top of the post to match the truss's $\frac{4}{12}$ pitch.

Prepare trusses for installation. Unload preassembled trusses designed to meet your building's specifications and approved by an engineer for use with your building. Before installation, mark the location of the first purlins on each side of the peak at 8½".

Raise the first truss. Place the first truss on the outside of the gable-wall posts, with the bottom edge of the bottom chord flush with the top edge of the ribbon board on the eave sides. Check for level and nail the truss to all posts with four or five 4" steel pole barn nails at each post location. *Option: Equip a skid steer with a lift to help raise each truss to the top of the posts (below), where one person on each side of the building should be waiting to guide it into place.*

Raise the second truss to rest in the notches of the second set of posts. Check for level and nail to the remaining plies of the posts with four or five 4" steel pole barn nails. The notched end of the posts should extend up on one side of the truss.

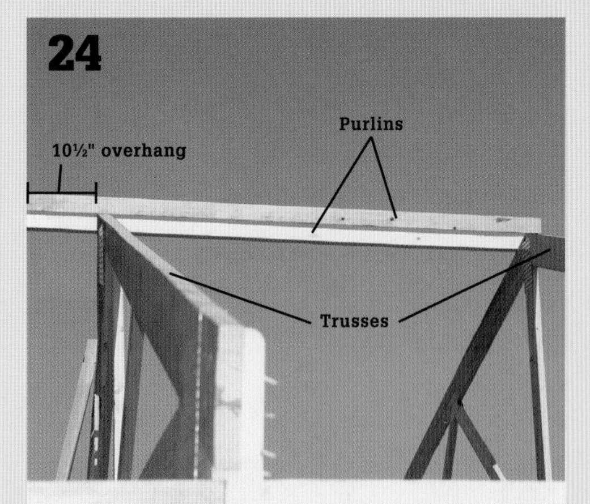

Tie the trusses together with purlins. Before you disconnect the second truss from the lift, position the first 2 × 4 purlin on-edge with its bottom edge flush to the mark made in step 21 and its end extended 10½" past the gable side of the truss. Nail the purlin to the truss with one 6" steel pole barn nail; repeat with a second purlin on the opposite side of the peak. Attach the first two purlins to the second truss in the same manner and disconnect the second truss from the lift.

Install the remaining trusses, repeating steps 23 and 24 for each truss. A section of three trusses makes up a bay. After each bay is completed, slide a new purlin down the length of the previous one for the next bay. Although they do not need to be fastened together, overlap the purlins approximately 12" between bays. When completed, the purlins will appear to be staggered slightly because of this overlap.

Trim posts on the gable end to follow the 4/12 pitch of the trusses using a chainsaw. Install skirt boards, wall girts, and 2 × 4 corner bracing on the gable walls at this time (see steps 15 to 17 and 19). If a door or window opening impedes on the corner space for the bracing, attach a brace in the next available opening between posts.

(continued)

Attach the remaining purlins on both sides of the peak, spaced every 24" on-center down the length of the roof. All purlins should extend 10½" past both gable ends and should be installed to cover a bay. This building has one extra truss-to-truss space not included in a bay; select a space after a full bay and use 9-ft. purlins for this space, overlapping the purlins on either side by 6".

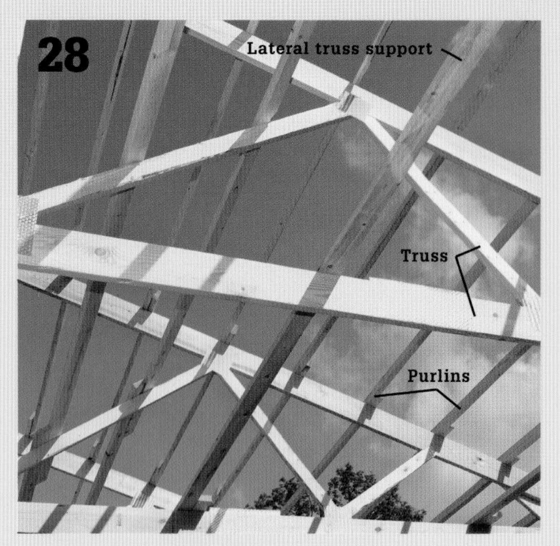

Install lateral truss support according to your truss design's specification sheet. The trusses used in this project required two 2 × 4 braces on the webbing and four 2 × 4 braces running on top of the bottom chord (pictured here, see LATERAL TRUSS DETAIL). Cut 2 × 4 bracing boards to 18-ft. and overlap them about 12" after each bay. Attach to trusses with 4" steel pole barn nails.

Install 12-ft.-long 2 × 6 knee bracing to each interior truss/post to keep the walls straight during roofing material installation. Cut a 45° angle into each end of the brace and install so its length is split equally between the truss and the inside of the building; attach the top 6 ft. to the truss, and attach the bottom of the brace to the corresponding interior post with 4" steel pole barn nails. Repeat on the other side.

Install the ledger board for the lean-to rafters. Cut a 2 × 10 and attach it flush with the top chords of the trusses. Nail to the ends of the trusses with four 4" steel pole barn nails at each location. Notch 1½" from the top of the exterior lean-to posts as in step 20 to a depth of 8".

Fasten the lean-to rafters to the ledger board using joist hangers. Use a framing square to mark one end of each of the four 2 × 10 rafters to match the roof's 4⁄12 pitch. Cut this angle. Attach the angled end to the ledger using joist hangers and joist hanger nails, and attach the outer end to the notched portion of the lean-to post with 4" steel pole barn nails. Cut the end of the rafter at the same 4⁄12 pitch 1½" past the outer edge of the post.

Attach a 90° return board to the outside of the posts at both gable ends of the lean-to roof, even with the top wall girt on the building. Nail the return boards to the posts with 4" steel pole barn nails. Cut the return boards to be flush with the outside of the posts. Install a 2 × 4 ribbon board on the eave side of the lean-to flush with the 90° return boards with two 4" steel pole barn nails.

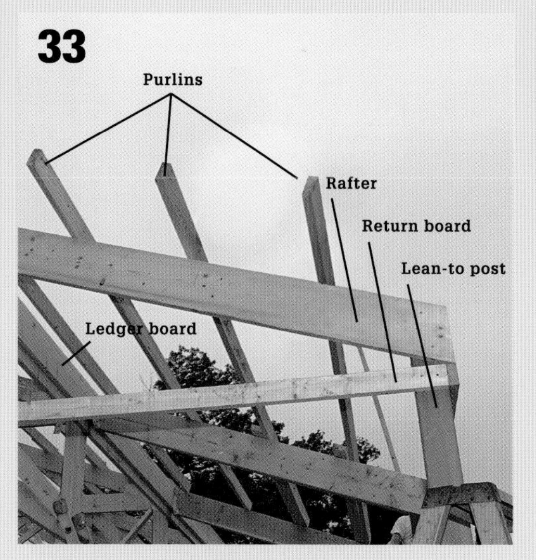

Install purlins on the lean-to rafters, staying consistent with the purlin spacing on the roof. Position the first lean-to roof purlin 24" on-center from the last purlin on the building roof, and attach it with one 6" steel pole barn nail at each rafter. Continue until all purlins on the lean-to roof are installed.

Cut openings for service doors, windows, and a garage door according to these components' specifications. Install service doors and windows as directed. The garage door will be installed later. Frame the openings.

(continued)

Cut and attach the first 2 × 4 tail to a gable wall truss to create the boxed overhang. Use a framing square to measure and cut one end of a 2 × 4 to match the roof's 4⁄12 pitch. Set the tail so the angled end is 10½" from the wall; a few feet of 2 × 4 will extend along the truss chords. Attach the tail to the truss with five 16d common nails, spaced about 10" apart. Next, install the tail on the opposite gable wall truss on the other end of the wall in the same manner.

To install the tails on the rest of the wall, pound a 16d common nail halfway into the end of both the first and last tails and connect the nails with a mason's string spanning from one end of the building to the other. Install the remaining tails flush to this string. Install tails on the opposite side of the building in this same manner.

Install 2 × 4 spacers between the rafters on the eave edge of the lean-to roof (these are nailers for subfascia). Mark a consistent spacer location on each rafter end and then measure and cut 2 × 4s to fit between the rafters. Attach the spacers to the rafters and posts with 4" pole barn nails. Before you install the subfascia, install temporary 2 × 4 bracing for each truss; wedge one end where the truss meets the post and rest the other on the ground to create a firm brace.

Attach 2 × 8 subfascia to the outside of the truss tails with 4" steel pole barn nails. Attach the first piece on the eave side with its bottom edge flush to the bottom edge of the tail, flush with the gabled end. Measure and cut the subfascia as you go. Cut and position the subfascia so that each board covers only half the tail face and the next piece can be easily attached.

Install 2 × 2 spacers on top of the ribbon between each truss. Measure each opening and cut the 2 × 2 to fit, then fasten it to the top of the ribbon with pole barn nails.

Install the soffit starter. This two-in-one trim has a J-channel side to cover the top of the wall and a C-channel side to receive the soffit. Mark the location of the soffit starter by marking a level line on the framing members of the building, leveling over from the bottom of the subfascia. Attach to the ribbons and 2 × 2 spacers with stainless steel screws on all four walls and on the front side of the lean-to (the back gable and eave side of the lean-to do not have soffits).

Install base trim (left photo) to the 2 × 6 skirt board, with its bottom edge flush with the joint between the skirt board and the first 2 × 4 wall girt. Install door jamb trim (right photo) and J-channel around the garage door opening and install J-channel around the service door. Install all doors and windows.

(continued)

Install the steel siding. Steel wall siding is made up of precut 36"-wide panels with 9"-on-center ribs. Attach the siding flush to the first corner, and hold it to the wall girts temporarily with two or three 1½" gasket washer bell cap screws per panel. Overlap the first rib of the next sheet with the last rib of the previous sheet and tack it in place. Cut steel panels to fit around door and window openings with a tin snips or a nibbler (right photo). Remember to side the gable ends of the lean-to roof as well. Finish installing siding around the building.

Install soffit panels and eave molding trim. Fit soffit panels into the C-channel and attach the other end to the bottom of the subfascia (left photo). Install steel eave molding trim to cover the top of the subfascia and slip it under the roof seal on the eave sides of the roof using galvanized trim nails (right photo).

Install the steel roof. Order steel to the correct dimensions for your project; for this building, steel panels are 36" wide. Begin on one gable side, and attach steel roofing to each purlin in the flats with 1½" gasket washer bell cap screws. Position the first piece so you have an overhang of between 2 to 3". Keep this overhang consistent for the remainder of the roofing installation. Overlap the ridge of each panel over the previous panel as you move down the roof. On the side of the roof with the lean-to, start installing the roofing on the lean-to roof, and then move up to the building roof, overlapping panels as you go. When you get to the end, cut the last piece of roofing to fit.

Install the ridge cap. First, install the closure between the roof steel and the ridge cap, composite material that allows for excellent airflow, but prevents dust and debris from entering the building. One side of this porous, thick material will be pre-glued; its profile will match the steel roofing material. Set the ridge cap on top of the composite material and attach with 2" stainless steel screws at each purlin.

Complete the siding installation by adding the remaining screws. Use a chalk line to mark a long line onto the siding at each wall girt location from one end of the building to the other. Use this chalk line to drive screws into the 9"-on-center flats in the siding, one at each wall girt location.

Attach finishing trim. Install corner trim on all corners with 2" stainless steel screws. Attach fascia to cover the edge of the soffit and the remaining exposed subfascia with 2" stainless steel screws. Remove temporary bracing. Fill the gap beneath the skirt board around the building with excess soil, if desired. *Option: Finish this building with a concrete floor. Unless you are experienced working with concrete, call in the professionals to finish a floor of this size.*

Mini Gambrel Barn

This 12 × 12-foot mini barn has several features that make it a versatile space, perfect for storing equipment or setting up a workshop. The barn's 144-square-foot floor is a poured concrete slab with a thickened edge that allows it to serve as the building's foundation. Designed for economy and durability, the floor can easily support heavy machinery, woodworking tools, and recreational vehicles.

The barn's sectional overhead door makes for quick access to equipment and supplies and provides plenty of air and natural light for working inside. The door opening is sized for an eight-foot-wide by seven-foot-tall door, but you can buy any size or style of door you like—just make your door selection before you start framing the barn.

Another important design feature of this building is its gambrel roof, which maximizes the usable interior space (see The Gambrel Roof, next page). Beneath the roof is a sizeable storage attic with 315 cubic feet of space and its own double doors above the garage door. *Note: We added a patio section to the front of this barn. This optional slab appears throughout the how-to photos.*

The gambrel roof creates space for lofted storage behind these classic hayloft doors.

Inside, the mini barn has ample space for equipment, tools, or a workshop.

The Gambrel Roof ▸

The gambrel roof is the defining feature of two structures in American architecture: the barn and the Dutch Colonial house. Adopted from earlier English buildings, the gambrel style became popular in America during the early 17th century and was used on homes and farm buildings throughout the Atlantic region. Today, the gambrel roof remains a favorite detail for designers of sheds, garages, and carriage houses.

The basic gambrel shape has two flat planes on each side, with the lower plane sloped much more steeply than the upper. More elaborate versions incorporate a flared eave, known as a "Dutch kick," that was often extended to shelter the front and rear facades of the building. Barns typically feature an extended peak at the front, sheltering the doors of the hayloft. The main advantage of the gambrel roof is the increased space underneath the roof, providing additional headroom for upper floors in homes or extra storage space in outbuildings.

Cutting List

Description	Quantity/Size	Material
Foundation		
Drainage material	1.75 cu. yds.	Compactable gravel
Concrete slab	2.5 cu. yds.	3,000 psi concrete
Mesh	144 sq. ft.	6 × 6", W1.4 × W1.4 welded wire mesh
Wall Framing		
Bottom plates	4 @ 12'	2 × 4 pressure-treated
Top plates	8 @ 12'	2 × 4
Studs	47 @ 92⅝"	2 × 4
Headers	2 @ 10', 2 @ 6'	2 × 8
Header spacers	1 @ 9', 1 @ 6'	½" plywood — 7" wide
Angle braces	1 @ 4'	2 × 4
Gable Wall Framing		
Plates	2 @ 10'	2 × 4
Studs	7 @ 10'	2 × 4
Header	2 @ 6'	2 × 6
Header spacer	1 @ 5'	½" plywood — 5" wide
Attic Floor		
Joists	10 @ 12'	2 × 6
Floor sheathing	3 sheets @ 4 × 8'	¾" tongue-&-groove ext.-grade plywood
Kneewall Framing		
Bottom plates	2 @ 12'	2 × 4
Top plates	4 @ 12'	2 × 4
Studs	8 @ 10'	2 × 4
Nailers	2 @ 14'	2 × 8
Roof Framing		
Rafters	28 @ 10'	2 × 4
Metal framing connectors — rafters	20, with nails	Simpson H2.5
Collar ties	2 @ 6'	2 × 4
Ridge board	1 @ 14'	2 × 6
Lookouts	1 @ 10'	2 × 4
Soffit ledgers	2 @ 14'	2 × 4
Soffit blocking	6 @ 8'	2 × 4
Exterior Finishes		
Plywood siding	14 sheets @ 4 × 8'	⅝" texture 1-11 plywood, grooves 8" O. C.
Z-flashing — siding	2 pieces @ 12'	Galvanized 18-gauge
Horizontal wall trim	2 @ 12'	1 × 4 cedar
Corner trim	8 @ 8'	1 × 4 cedar
Fascia	6 @ 10', 2 @ 8'	1 × 6 cedar
Subfascia	4 @ 8'	1 × 4 pine
Plywood soffits	1 sheet @ 10'	⅜" cedar or fir plywood
Soffit vents	4 @ 4 × 12"	Louver w/ bug screen
Z-flashing — garage door	1 @ 10'	Galvanized 18-gauge

Description	Quantity/Size	Material
Roofing		
Roof sheathing	12 sheets @ 4 × 8'	½" plywood
Shingles	3 squares	250# per square (min.)
15# building paper	300 sq. ft.	
Metal drip edge	2 @ 14', 2 @ 12'	Galvanized metal
Roof vents (optional)	2 units	
Window		
Frame	3 @ 6'	¾ × 4" (actual) S4S cedar
Stops	4 @ 8'	1 × 2 S4S cedar
Glazing tape	30 linear ft.	
Glass	1 piece — field measure	¼" clear, tempered
Exterior trim	3 @ 6'	1 × 4 S4S cedar
Interior trim (optional)	3 @ 6'	1 × 2 S4S cedar
Door		
Frame	2 @ 8'	1 × 6 S4S cedar
Doorsill	1 @ 6'	1 × 6 S4S cedar
Stops	1 @ 8', 1 @ 6'	1 × 2 S4S cedar
Panel material	4 @ 8'	1 × 8 T&G V-joint S4S cedar
Door X-brace/panel trim	4 @ 6', 2 @ 8'	1 × 4 S4S cedar
Exterior trim	1 @ 8', 1 @ 6'	1 × 4 S4S cedar
Interior trim (optional)	1 @ 8', 1 @ 6'	1 × 2 S4S cedar
Strap hinges	4	
Garage Door		
Frame	3 @ 8'	1 × 8 S4S cedar
Door	1 @ 8' × 6' - 8"	Sectional flush door w/2" track
Rails	2 @ 8'	2 × 6
Trim	3 @ 8'	1 × 4 S4S cedar
Fasteners		
Anchor bolts	16	⅜" × 8", with washers & nuts, galvanized
16d galvanized common nails	2 lbs.	
16d common nails	17 lbs.	
10d common nails	2 lbs.	
10d galvanized casing nails	1 lb.	
8d common nails	3 lbs.	
8d galvanized finish nails	6 lbs.	
8d box nails	6 lbs.	
6d galvanized finish nails	20 nails	
3d galvanized box nails	½ lb.	
⅞" galvanized roofing nails	2½ lbs.	
2½" deck screws	24 screws	
1¼" wood screws	48 screws	
Construction adhesive	2 tubes	
Silicone-latex caulk	2 tubes	

Building Section

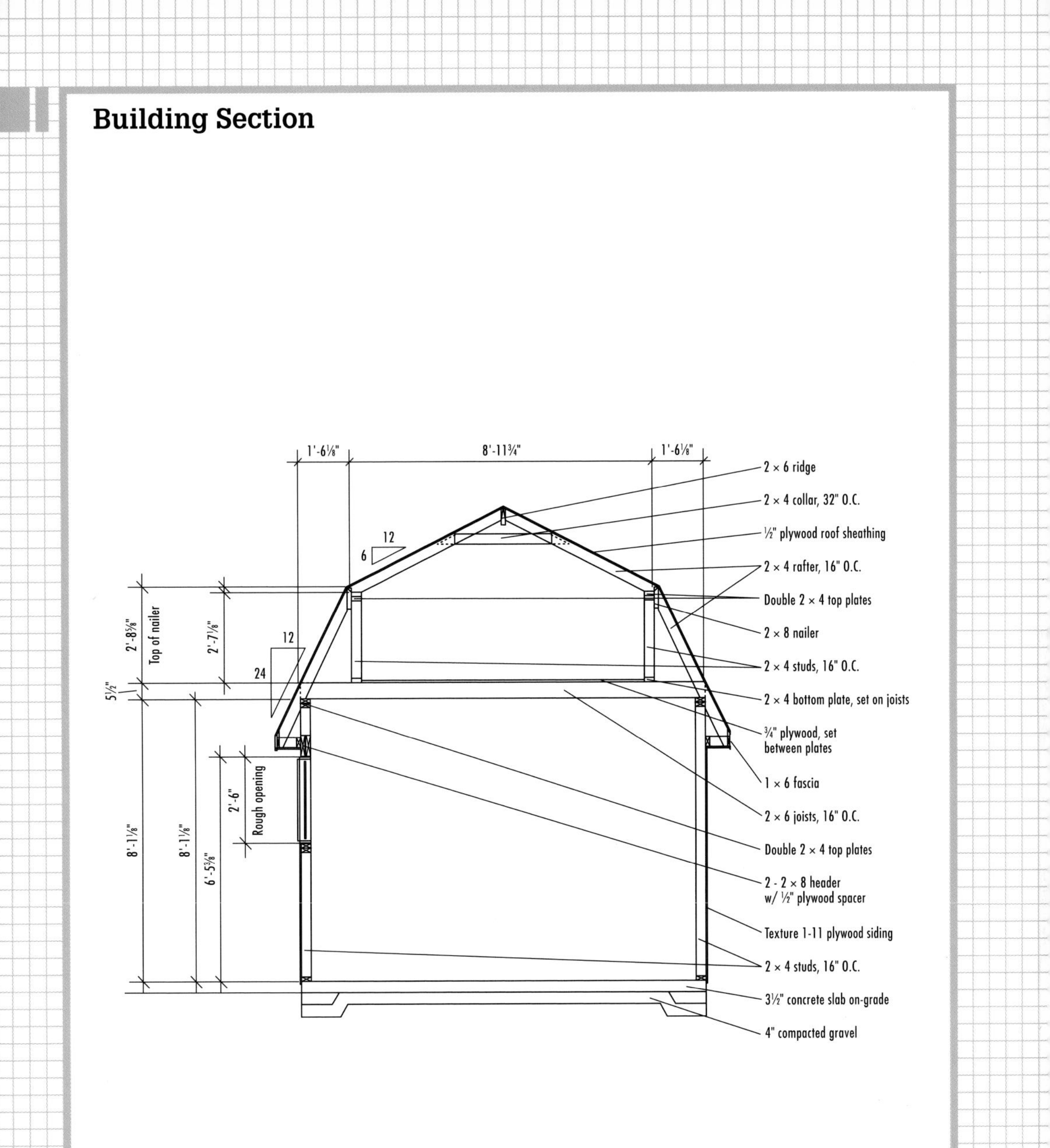

Floor Plan

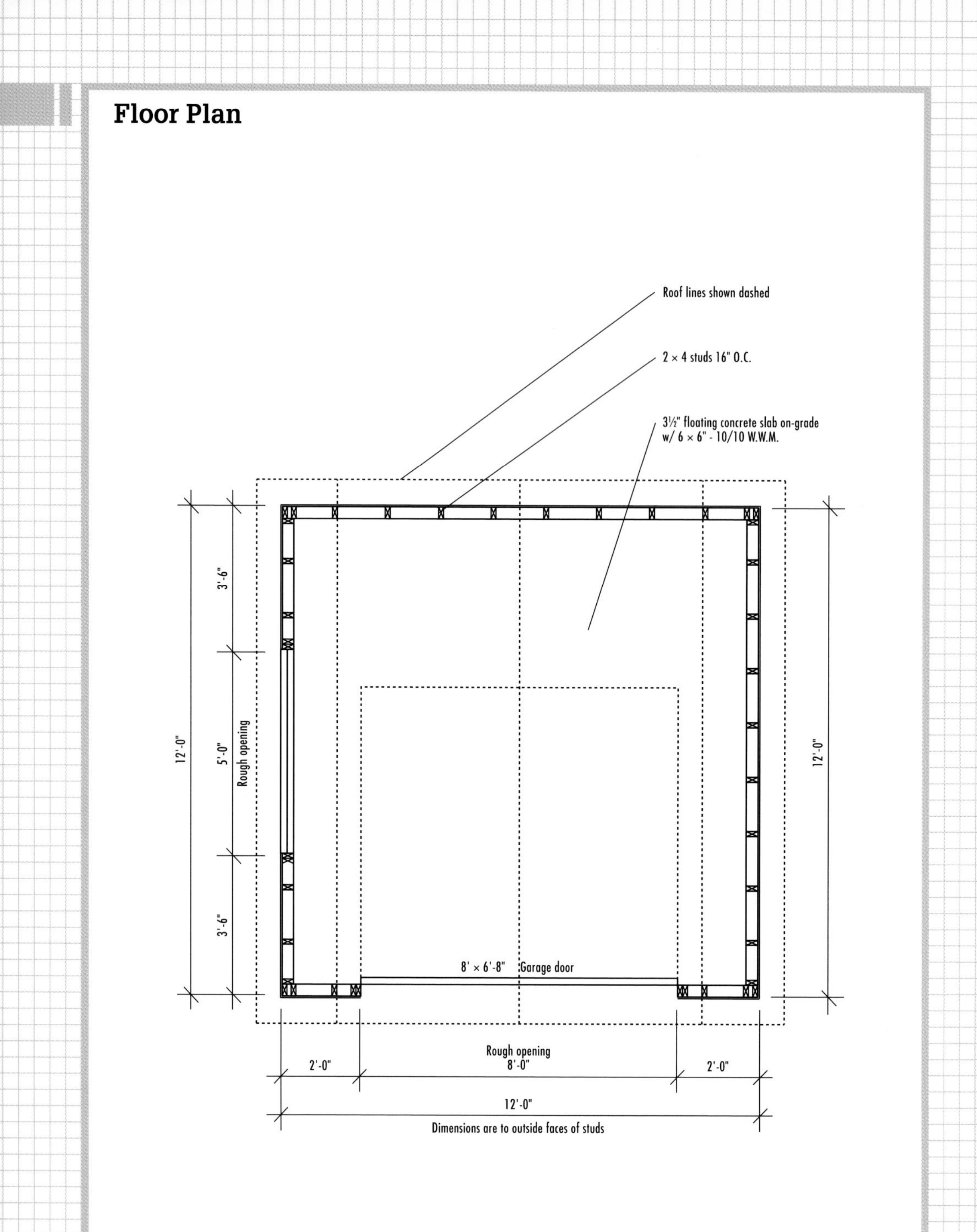

Roof lines shown dashed
2 × 4 studs 16" O.C.
3½" floating concrete slab on-grade w/ 6 × 6" - 10/10 W.W.M.
3'-6"
5'-0"
Rough opening
3'-6"
12'-0"
12'-0"
8' × 6'-8" Garage door
2'-0"
Rough opening 8'-0"
2'-0"
12'-0"
Dimensions are to outside faces of studs

Rafter Templates

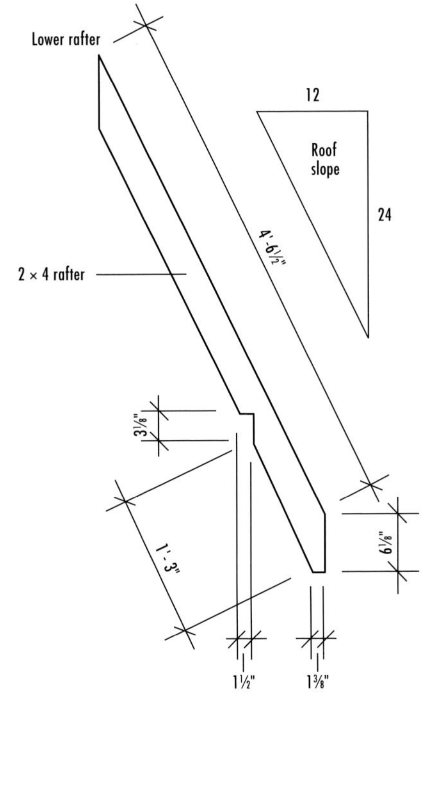

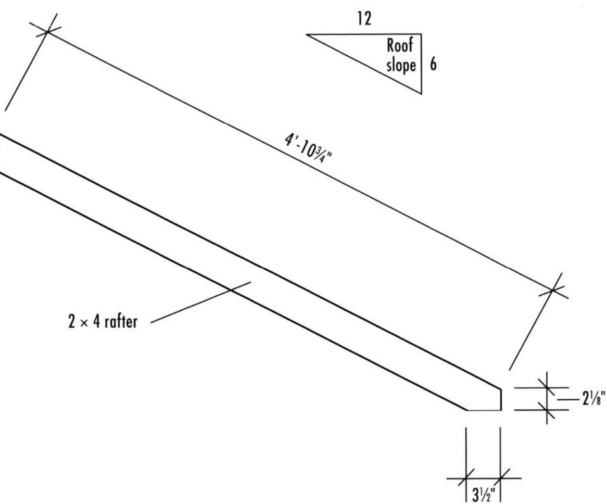

Front Elevation

12
6
12
24
Double door - see detail
1 × 4 trim
1 × 6 fascia
Pork chop
Flashing
1 × 4 trim, mitered corners
Texture 1-11 plywood siding
Flush overhead garage door
1 × 4 trim

Left Side Elevation

Roof vent
Asphalt shingles
1 × 6 fascia
1 × 4 trim, mitered corners
Texture 1-11 plywood siding
1 × 4 trim
Window with ¼" clear tempered glass

Rear Elevation

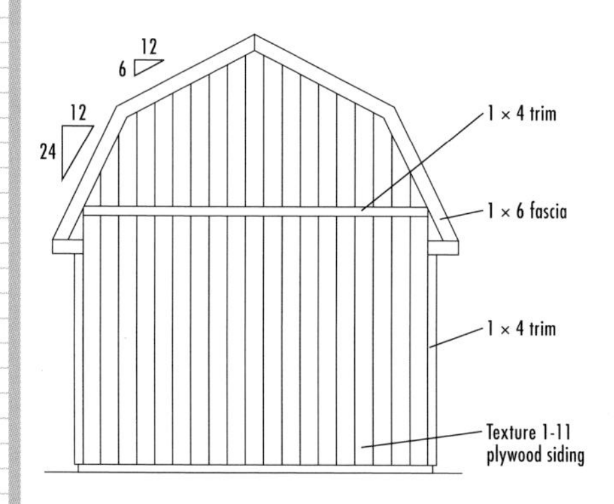

Right Side Elevation

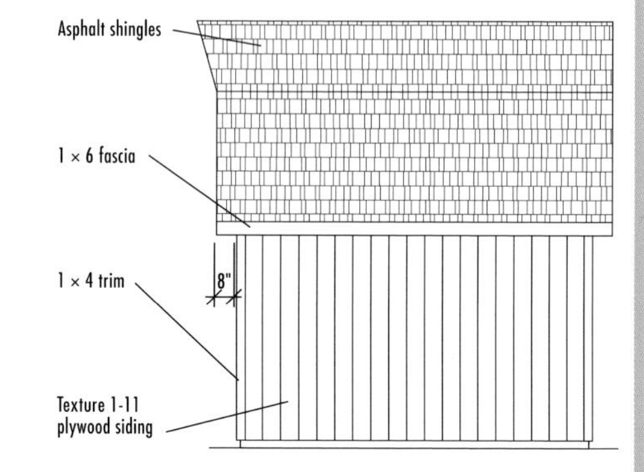

Gable Overhang Detail

Gable Overhang Rafter Details

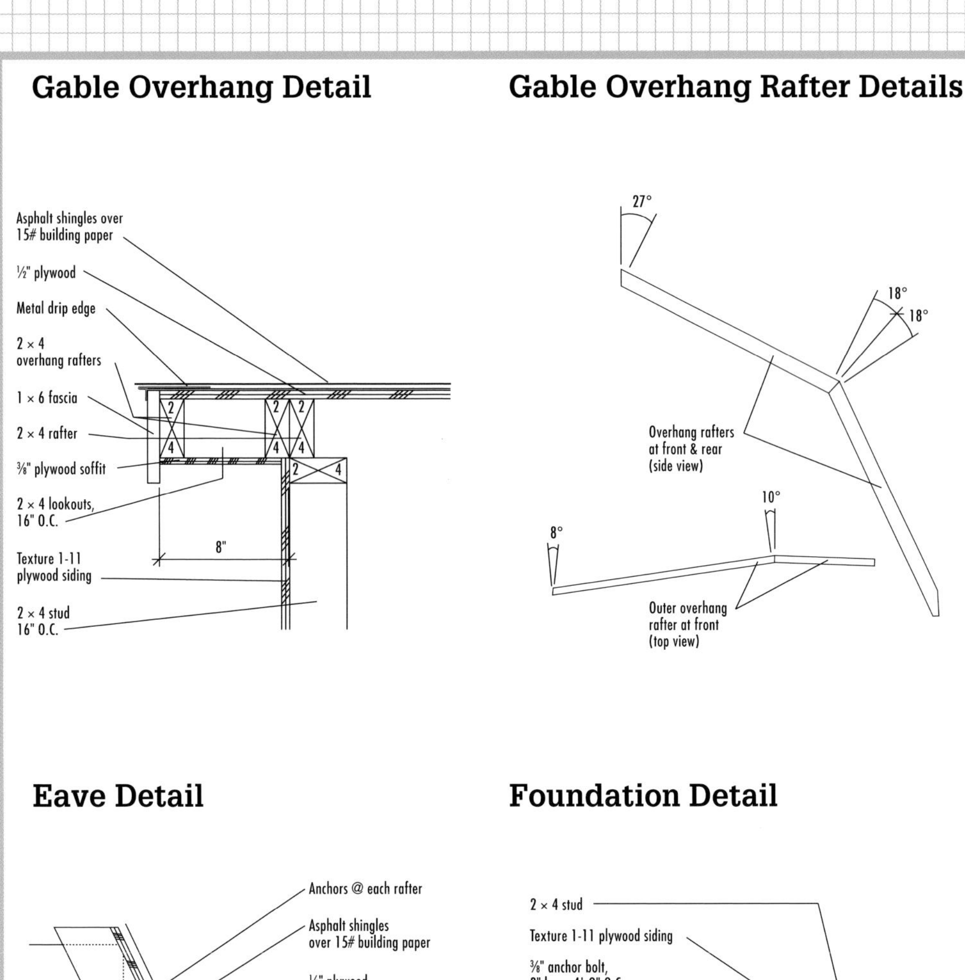

Eave Detail

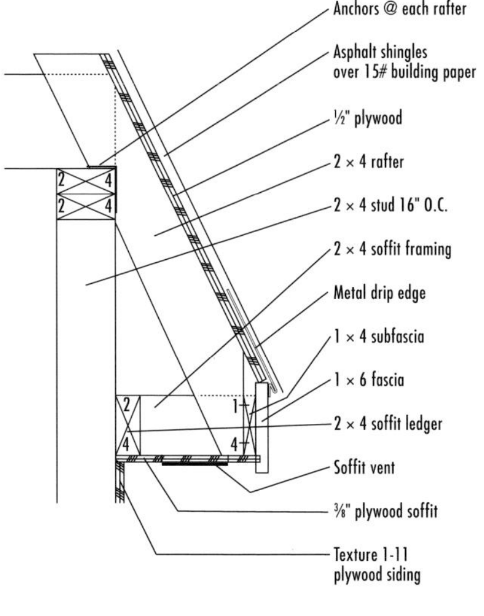

Foundation Detail

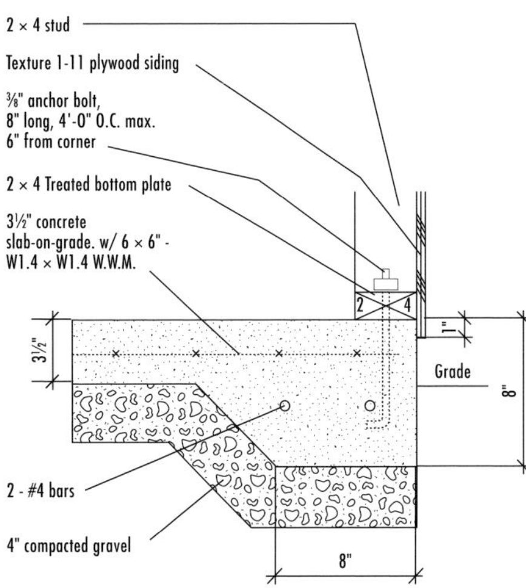

Attic Door Elevation

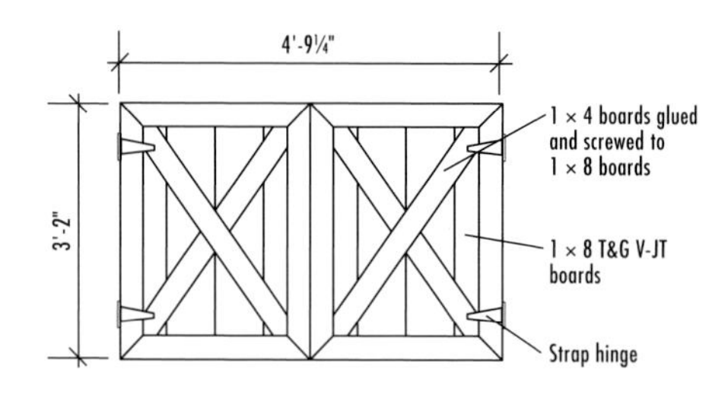

Attic Door Jamb Detail

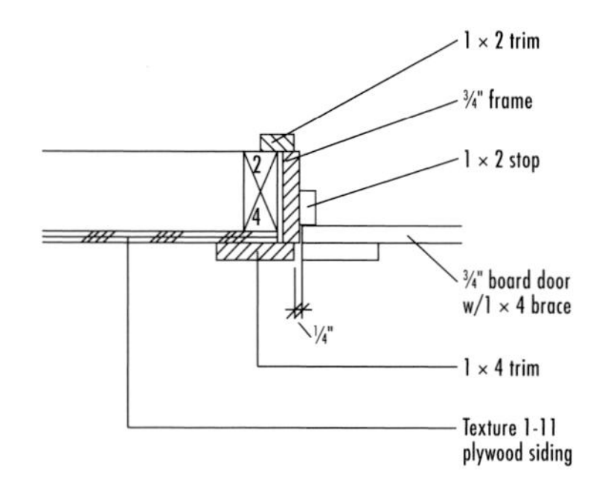

Garage Door Trim Detail

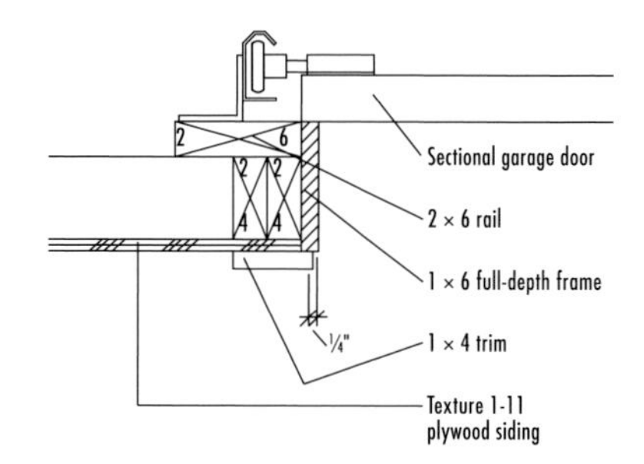

Attic Door Sill Detail

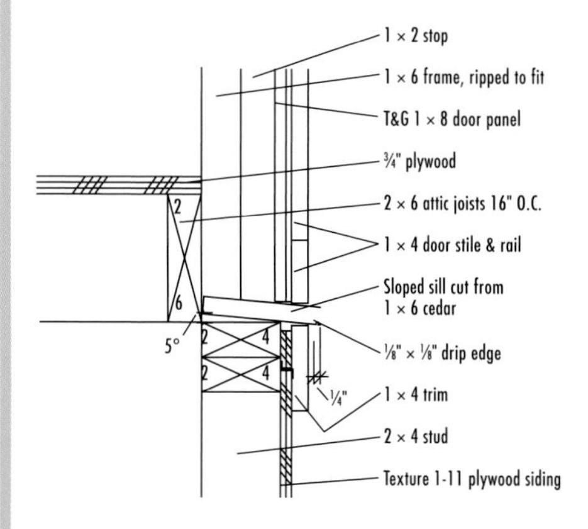

Window Jamb Detail

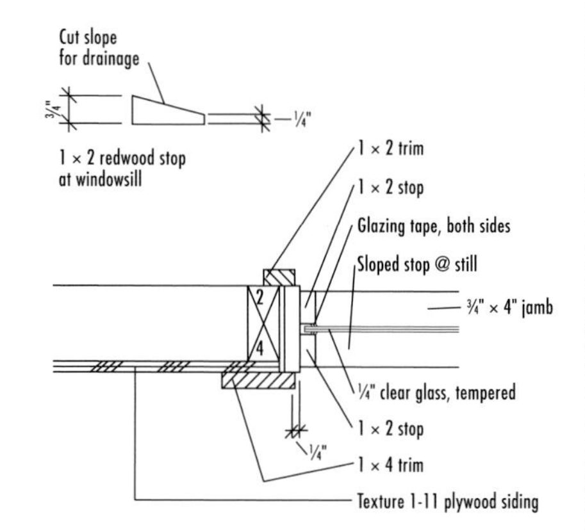

Front Framing Elevation

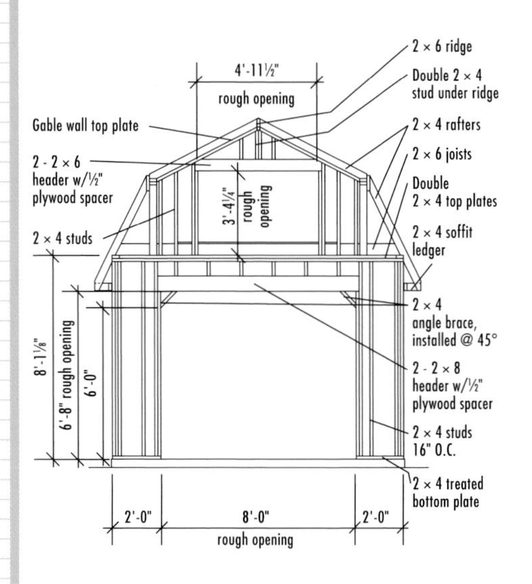

Left Side Framing Elevation

2 × 6 ridge

2 × 4 collar tie - 32" O.C.

2 × 8 nailer

2 × 4 rafters

2 × 6 joists

Double 2 × 4 top plates

2 × 4 soffit ledger

2 - 2 × 8 header w/½" plywood spacer

Double 2 × 4 sill

2 × 4 studs 16" O.C.

2 × 4 treated bottom plate

8"

8"

2'-6" rough opening

3'-11⅜"

Rear Framing Elevation

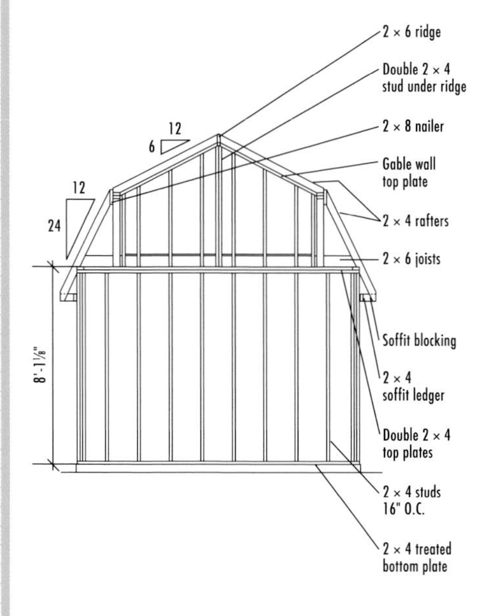

Right Side Framing Elevation

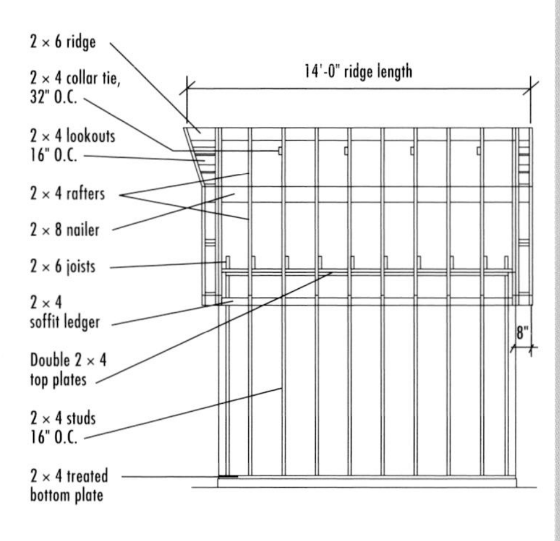

How to Build the Mini Gambrel Barn

Build the slab foundation at 144" × 144", following the basic procedure outlined on pages 28 to 33. Set J-bolts into the concrete 1¾" from the outer edges and extending 2½" from the surface. Set a bolt 6" from each corner and every 48" in between (except in the door opening). Let the slab cure for at least three days before you begin construction. *Note: Add the optional slab now, as desired.*

Snap chalk lines on the slab for the wall plates. Cut two bottom plates and two top plates at 137" for the side walls. Cut two bottom and two top plates at 144" for the front and rear walls. Use pressure-treated lumber for all bottom plates. Cut 38 studs at 92⅝", plus two jack studs for the garage door at 78½" and two window studs at 75⅞".

Construct the built-up 2 × 8 headers at 99" (garage door) and 63" (window). Frame, install, and brace the walls with double top plates one at a time, following the FLOOR PLAN (page 118) and FRAMING ELEVATION drawings (page 123). Use galvanized nails to attach the studs to the bottom plates. Anchor the walls to the J-bolts in the slab with galvanized washers and nuts.

Frame the attic floor. Cut ten 2 × 6 joists to 144" long, then clip each top corner with a 1½"-long, 45° cut. Install the joists as shown in the FRAMING ELEVATIONS, leaving a 3½" space at the front and rear walls for the gable wall studs. Fasten the joists with three 8d nails at each end.

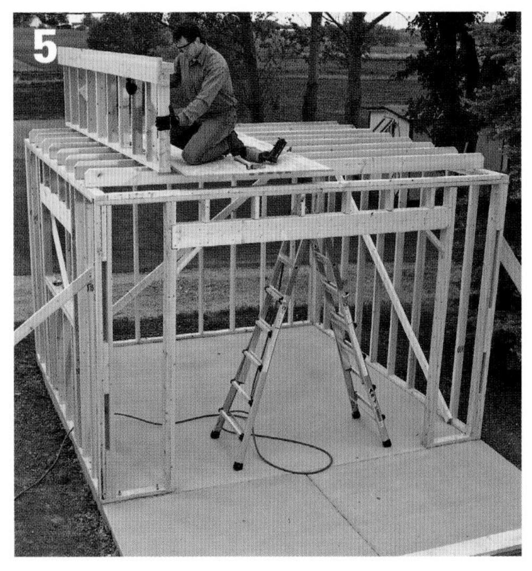

Frame the attic kneewalls: Cut four top plates at 144" and two bottom plates at 137". Cut 20 studs at $26\frac{5}{8}$" and four end studs at $33\frac{5}{8}$". Lay out the plates so the studs fall over the attic joists. Frame the walls and install them $18\frac{1}{8}$" from the ends of the joists, then add temporary bracing. *Option: You can begin building the roof frame by cutting two 2 × 8 nailers to 144" long. Fasten the nailers to the kneewalls so their top edges are $32\frac{5}{8}$" above the attic joists.*

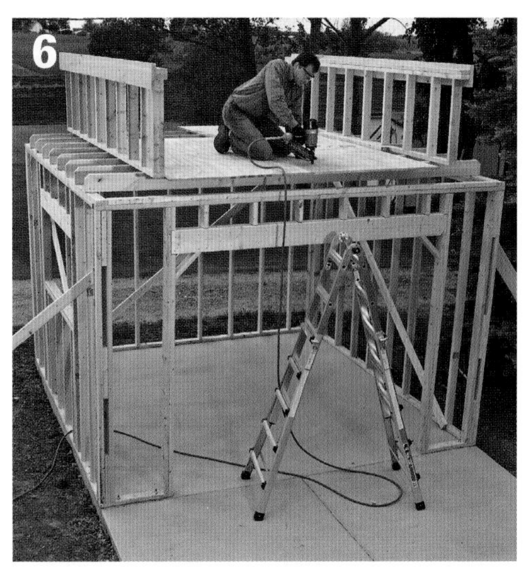

Cover the attic floor between the kneewalls with $\frac{3}{4}$" plywood. Run the sheets perpendicular to the joists, and stop them flush with the outer joists. Fasten the flooring with 8d box or ring-shank nails every 6" along the edges and every 12" in the field of the sheets.

Mark the rafter layouts onto the top and outside faces of the 2 × 8 nailers; see the FRAMING ELEVATIONS.

Cut the 2 × 6 ridge board at 168", mitering the front end at 16°. Mark the rafter layout onto the ridge. The outer common rafters should be 16" from the front end and 8" from the rear end of the ridge.

(continued)

Use the RAFTER TEMPLATES (page 119) to mark and cut two upper pattern rafters and one lower pattern rafter. Use the patterns to mark the remaining common rafters. For the gable overhangs, cut an additional eight lower and six upper rafters following the GABLE OVERHANG RAFTER DETAILS (page 121).

Install the common rafters. Nail the upper rafters to the ridge and kneewalls. Toenail the lower rafters to the nailers and wall plates. Reinforce the wall-plate connection with metal framing connectors. Nail the attic joists to the sides of the lower rafters. Cut four 2 × 4 collar ties at 34", mitering the ends at 26.5°. Fasten them between pairs of upper rafters, as shown in the BUILDING SECTION (page 117) and FRAMING ELEVATIONS (page 123).

Snap a chalk line across the sidewall studs, level with the ends of the rafters. Cut two 2 × 4 soffit ledgers at 160" and fasten them to the studs on top of the chalk lines, with their ends overhanging the walls by 8". Cut 24 2 × 4 blocks to fit between the ledger and rafter ends, as shown in the EAVE DETAIL (page 121). Install the blocks.

Frame the gable overhangs. Cut 12 2 × 4 lookouts at 5" and nail them to the inner overhang rafters as shown in the LEFT and RIGHT SIDE FRAMING ELEVATIONS. Install the inner overhang rafters over the common rafters, using 10d nails. Cut the two front (angled) overhang rafters; see the GABLE OVERHANG RAFTER DETAILS. Install those rafters; then add two custom-cut lookouts for each rafter.

To complete the gable walls, cut top plates to fit between the ridge and the attic kneewalls. Install the plates flush with the outer common rafters. Mark the stud layout onto the walls and gable top plate; see the FRONT and REAR FRAMING ELEVATIONS. Cut the gable studs to fit and install them. Construct the built-up 2 × 6 attic door header at 62½"; then clip the top corners to match the roof slope. Install the header with jack studs cut at 40¼".

Install siding on the walls, holding it 1" below the top of the concrete slab. Add Z-flashing along the top edges, and then continue the siding up to the rafters. Below the attic door opening, stop the siding about ¼" below the top wall plate, as shown in the ATTIC DOOR SILL DETAIL (page 122). Don't nail the siding to the garage door header until the flashing is installed (Step 20).

Mill ⅜"-wide × ¼"-deep grooves into the 1 × 6 boards for the horizontal fascia along the eaves and gable ends (about 36 linear ft.); see the EAVE DETAIL. Use a router or table saw with a dado-head blade to mill the groove, and make the groove ⅞" above the bottom edge of the fascia.

Install the 1 × 4 subfascia along the eaves, keeping the bottom edge flush with the ends of the rafters and the ends flush with the outsides of the outermost rafters; see the EAVE DETAIL. Add the milled fascia at the eaves, aligning the top of the groove with the bottom of the subfascia. Cut fascia to wrap around the overhangs at the gable ends but don't install them yet.

(continued)

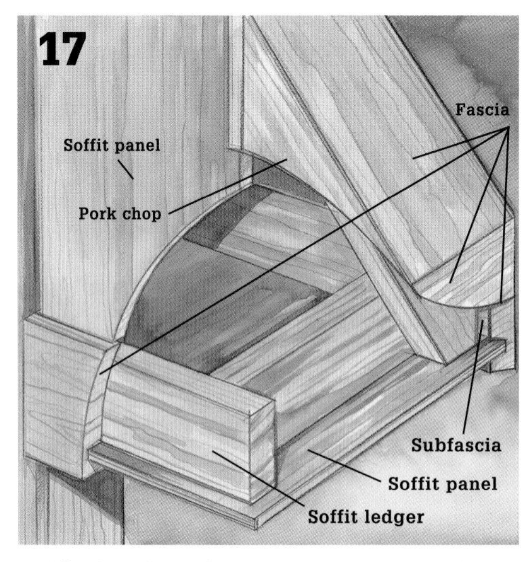

Add fascia at the gable ends, holding it up ½" to be flush with the roof sheathing. Cut soffit panels to fit between the fascia and walls, and fasten them with 3d galvanized nails. Install the end and return fascia pieces at the gable overhangs. Enclose each overhang at the corners with a triangular piece of grooved fascia (called a pork chop) and a piece of soffit material. Install the soffit vents as shown in the EAVE DETAIL.

Cover the roof, starting at one of the lower corners. Add metal drip edge along the eaves, followed by building paper; then add drip edge along the gable ends, over the paper. Install the asphalt shingles (see pages 60 to 67). Plan the courses so the roof transition occurs midshingle, not between courses; the overlapping shingles will relax over time. If desired, add roof vents or a continuous ridge vent (pages 66 to 67).

Cover the Z-flashing at the rear wall with horizontal 1 × 4 trim. Finish the four wall corners with overlapping vertical 1 × 4 trim. Install the 2 × 6 rails that will support the garage door tracks, following the door manufacturer's instructions to determine the sizing and placement; see the GARAGE DOOR TRIM DETAIL (page 122).

For the garage door frame, rip 1 × 8 trim boards to width so they cover the front wall siding and 2 × 6 rails, as shown in the GARAGE DOOR TRIM DETAIL. Install the trim, mitering the pieces at 22.5°. Install the 1 × 4 trim around the outside of the opening, adding flashing along the top; see the FRONT ELEVATION (page 120).

Build the window frame, which should be ½" narrower and shorter than the rough opening. Install the frame using shims and 10d galvanized casing nails, as shown in the WINDOW JAMB DETAIL (page 122). Cut eight 1 × 2 stop pieces to fit the frame. Bevel the outer sill stop for drainage. Order glass to fit, or cut your own plastic panel. Install the glazing and stops using glazing tape for a watertight seal. Add the window trim.

For the attic doorframe, rip 1 × 6s to match the depth of the opening and cut the head jamb and side jambs. Cut the sill from full-width 1 × 6 stock; then cut a kerf for a drip edge (see the ATTIC DOOR SILL DETAIL). Fasten the head jamb to the side jambs and install the sill at a 5° slope between the side jambs. Install the doorframe using shims and 10d casing nails. Add shims or cedar shingles along the length of the sill to provide support underneath. The front edge of the frame sides and top should be flush with the face of the siding. Add 1 × 2 stops at the frame sides and top, ¾" from the front edges.

Build the attic as shown in the ATTIC DOOR ELEVATION (page 122), using glue and 1¼" screws. Each door measures 28⅝" × 38". Cut the 1 × 8 panel boards about ⅛" short along the bottom to compensate for the sloping sill. Install the door with two hinges each. Add 1 × 4 horizontal trim on the front wall, up against the doorsill; then trim around both sides of the doorframe. Prime and paint the barn as desired.

Install the garage door in the door opening, following the manufacturer's directions.

Simple Storage Shed

The name of this practical outbuilding says it all. It's an easy-to-build, sturdy, 8 × 10-ft. shed with plenty of storage space. With no windows it also offers excellent security. The clean, symmetrical interior and centrally located double doors make for easy access to your stuff. The walls are ready to be lined with utility shelves, and you can quickly add a ramp to simplify parking the lawnmower, wheelbarrow, and other yard equipment.

This shed is indeed basic, but it's also a nicely proportioned building with architecturally appropriate features like overhanging eaves and just enough trim to give it a quality, hand-built appearance. Without getting too fancy—remember, simplicity is the central design idea—you might consider finishing the exterior walls and roof of the shed with the same materials used on your house. This easy modification visually integrates the shed with the rest of the property and provides a custom look that you can't get with kit buildings.

Inside the shed, you can maximize storage space by building an attic: Install full-length 2 × 4 or 2 × 6 joists (which also serve as rafter ties) and cover them with ½" plywood. Include one or more framed-in access openings that you can easily reach with a stepladder. This type of storage space is ideal for seldom-used household items—like winter clothing and holiday decorations—that you can stow in covered plastic bins.

The simplicity and economy of this shed design also make it a great choice for cabins, vacation homes, and other remote locations. A heavy-duty hasp latch and padlock on the door, along with head and foot slide bolts inside, will provide the security you need when you're away for long periods.

This simple structure is the perfect storage space for frequently used lawn equipment or seldom-used items, such as holiday decorations.

Add a padlock to the heavy-duty clasp latch for security. Paired with the head and foot slide bolts inside, you can feel confident your stuff inside is safe and secure.

Cutting List

Description	Quantity/Size	Material
Foundation		
Drainage material	1.25 cu. yd.	Compactable gravel
Skids	2 @ 10'	4 × 6 pressure-treated landscape timbers
Floor		
Rim joists	2 @ 10'	2 × 8 pressure-treated, rated for ground contact
Joists	9 @ 8'	2 × 8 pressure-treated, rated for ground contact
Floor sheathing	3 sheets @ 4 × 8'	¾" tongue-&-groove ext.-grade plywood
Wall Framing		
Bottom plates	2 @ 10', 2 @ 8'	2 × 4
Top plates	4 @ 10', 4 @ 8'	2 × 4
Studs	36 @ 8'	2 × 4
Door header	1 @ 10'	2 × 6
Roof Framing		
Rafters	6 @ 12'	2 × 6
Rafter blocking	2 @ 10'	2 × 6
Ridge board	1 @ 10'	1 × 8
Collar ties	2 @ 12'	2 × 4
Exterior Finishes		
Siding	11 sheets @ 4 × 9'	½" Texture 1-11 plywood siding
Fascia	4 @ 12'	1 × 8
Corner trim	8 @ 8'	1 × 2
Gable wall trim	2 @ 8'	1 × 4
Siding flashing	16 linear ft.	Metal Z-flashing

Description	Quantity/Size	Material
Roofing		
Sheathing (& door header spacer)	5 sheets @ 4 × 8'	½" exterior-grade plywood roof sheathing
15# building paper	1 roll	
Shingles	1¼ squares	Asphalt shingles
Drip edge	45 linear ft.	Metal drip edge
Door		
Frames	7 @ 8'	2 × 4 pressure-treated
Panels	1 sheet @ 4 × 8'	½" Texture 1-11 plywood siding
Stops & overlap trim	4 @ 8'	1 × 2 pressure-treated
Fasteners & Hardware		
16d galvanized common nails	4 lbs.	
16d common nails	10 lbs.	
10d common nails	2 lb.	
8d galvanized common nails	3 lbs.	
8d box nails	3 lbs.	
8d galvanized siding or finish nails	9 lbs.	
1" galvanized roofing nails	5 lbs.	
Door hinges with screws	6 @ 3½"	Galvanized metal hinges
Door handle	1	
Door lock (optional)	1	
Door head bolt	1	
Door foot bolt	1	
Construction adhesive		

Elevation

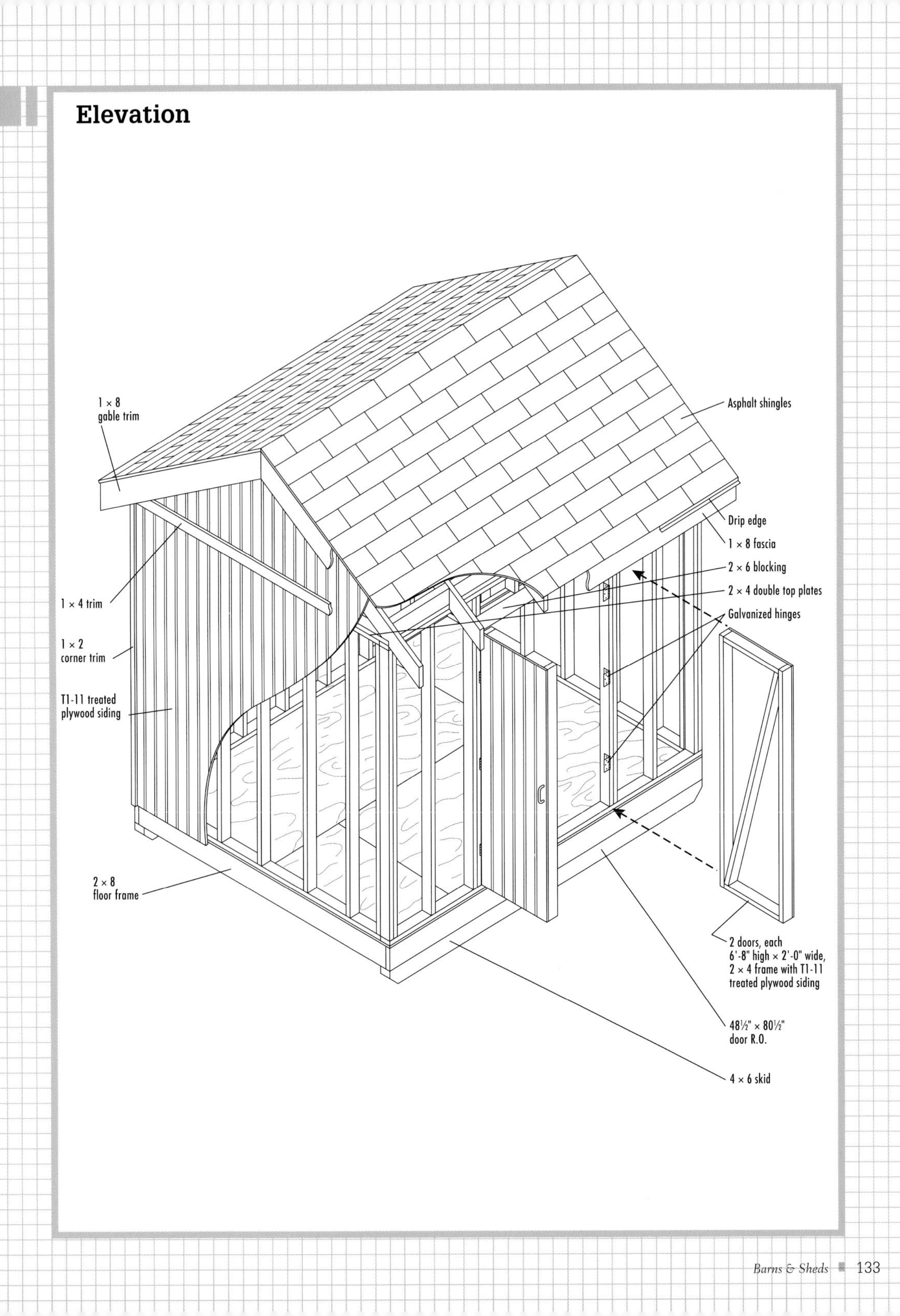

1 × 8
gable trim
Asphalt shingles
1 × 4 trim
Drip edge
1 × 8 fascia
2 × 6 blocking
2 × 4 double top plates
Galvanized hinges
1 × 2
corner trim
T1-11 treated
plywood siding
2 × 8
floor frame
2 doors, each
6'-8" high × 2'-0" wide,
2 × 4 frame with T1-11
treated plywood siding
48½" × 80½"
door R.O.
4 × 6 skid

Framing Elevation

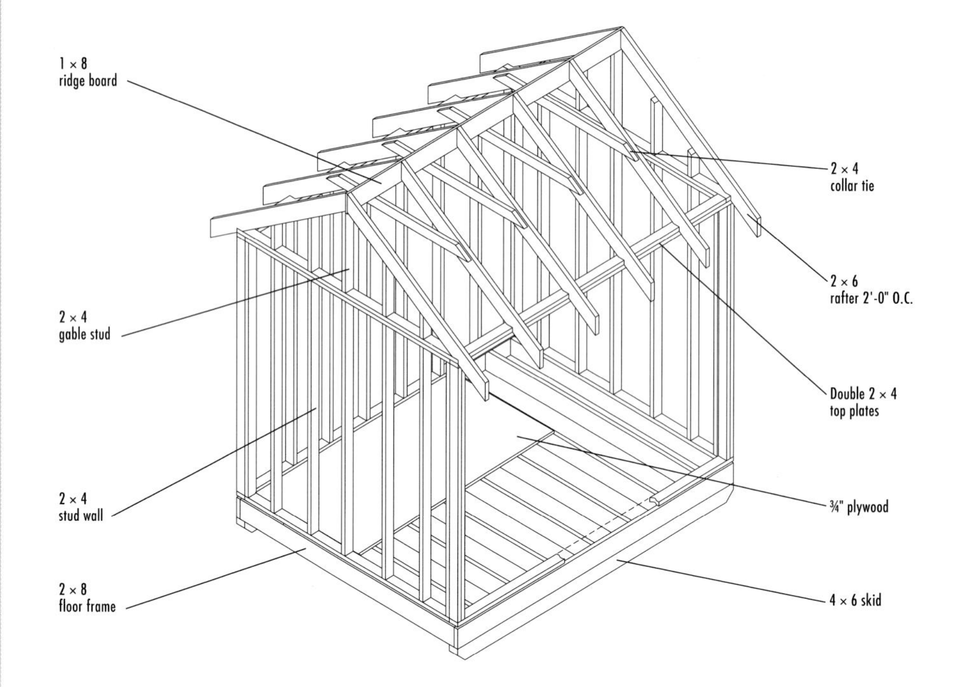

Side Framing

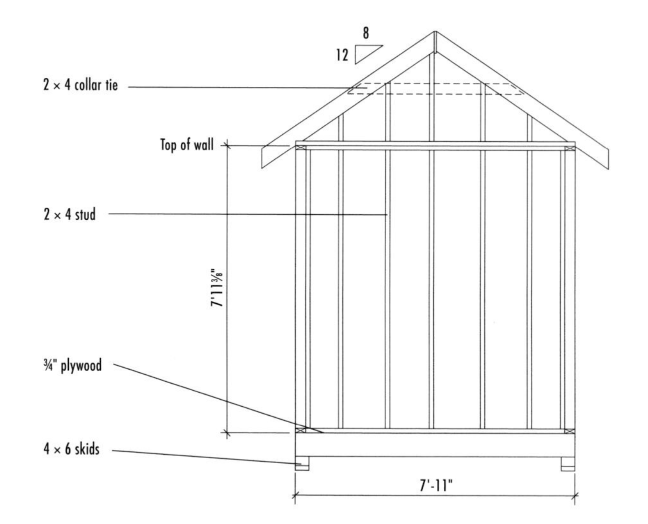

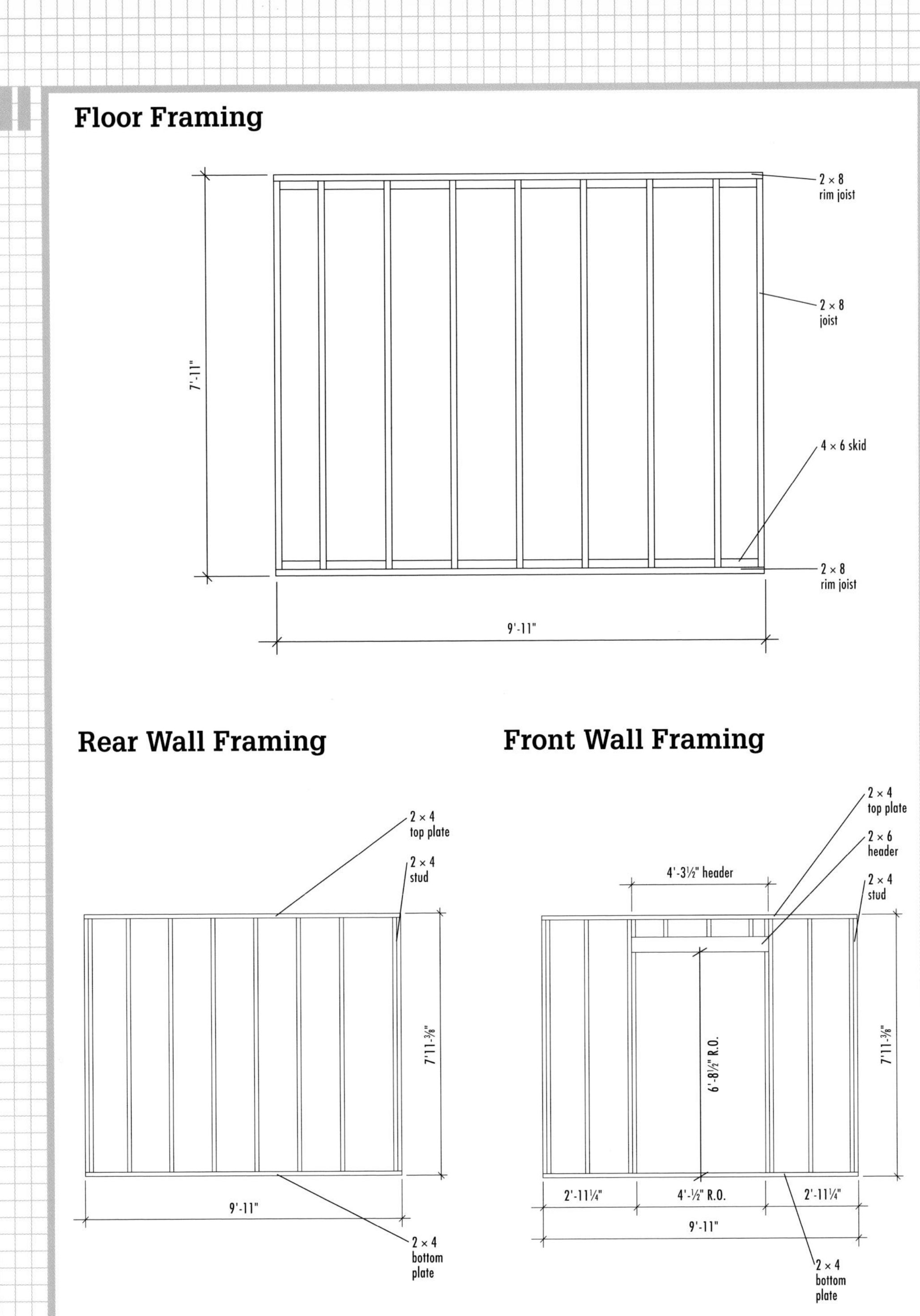

Floor Framing
2 × 8
rim joist
2 × 8
joist
4 × 6 skid
2 × 8
rim joist
7'-11"
9'-11"
Rear Wall Framing
2 × 4
top plate
2 × 4
stud
7'11-3/8"
9'-11"
2 × 4
bottom
plate
Front Wall Framing
2 × 4
top plate
2 × 6
header
2 × 4
stud
4'-3½" header
6'-8½" R.O.
7'11-3/8"
2'-11¼"
4'-½" R.O.
2'-11¼"
9'-11"
2 × 4
bottom
plate

Roof Plan

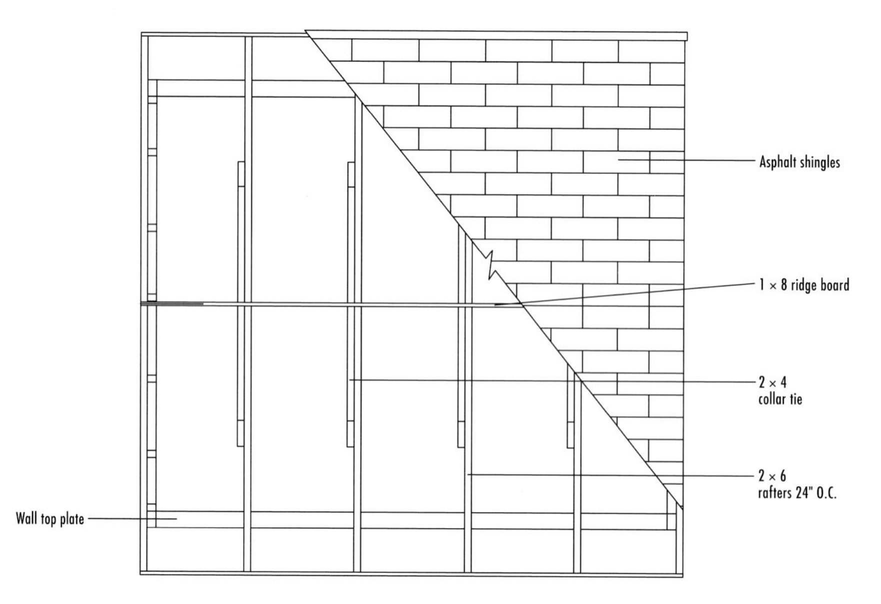

Floor Plan

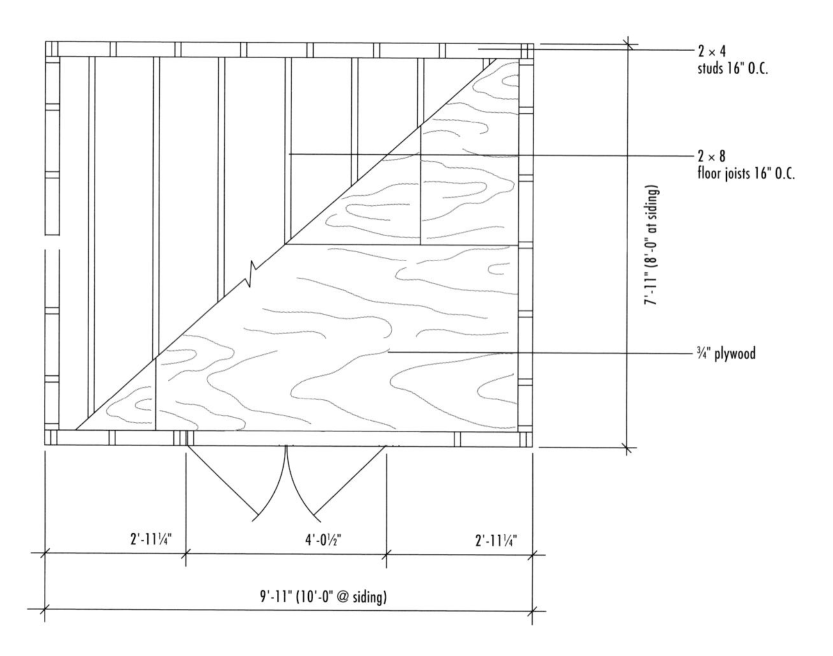

Rafter Template

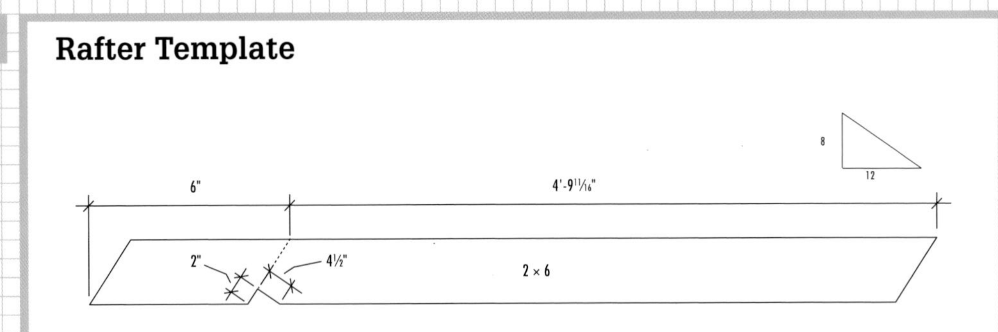

Door Detail

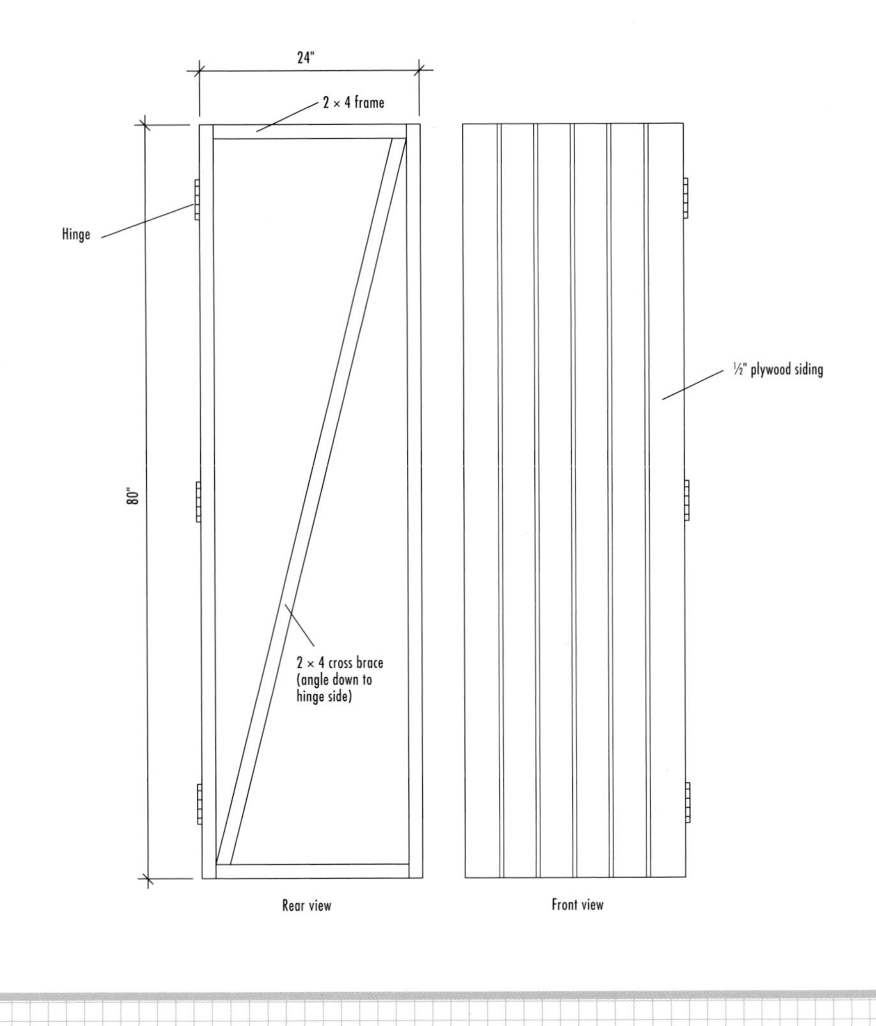

How to Build a Simple Storage Shed

Prepare the foundation site with a 4" layer of compacted gravel. Cut the two 4 × 6 timber skids at 119". Position the skids on the gravel so their outside edges are 95" apart, making sure they are level and parallel.

Cut two 2 × 8 rim joists at 119". Cut nine 2 × 8 joists at 92". Assemble the floor frame following FLOOR FRAMING (page 135), then set it on the skids and measure the diagonals to make sure the frame is square. Fasten the joists to the skids with 16d galvanized common nails.

Attach tongue-and-groove plywood flooring to the floor frame, starting at the left front corner of the shed. Begin the second row of plywood with a full sheet in the right rear corner to stagger end joints. Make sure the tongues are fully seated in the mating grooves. Fasten the sheathing with 8d galvanized common nails.

Frame the rear wall: Cut one 2 × 4 bottom plate and one top plate at 119". Cut ten 2 × 4 studs at 92⅜". Assemble the wall using 16" on-center spacing, as shown in REAR WALL FRAMING (page 135). Raise the wall and fasten it flush to the rear edge of the floor, then brace the wall in position with 2 × 4 braces.

Build the side walls following SIDE FRAMING (page 134). The two side walls are identical. Each has a bottom and top plate at 88" and seven studs at 92$\frac{3}{8}$". Assemble each wall, then install it and brace it in position.

Frame the front wall following FRONT WALL FRAMING (page 135): Cut two plates at 119", cut eight studs at 92$\frac{3}{8}$", and cut two jack studs at 79". Install the 2 × 6 built-up header (add a layer of $\frac{1}{2}$" plywood as a spacer between the 2 × 6s), then add three cripple studs. Raise and fasten the front wall, then install the double top plates along all four walls.

Cut the 1 × 8 ridge board at 119". Mark the rafter layout onto the ridge and the front and rear wall plates following the ROOF PLAN (page 136).

Cover the shed exterior with $\frac{1}{2}$" siding, starting at the left end of the rear wall. Align the sheets with the tops of the wall plates, letting the bottom edges overhang the floor frame by at least 1". Complete the front wall, and then the side walls, keeping the bottom edges even with the sheets on the front and side walls. Cut two 2 × 6 pattern rafters following the RAFTER TEMPLATE (page 137). Test-fit the rafters and make any necessary adjustments. Use one of the patterns to mark and cut the remaining 10 rafters.

(continued)

Install the rafters and ridge board. Cut four 2 × 4 collar ties at 64", mitering the ends at 33.5°. Fasten the collar ties between each set of the four inner rafters, using 10d common nails. Make sure the ties are level and extend close to but not above the top edges of the rafters.

Mark the gable wall stud layout onto the sidewall top plates. Use a level to transfer the marks to the end rafters. Cut each of the 10 2 × 4 studs to fit, mitering the top ends at 33.5°. Install the studs. *Note: The center stud on each wall is located to the rear side of the ridge board. If desired, frame in the attic floor at this time (see Adding an Attic, below).*

Adding an Attic ▸

To build an attic floor for storage, cut six 2 × 4 or 2 × 6 floor joists at 95" (use 2 × 6s if you plan to store heavy items in the attic). If necessary, clip the top corners of the joists so they won't extend above the tops of the rafters. Fasten the joists to the rafters and wall plates with 10d common nails (photo A). At the end rafters, install 2" blocking against the rafters, then attach the joists to the blocking and gable wall studs.

Frame access openings with two header joists spanning neighboring floor joists (photo B). For heavier storage, double up the floor joists on either side of the opening, then use doubled headers to frame the opening. Join doubled members with pairs of 10d common nails every 16". Cover the joists with ½" plywood fastened with 8d nails to complete the attic floor.

Enclose the rafter bays over the walls with 2 × 6 blocking. Bevel the top edge of the blocking at 33.5° so it will be flush with the rafters. Cut the blocks to fit snugly between pairs of rafters and install them. Install 1 × 8 fascia boards at the ends of the rafters along the eaves and over the siding on the gable ends. Keep the fascia ½" above the tops of the rafters. Add Z-flashing and complete the siding on the gable ends.

Apply ½" roof sheathing, starting at the bottom corner of either roof plane. The sheathing should be flush with the tops of the fascia boards. Add the metal drip edge, building paper, and asphalt shingle roofing following the steps on pages 60 to 67.

Construct the two doors from 2 × 4 bracing and ½" siding, as shown in the DOOR DETAIL (page 137). The doors are identical. Each measures 24" × 80". Add hinges and install the doors leaving a ¼" gap between the doors and along the top and bottom.

Trim the corners of the shed with 1 × 2s. Also add a piece of 1 × 2 trim on one of the doors to cover the gap between the doors. Install 1 × 4 trim horizontally to cover the Z-flashing at the side walls. Install door locks and hardware as desired.

Metal & Wood Kit Sheds

The following pages walk you through the steps of building two new sheds from kits. The metal shed measures 8 × 9 ft. and comes with every piece in the main building precut and predrilled. All you need is a ladder and a few hand tools for assembly. The woodshed is a cedar building with panelized construction—most of the major elements come in preassembled sections. The wall panels have exterior siding installed, and the roof sections are already shingled. For both sheds, the pieces are lightweight and maneuverable, but it helps to have at least two people for fitting everything together.

As with most kits, these sheds do not include foundations as part of the standard package. The metal shed can be built on top of a patio surface or out in the yard, with or without an optional floor. The woodshed comes with a complete wood floor, but the building needs a standard foundation, such as wooden skid, concrete block, or concrete slab foundation. To help keep either type of shed level and to reduce moisture from ground contact, it's a good idea to build it over a bed of compacted gravel. A 4"-deep bed that extends about 6" beyond the building footprint makes for a stable foundation and helps keep the interior dry throughout the seasons.

Before you purchase a shed kit, check with your local building department to learn about restrictions that affect your project. It's recommended—and often required—that lightweight metal sheds be anchored to the ground. Shed manufacturers offer different anchoring systems, including cables for tethering the shed into soil and concrete anchors for tying into a concrete slab.

Metal kit sheds like this one come with all the building components—just lay the foundation and then you can set it up with just a few hand tools and a little know-how.

This attractive kit shed looks like a custom-built unit, but, in this case, all the pieces come in preassembled sections.

Shopping for a Kit Shed ▸

If you need an outbuilding but don't have the time or inclination to build one from scratch, a kit shed is the answer. Today's kit sheds are available in a wide range of materials, sizes, and styles—from snap-together plastic lockers to Norwegian pine cabins with divided-light windows and loads of architectural details. Equally diverse is the range of quality and prices for shed kits. One thing to keep in mind when choosing a shed is that much of what you're paying for are the materials and the ease of installation. Better kits are made with quality, long-lasting materials, and many come largely preassembled. Most of the features discussed here will have an impact on a shed's cost.

The best place to start shopping for shed kits is on the Internet. Large manufacturers and small-shop custom designers alike have websites featuring their products and available options. A quick online search should help you narrow your choices to sheds that fit your needs and budget. From there, you can visit local dealers or builders to view assembled sheds firsthand. When figuring cost, be sure to factor in all aspects of the project, including the foundation, extra hardware, tools you don't already own, and paint and other finishes not included with your kit.

High-tech plastics, like polyethylene and vinyl are often combined with steel and other rigid materials to create tough, weather-resistant—and washable—kit buildings.

If you're looking for something special, higher-end shed kits allow you to break with convention without breaking your budget on a custom-built structure.

FEATURES TO CONSIDER

Here are some of the key elements to check out before purchasing a kit shed:

MATERIALS

Shed kits are made of wood, metal, vinyl, various plastic compounds, or any combination thereof. Consider aesthetics, of course, but also durability and appropriateness for your climate. For example, check the snow load rating on the roof if you live in a snowy climate, or inquire about the material's UV resistance if your shed will receive heavy sun exposure. The finish on metal sheds is important for durability. Protective finishes include paint, powder-coating, and vinyl. For wood sheds, consider all of the materials, from the framing to the siding, roofing, and trimwork.

EXTRA FEATURES

Do you want a shed with windows or a skylight? Some kits come with these features, while others offer them as optional add-ons. For a shed workshop, office, or other workspace where you'll be spending a lot of time, consider the livability and practicality of the interior space, and shop accordingly for special features.

WHAT'S INCLUDED?

Many kits do not include foundations or floors, and floors are commonly available as extras. Other elements that may not be included:

- Paint, stain, etc.—Also, some sheds come pre-painted (or pre-primed), but you won't want to pay extra for a nice paint job if you plan to paint the shed to match your house
- Roofing—Often the plywood roof sheathing is included but not the building paper, drip edge, or shingles.

Most shed kits include hardware (nails, screws) for assembling the building, but always check this to make sure.

ASSEMBLY

Many kit manufacturers have downloadable assembly instructions on their websites, so you can really see what's involved in putting their shed together. Assembly of wood sheds varies considerably among manufacturers—the kit may arrive as a bundle of precut lumber or with screw-together prefabricated panels. Easy-assembly models may have wall siding and roof shingles already installed onto panels.

EXTENDERS

Some kits offer the option of extending the main building with extenders, or expansion kits, making it easy to turn an 8 × 10-ft. shed into a 10 × 12-ft. shed, for example.

FOUNDATION

Check with the manufacturer for recommended foundation types to use under their sheds. The basic foundations designs shown in this book should be appropriate for most kit sheds.

Shed hardware kits make it easy to build a shed from scratch. Using the structural gussets and framing connectors, you avoid tricky rafter cuts and roof assembly. Many hardware kits come with lumber cutting lists so you can build the shed to the desired size without using plans.

How to Assemble a Metal Kit Shed

Prepare the building site by leveling and grading as needed and then excavating and adding a 4"-thick layer of compactable gravel. If desired, apply landscape fabric under the gravel to inhibit weed growth. Compact the gravel with a tamper and use a level and a long, straight 2 × 4 to make sure the area is flat and level.

Note: *Always wear work gloves when handling shed parts—the metal edges can be very sharp.* Assemble the floor kit according to the manufacturer's directions—these will vary quite a bit among models, even within the same manufacturer. Be sure that the floor system parts are arranged so the door is in the desired location. Do not fasten the pieces at this stage.

Once you've laid out the floor system parts, check to make sure they're square before you begin fastening them. Measuring the diagonals to see if they're the same is a quick and easy way to check for square.

Fasten the floor system parts together with kit connectors once you've established that the floor is square. Anchor the floor to the site if applicable. Some kits are designed to be anchored after full assembly is completed.

Begin installing the wall panels according to the instructions. Most panels are predrilled for fasteners, so the main trick is to make sure the fastener holes align between panels and with the floor.

Tack together mating corner panels on at least two adjacent corners. If your frame stiffeners require assembly, have them ready to go before you form the corners. With a helper, attach the frame stiffener rails to the corner panels.

Install the remaining fasteners at the shed corners once you've established that the corners are square.

Lay out the parts for assembling the roof beams and the upper side frames and confirm that they fit together properly. Then, join the assemblies with the fasteners provided.

(continued)

Attach the moving and nonmoving parts for the upper door track to the side frames, if your shed has sliding doors.

Fasten the shed panels to the top frames, making sure to check that any fastener holes are aligned and that crimped tabs are snapped together correctly.

Fill in the wall panels between the completed corners, attaching them to the frames with the provided fasteners. Take care not to overdrive the fasteners.

Fasten the door frame trim pieces to the frames to finish the door opening. If the fasteners are colored to match the trim, make sure you choose the correct ones.

Insert the shed gable panels into the side frames and the door track and slide them together so the fastener holes are aligned. Attach the panels with the provided fasteners.

Fit the main roof beam into the clips or other fittings on the gable panels. Have a helper hold the free end of the beam. Position the beam and secure it to both gable ends before attaching it.

Drive fasteners to affix the roof beam to the gable ends and install any supplementary support hardware for the beam, such as gussets or angle braces.

(continued)

Begin installing the roof panels at one end, fastening them to the roof beam and to the top flanges of the side frames.

Apply weatherstripping tape to the top ends of the roof panels to seal the joints before you attach the overlapping roof panels. If your kit does not include weatherstripping tape, look for adhesive-backed foam tape in the weatherstripping products section of your local building center.

As the overlapping roof panels are installed and sealed, attach the roof cap sections at the roof ridge to cover the panel overlaps. Seal as directed. *Note: Completing one section at a time allows you to access subsequent sections from below so you don't risk damaging the roof.*

Attach the peak caps to cover the openings at the ends of the roof cap and then install the roof trim pieces at the bottoms of the roof panels, tucking the flanges or tabs into the roof as directed. Install the plywood floor (if applicable), according to manufacturer instructions.

Assemble the doors, paying close attention to right/left differences on double doors. Attach hinges for swinging doors and rollers for sliding doors.

Install door tracks and door roller hardware (as applicable) on the floor as directed, and then install the doors according to the manufacturer's instructions. Test the action of the doors and make adjustments so the doors roll or swing smoothly and are aligned properly.

Tips for Maintaining a Metal Shed ▸

- Touch up scratches or any exposed metal as soon as possible to prevent rust. Clean the area with a wire brush, and then apply a paint recommended by the shed's manufacturer.
- Inspect your shed once or twice a year and tighten loose screws, bolts, and other hardware. Loose connections lead to premature wear.
- Sweep off the roof to remove wet leaves and debris, which can be hard on the finish. Also clear the roof after heavy snowfall to reduce the risk of collapse.
- Seal open seams and other potential entry points for water with silicone caulk. Keep the shed's doors closed and latched to prevent damage from wind gusts.

Anchor the Shed ▸

Metal sheds tend to be light in weight and require secure anchoring to the ground, generally with an anchor kit that may be sold separately by your kit manufacturer. There are many ways to accomplish this. The method you choose depends mostly on the type of foundation you've built on, be it concrete or wood or gravel. On concrete and wood foundations, look for corner gusset anchors that are attached directly to the floor frame and then fastened with landscape screws (wood) or masonry anchors driven into concrete. Sheds that are built on a gravel or dirt base can be anchored with auger-type anchors that are driven into the ground just outside the shed. You'll need to anchor the shed on at least two sides. Once the anchors are driven, string cables through the shed so they are connected to the roof beam. The ends of the cables should exit the shed at ground level and then be attached to the anchors with cable clamps.

How to Build a Wood Kit Shed

Prepare the base for the shed's wooden skid foundation with a 4" layer of compacted gravel. Make sure the gravel is flat, smooth, and perfectly level. *Note: For a sloping site, a concrete block foundation may be more appropriate (check with your shed's manufacturer).*

Cut two 4 × 4 (or 6 × 6) pressure-treated timbers to match the length of the shed's floor frame. Position these two outer skids so they will be flush with the outside edges of the frame. Cut five timbers to fit between the two outer skids. Make sure that each skid is perfectly level and the skids are level with one another. Fasten the skids together.

Prepare for the Delivery ▸

Panelized shed kits are shipped on pallets. The delivery truck should have a forklift, so the driver can take off the load by whole pallets. Otherwise, you'll have to unload the pieces one at a time. Make sure to have two helpers on hand to help you unload (often drivers aren't allowed to help due to insurance liability).

Once the load is on the ground, carry the pieces to the building site and stack them on pallets or scrap-wood skids to keep them clean and dry. Look through the manufacturer's instructions and arrange the stacks according to the assembly steps.

Assemble the floor frame pieces with screws. First, join alternating pairs of large and small pieces to create three full-width sections. Fasten the sections together to complete the floor frame.

Attach the floor runners to the bottom of the floor frame using exterior screws. Locate the side runners flush to the outsides of the frame, and center the middle runner in between. Set the frame on the skids with the runners facing down. Check the frame to make sure it is level. Secure the floor to the skids following the manufacturer's specifications.

Cover the interior portion of the floor frame with plywood, starting with a large sheet at the left rear corner of the frame. Fasten the plywood with screws. Install the two outer deck boards in the deck area. Lay out all of the remaining boards in between, then set even gapping for each board. Fasten the remaining deckboards.

Lay out the shed's wall panels in their relative positions around the floor. Make sure you have them right-side-up: the windows are on the top half of the walls; on the windowless panels, the siding tells you which end is up.

(continued)

Position the two rear corner walls upright onto the floor so the wall framing is flush with the floor's edges. Fasten the wall panels together. Raise and join the remaining wall panels one at a time. Do not fasten the wall panels to the shed floor in this step.

Place the door header on top of the narrow front wall panel so it's flush with the wall framing. Fasten the header with screws. Fasten the door jamb to the right-side wall framing to create a ½" overhang at the end of the wall. Fasten the header to the jamb with screws.

Confirm that all wall panels are properly positioned on the floor: The wall framing should be flush with the edges of the floor frame; the wall siding overhangs the outsides of the floor. Fasten the wall panels by screwing through the bottom wall plate, through the plywood flooring, and into the floor framing.

Install the wall's top plates starting with the rear wall. Install the side wall plates as directed—these overhang the front of the shed and will become part of the porch framing. Finally, install the front wall top plates.

Assemble the porch rail sections using the screws provided for each piece. Attach the top plate extension to the 4 × 4 porch post, and then attach the wall trim/support to the extension. Fasten the corner brackets, centered on the post and extension. Install the handrail section 4" up from the bottom of the post.

Install each of the porch rail sections: Fasten through the wall trim/support and into the side wall, locating the screws where they will be least visible. Fasten down through the wall top plate at the post and corner bracket locations to hide the ends of the screws. Anchor the post to the decking and floor frame with screws driven through angled pilot holes.

Hang the Dutch door using two hinge pairs. Install the hinges onto the door panels. Use three pairs of shims to position the bottom door panel: ½" shims at the bottom, ⅜" shims on the left side, and ⅛" shims on the right side. Fasten the hinges to the wall trim/support. Hang the top door panel in the same fashion using ¼" shims between the door panels.

Join the two pieces to create the rear wall gable, screwing through the uprights on the back side. On the outer side of the gable, slide in a filler shingle until it's even with the neighboring shingles. Fasten the filler with two finish nails located above the shingle exposure line, two courses up. Attach the top filler shingle with two (exposed) galvanized finish nails.

(continued)

Position the rear gable on top of the rear wall top plates and center it from side to side. Use a square or straightedge to align the angled gable supports with the angled ends of the outer plates. Fasten the gable to the plates and wall framing with screws. Assemble and install the middle gable wall.

Arrange the roof panels on the ground according to their installation. Flip the panels over and attach framing connectors to the rafters at the marked locations using screws.

With one or two helpers, set the first roof panel at the rear of the shed, then set the opposing roof panel in place. Align the ridge boards of the two panels, and then fasten them together with screws. Do not fasten the panels to the walls at this stage.

Position one of the middle roof panels, aligning its outer rafter with that of the adjacent rear roof panel. Fasten the rafters together with screws. Install the opposing middle panel in the same way. Set the porch roof panels into place one at a time—these rest on a ½" ledge at the front of the shed. From inside the shed, fasten the middle and porch panels together along their rafters.

Check the fit of all roof panels at the outside corners of the shed. Make any necessary adjustments. Fasten the panels to the shed with screws, starting with the porch roof. Inside the shed, fasten the panels to the gable framing, then anchor the framing connectors to the wall plates.

Install the two roof gussets between the middle rafters of the shed roof panels (not the porch panels): First measure between the side walls—this should equal 91" for this kit. If not, have two helpers push on the walls until the measurement matches your requirement. Hold the gussets level, and fasten them to the rafters with screws.

(continued)

Add filler shingles at the roof panel seams. Slide in the bottom shingle and fasten it above the exposure line two courses up using two screws. Drive the screws into the rafters. Install the remaining filler shingles the same way. Attach the top shingle with two galvanized finish nails.

Cover the underside of the rafter tails (except on the porch) with soffit panels, fastening to the rafters with finish nails. Cover the floor framing with skirting boards, starting at the porch sides. Hold the skirting flush with the decking boards on the porch and with the siding on the walls, and fasten it with screws.

Add interior trim. Add vertical trim boards to cover the wall seams and shed corners. The rear corners get a filler trim piece, followed by a wide trim board on top. Add horizontal trim boards at the front wall and along the top of the door. Fasten all trim with finish nails.

Add exterior trim. At the rear of the shed, fit the two fascia boards over the ends of the roof battens so they meet at the roof peak. Install the side fascia pieces over the rafter tails with finish nails. The rear fascia overlaps the ends of the side fascia. Cover the fascia joints and the horizontal trim joint at the front wall with decorative plates. Add corner boards.

Place the two roof ridge caps along the roof peak, overlapping the caps' roofing felt in the center. Fasten the caps with screws. Install the decorative gusset gable underneath the porch roof panels using mounting clips. Finish the gable ends with two fascia pieces installed with screws.

Complete the porch assembly by fastening each front handrail section to a deck post using screws. Fasten the handrail to the corner porch post. The handrail should start 4" above the bottoms of the posts, as with the side handrail sections. Anchor each deck post to the decking and floor frame with screws (see *Drilling Counterbored Pilot Holes,* this page).

Drilling Counterbored Pilot Holes ▸

Use a combination piloting/ counterbore bit to pre-drill holes for installing posts. Angle the pilot holes at about 60°, and drive the screws into the framing below whenever possible. The counterbore created by the piloting bit helps hide the screw heads.

Clerestory Studio

The distinctive design of this outbuilding adds sophistication and intrigue to your property.

This easy-to-build shed is made distinctive by its three clerestory windows on the front side. In addition to their unique architectural effect, clerestory windows offer some practical advantages over standard windows. First, their position at the top of the building allows sunlight to spread downward over the interior space to maximize illumination. Most of the light is indirect, creating a soft glow without the harsh glare of direct sunlight. Clerestories also save on wall space and offer more privacy and security than windows at eye level. These characteristics make this shed design a great choice for a backyard office, artist's studio or even a remote spot for the musically inclined to get together and jam.

As shown, the Clerestory Studio has a 10 × 10-ft. floor plan. It can be outfitted with double doors that open up to a 5 ft.-wide opening, as seen here. But if you don't need a door that large, you can pick up about 2½ ft. of additional (and highly prized) wall space by framing the opening for a 30" wide door. The studio's striking roofline is created by two shed-style roof planes, which makes for deceptively easy construction.

The shed's walls and floor follow standard stick-frame construction. For simplicity, you can frame the square portions of the lower walls first, then piece in the framing for the four "rake," or angled, wall sections. To support the roof rafters, the clerestory wall has two large headers (beams) that run the full length of the building. These and the door header are all made with standard 2× lumber and a ½" plywood spacer.

You can increase the natural light in your studio—and add some passive solar heating—by including the two optional skylights. To prevent leaks, be sure to carefully seal around the glazing and the skylight frame. Flashing around the frame will provide an extra measure of protection.

The indirect light from the clerestory windows creates a soft glow inside the building.

Cutting List

Description	Quantity/Size	Material
Foundation		
Drainage material	1.5 cu. yd.	Compactable gravel
Skids	2 @ 10'	4 × 6 pressure-treated landscape timbers
Floor		
Rim joists	2 @ 10'	2 × 6 pressure-treated
Joists	9 @ 10'	2 × 6 pressure-treated
Floor sheathing	4 sheets, 4 × 8'	¾" tongue-&-groove ext.-grade plywood
Wall Framing		
Bottom plates	4 @ 10'	2 × 4
Top plates, front walls	5 @ 10'	2 × 4
Top plates, rear wall	2 @ 10'	2 × 4
Top plates, side walls	6 @ 10'	2 × 4
Studs, rear wall	11 @ 8'	2 × 4
Studs, front wall (& clerestory wall)	11 @ 8'	2 × 4
Studs, side walls	26 @ 8'	2 × 4
Header, above windows	2 @ 10'	2 × 6
Header, below windows	2 @ 10'	2 × 10
Header, door	2 @ 8'	2 × 6
Header & post spacers		See Sheathing, below
Roof Framing		
Rafters (& blocking)	20 @ 8'	2 × 6
Exterior Finishes		
Side wall fascia	4 @ 8'	2 × 6
Eave fascia	3 @ 12'	2 × 6
Fascia drip edge	8 @ 8'	1 × 2
Siding	10 sheets @ 4 × 8'	⅝" texture 1-11 plywood siding
Corner trim	10 @ 8'	1 × 4 cedar
Bottom siding trim	5 @ 12'	1 × 4 cedar
Vents	8	2"-dia. round metal vents
Roofing		
Sheathing (& header/post spacers)	6 sheets @ 4 × 8'	½" exterior-grade plywood roof sheathing

Description	Quantity/Size	Material
15# building paper	1 roll	
Shingles	1⅔ squares	Asphalt shingles—250# per sq. min.
Roof flashing	10'-6"	Aluminum
Windows		
Glazing	3 pieces @ 21 × 36"	¼"-thick acrylic or polycarbonate glazing
Window stops	5 @ 8'	1 × 2 cedar
Glazing tape	60 linear ft.	
Clear exterior caulk	1 tube	
Door		
Panels	2 sheets @ 4 × 8'	¾" exterior-grade plywood
Panel trim	8 @ 8'	1 × 4 cedar
Stops	3 @ 8'	1 × 2 cedar
Flashing	6 linear ft.	Aluminum
Skylights (optional)		
Glazing	2 pieces @ 13 × 22½"	¼"-thick plastic or polycarbonate glazing
Frame	2 @ 8'	1 × 4 cedar
Stops	2 @ 8'	1 × 2 cedar
Glazing tape	25 linear ft.	
Fasteners & Hardware		
16d galvanized common nails	4 lbs.	
16d common nails	16½ lbs.	
10d common nails	1 lb.	
8d galvanized common nails	3 lbs.	
8d box nails	3½ lbs.	
8d galvanized siding nails	7 lbs.	
1" galvanized roofing nails	5 lbs.	
8d galvanized casing nails	2 lbs.	
1¼" galvanized screws	1 lb.	
2" galvanized screws	1 lb.	
Door hinges with screws	6 @ 3½"	
Door handle	2	
Door lock (optional)	1	

Front Elevation

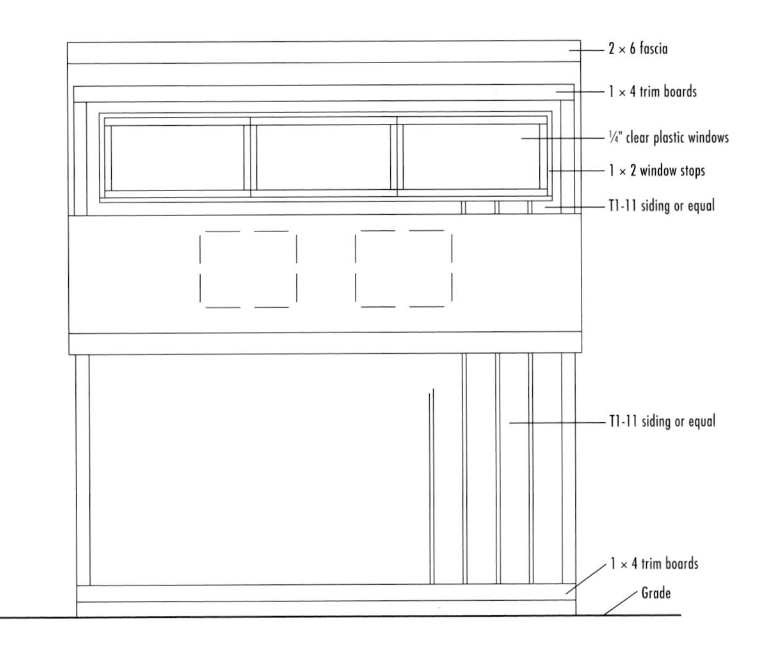

Rear Elevation

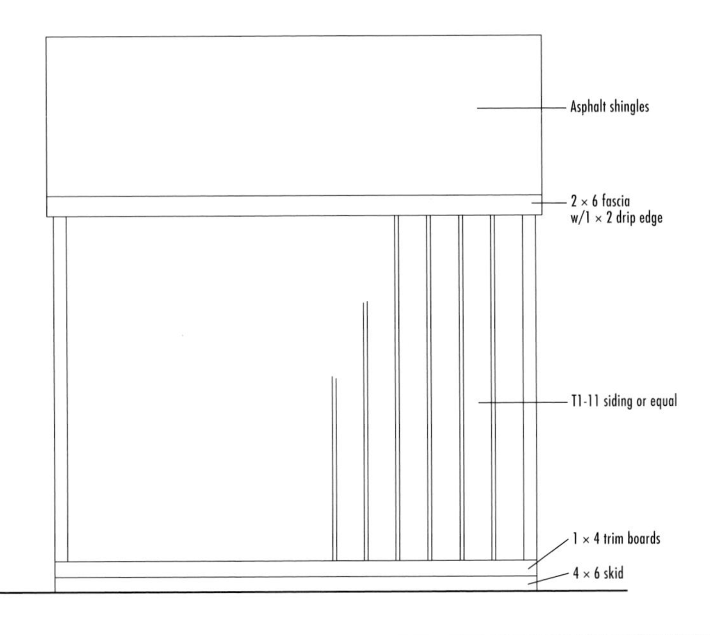

Building Section

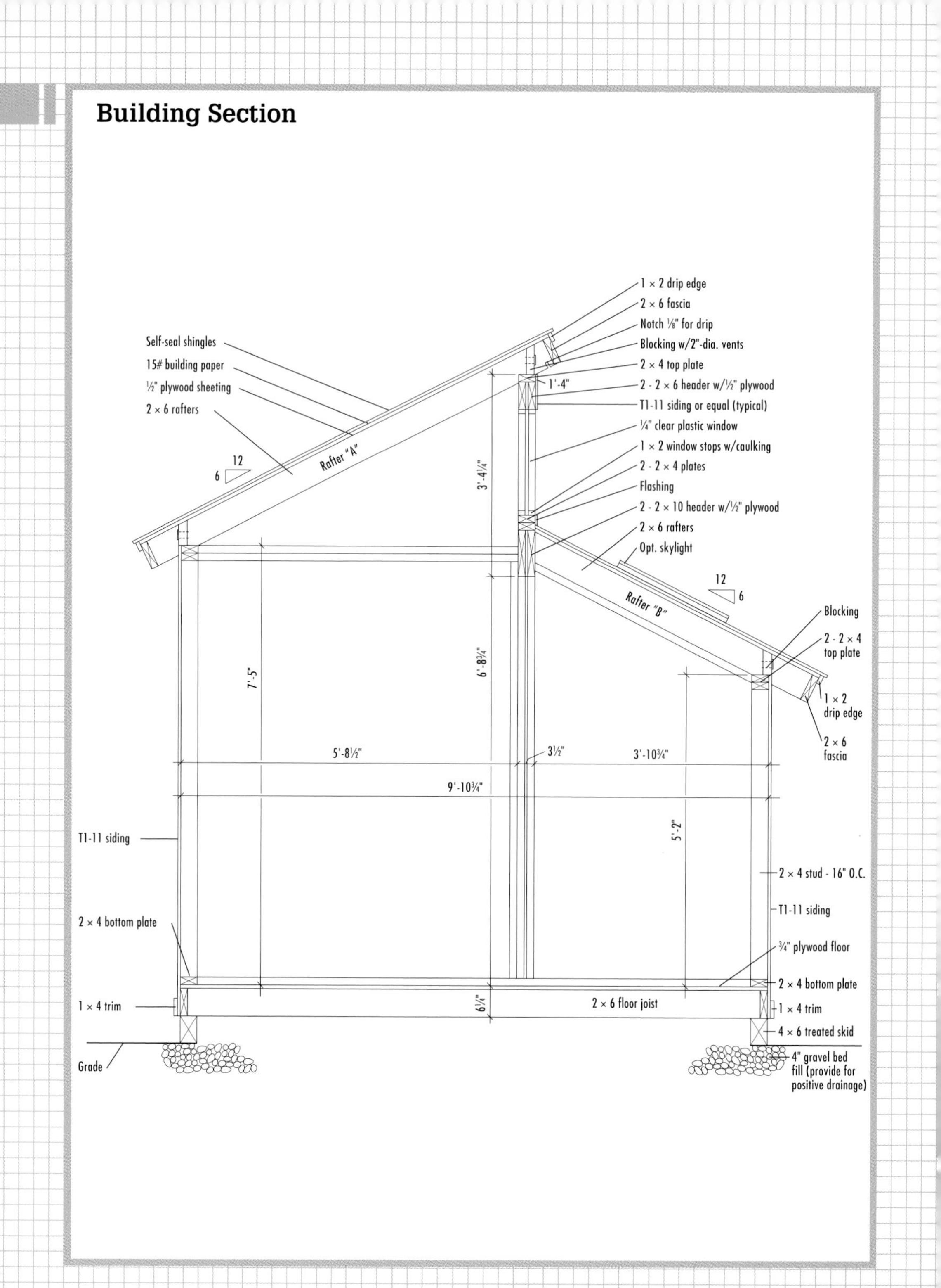

Front Framing

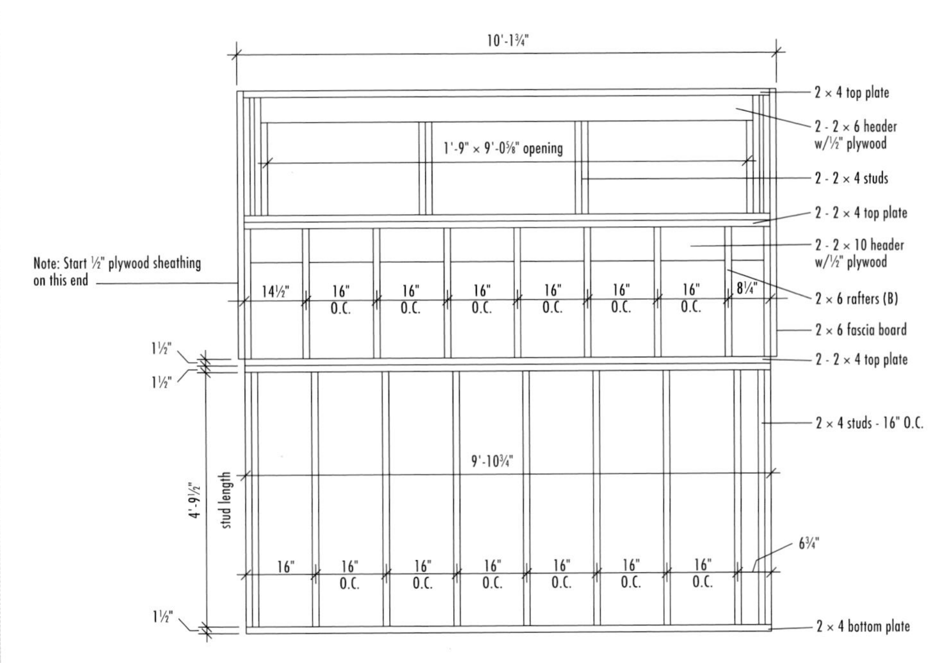

Rear Framing

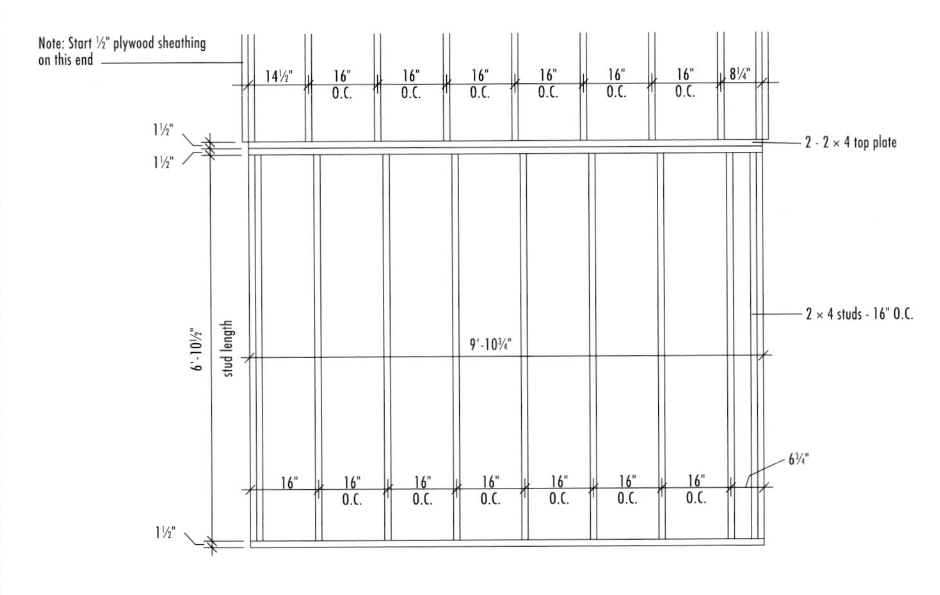

Left Side Wall Framing

Rake Detail

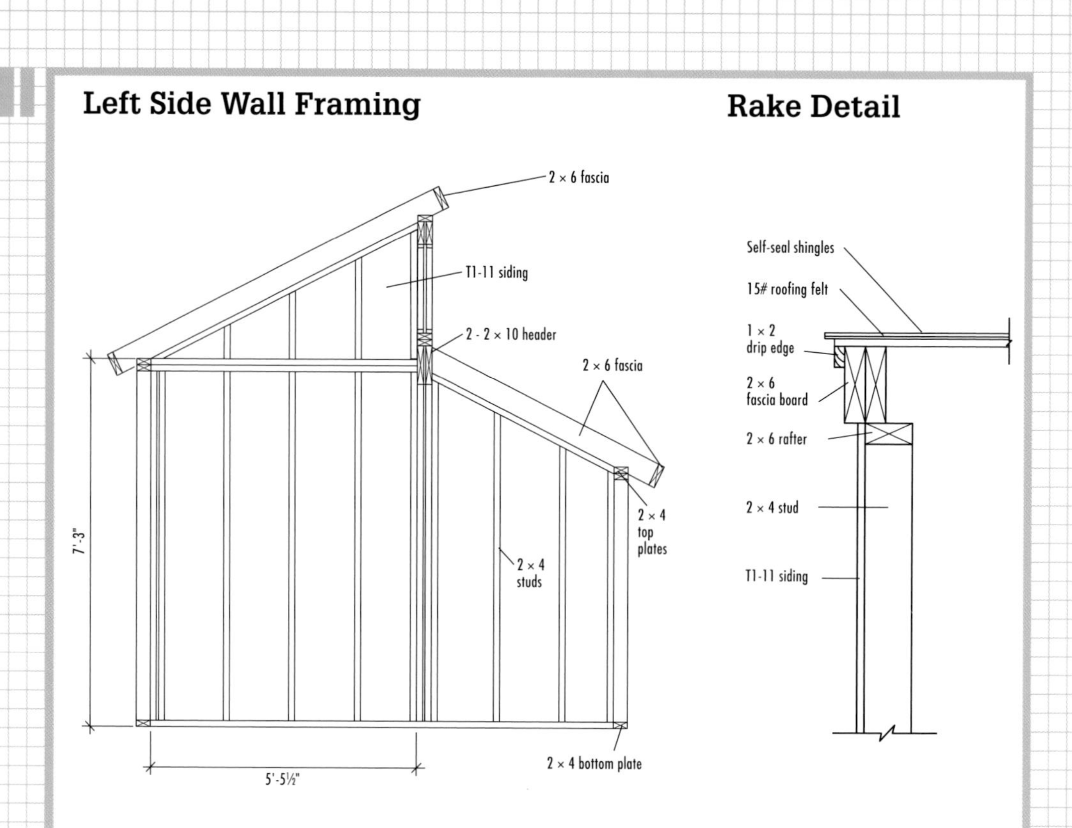

Door Detail

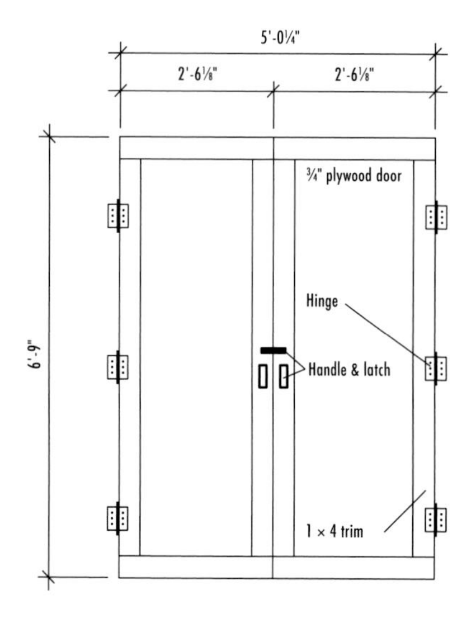

Jamb/Corner Detail

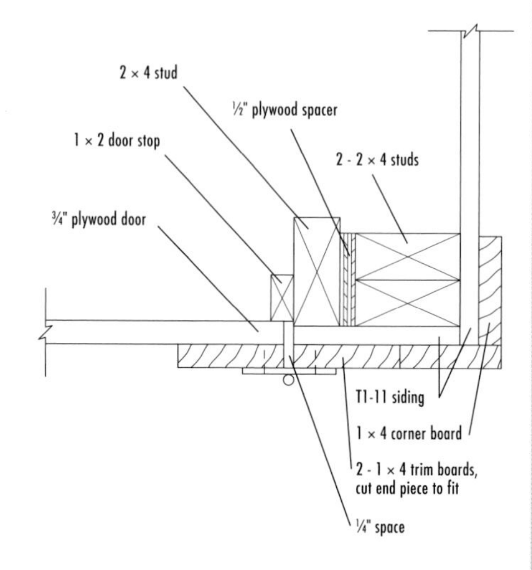

Right Side Elevation

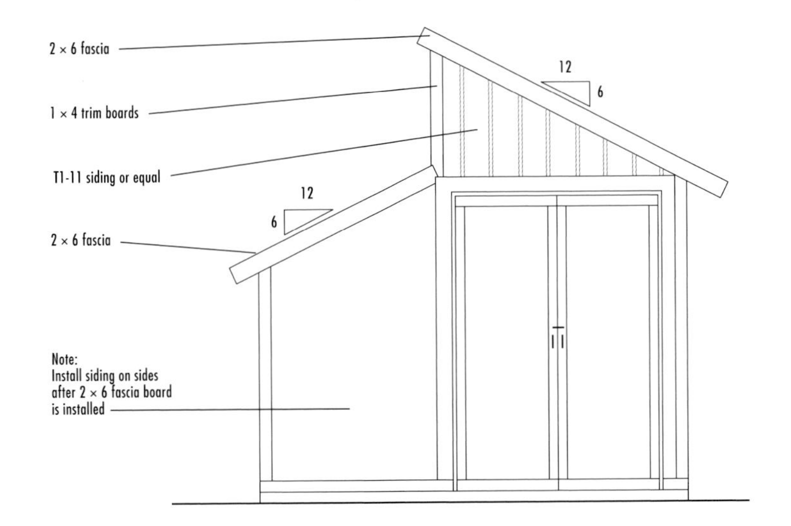

Floor Plan

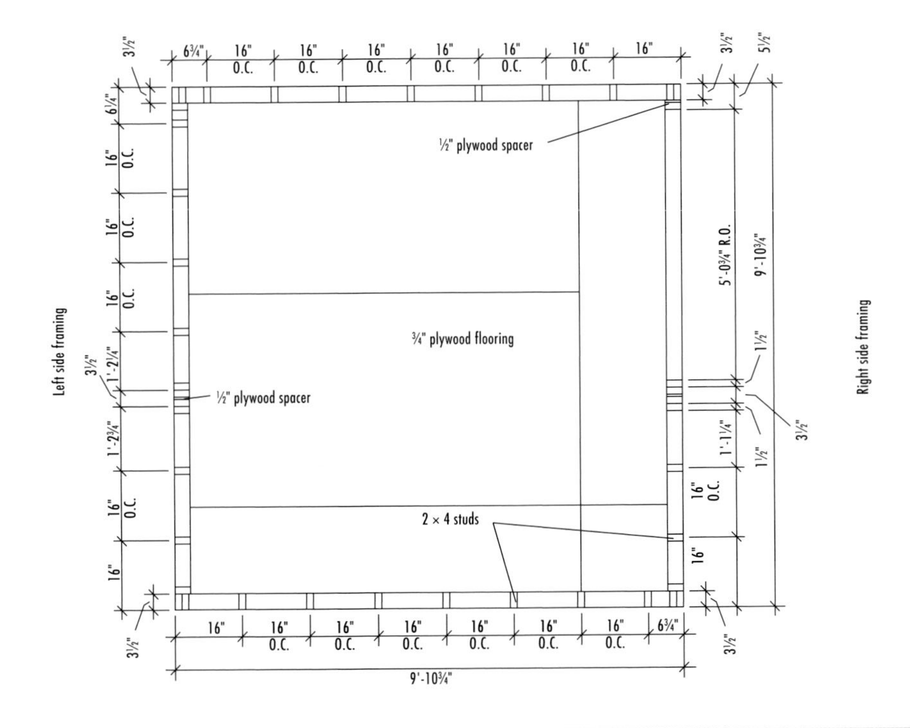

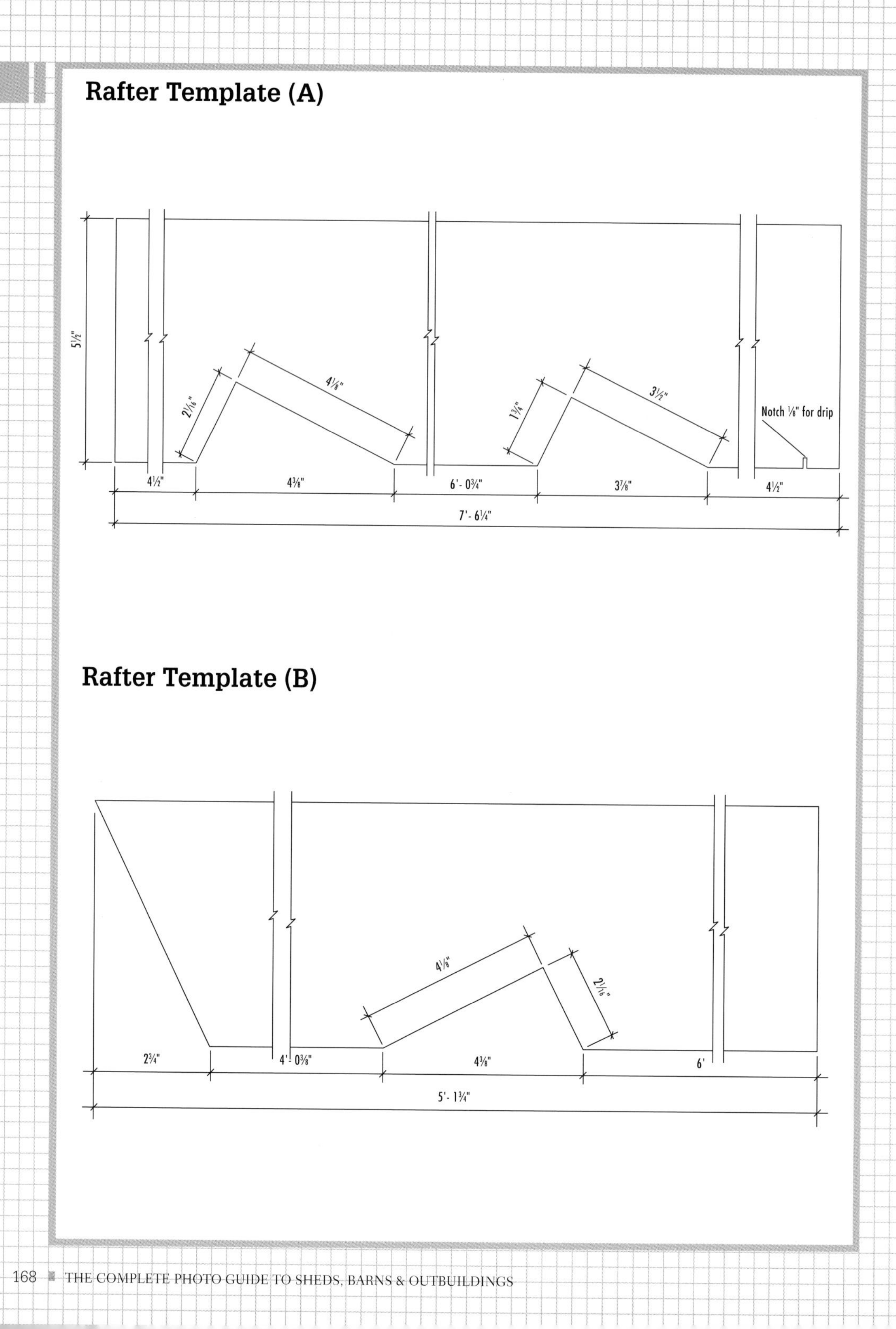
Rafter Template (A)
5 1/2"
2 7/16"
4 1/8"
1 3/4"
3 1/2"
Notch 1/8" for drip
4 1/2"
4 3/8"
6'- 0 3/4"
3 7/8"
4 1/2"
7'- 6 1/4"
Rafter Template (B)
4 1/8"
2 7/16"
2 3/4"
4'- 0 3/8"
4 3/8"
6'
5'- 1 3/4"

How to Build a Clerestory Studio

Prepare the foundation site with a 4"-deep layer of compacted gravel. Cut the two 4 × 6 timber skids at 118¾". Position the skids on the gravel bed so their outside edges are 118¾" apart, making sure they are level and parallel with one another.

Cut two 2 × 6 rim joists at 118¾". Cut nine 2 × 6 joists at 115¾". Build the floor frame on the skids and measure the diagonals to make sure the frame is square. Fasten the rim joists to the skids with 16d galvanized common nails driven toenail style through the joists and into the skids.

Install floor sheathing onto the floor frame, starting at the left rear corner of the shed, as shown in the FLOOR PLAN (page 167). Rip the two outer pieces and final corner piece so their outside edges are flush with the sides of the floor frame.

Cut the studs and top and bottom plates for the front wall and assemble the wall with 16d common nails. Position the wall on the floor deck and raise it. Fasten it by driving 16d common nails through the bottom plate and into the floor deck and frame.

(continued)

Complete the back wall framing with a bottom plate and double top plate at 118¾" and 82½" studs 16" o.c. Build the square portions of the left and right side walls. Attach the back wall and nail the side walls in place.

Take measurements to confirm the dimensions for the clerestory wall frame. Build the clerestory frame wall to match the dimensions.

Complete the sloped portions of the side walls. Install them by nailing them to the floor deck with 16d common nails. Also nail the ends to the front wall and the main header posts.

Create the headers by sandwiching a ½" plywood strip between two 2× dimensional framing members. Assemble the headers with 16d common nails driven through both faces.

Set the main header on top of the sidewall posts and toenail it in place with 16d common nails. The main header ends should be flush with the outsides of the side walls.

Lift the clerestory wall frame onto the main header. Orient the wall so it is flush with the front and ends of the header, and then attach it to the main header with 16d nails.

Install T1-11 siding on the front wall, starting at the left side (when facing the front of the shed). Cut the siding to length so it's flush with the top of the top plate and the bottom of the floor joists. Make sure any vertical seams fall at stud locations. Add strips of siding to cover the framing on the clerestory wall. Install siding on the rear wall, starting at the left side (when facing the rear side of the shed).

Cut one of each "A" and "B" pattern rafters from a single 16-ft. 2 × 6 using the RAFTER TEMPLATES (page 168). Both roof planes have a 6-in-12 slope. Test-fit the rafters and make any necessary adjustments, then use the patterns to cut eight more rafters of each type. Install the rafters as shown in REAR FRAMING and FRONT FRAMING (page 165). Toenail the top ends of the "B" rafters to the main header.

(continued)

Frame each of the upper rake walls following the same technique used for the side walls. Cut the top plate to fit between the clerestory header and the door header (on the right side wall) or the top plate (on the left side wall). Install four studs in each wall using 16" on-center spacing.

Install 2 × 6 fascia boards flush with the top edges and ends of the rafters. The upper roof gets fascia on all four sides; the lower roof on three sides. Miter the corner joints if desired. Install siding on the side walls, flush with the bottom of the fascia; see the RAKE DETAIL (page 166.)

Begin installing ½" plywood roof sheathing, starting at the bottom left side of the roof on both sides of the shed. Run the sheathing ¾" past the outside edge of the fascia. Add 1 × 2 trim to serve as a drip edge along all fascia boards, flush with the underside of the sheathing.

Fasten 1 × 2 stops inside the window rough openings, flush with the inside edges of the framing using 2" screws. Set each window panel into its opening using glazing tape as a sealant. Install the outer stops; see the BUILDING SECTION (page 164). Caulk around the windows and the bottom outside stops to prevent leaks. Add 2 × 6 blocking (and vents) or screen to enclose the rafter bays above the walls.

Add vertical trim at the wall corners. Trim and flash around the door opening and windows. Install flashing—and trim, if desired—along the joint where the lower roof plane meets the clerestory wall. *Note: The bottom flange of the flashing should sit on top of the building paper (step 18).*

Add 15# building paper and install the asphalt shingle roofing. The shingles should overhang the fascia drip edge by ½" along the bottom of the roof and by ⅜" along the sides. Install 1 × 4 horizontal trim boards flush with the bottom of the siding on all four walls.

Cut out the bottom plate inside the door's rough opening. Cut the two door panels at 30⅛" × 81". Install 1 × 4 trim around the panels, as shown in the DOOR DETAIL (page 166), using exterior wood glue and 1¼" screws or nails. Add 1 × 2 stops at the sides and top of the rough opening; see the JAMB/CORNER DETAIL (page 166). Also add a 1 × 4 stop to the back side of one of the doors. Hang the doors with galvanized hinges, leaving consistent gaps all around.

Finish the interior as desired. If you will be occupying the shed for activities, adding some wall covering, such as paneling, makes the interior much more pleasant. If you add wiring and insulation, the Clerestory Studio can function as a 3-season studio in practically any climate.

Timber-frame Shed

Timber-framing is a traditional style of building that uses a simple framework of heavy timber posts and beams connected with hand-carved joints. From the outside, a timber-frame building looks like a standard stick-frame structure, but on the inside, the stout, rough-sawn framing members evoke the look and feel of an 18th-century workshop. This 8 × 10-ft. shed has the same basic design used in traditional timber-frame structures but with joints that are easy to make.

In addition to the framing, some notable features of this shed are its simplicity and proportions. It's a nicely symmetrical building with full-height walls and an attractively steep-pitched roof, something you seldom find on manufactured kit sheds. The clean styling gives it a traditional, rustic look, but also makes the shed ideal for adding custom details. Install a skylight or windows to brighten the interior, or perhaps cut a crescent moon into the door in the style of old-fashioned backyard privies.

The materials for this project were carefully chosen to enhance the traditional styling. The 1 × 8 tongue-and-groove siding and all exterior trim boards are made from rough-sawn cedar, giving the shed a natural, rustic quality. The door is hand-built from rough cedar boards and includes exposed Z-bracing, a classic outbuilding detail. As shown here, the roof frame is made with standard 2 × 4s, but if you're willing to pay a little more to improve the appearance, you can use rough-cut 2 × 4s or 4 × 4s for the roof framing.

Another option to consider is traditional spaced sheathing instead of plywood for the roof deck. Spaced sheathing consists of 1 × 4 boards nailed perpendicular to the roof frame, with a 1½" gap between boards. The roof shingles are nailed directly to the sheathing without building paper in between, creating an attractive ceiling of exposed boards and shingles inside the shed.

Timber-framing impacts the inside of an outbuilding even more than the outside. The exposed posts and beams are an attractive design element as well as a smart technique.

The steep pitch of this building's roof adds distinctive styling not found in manufactured kit sheds.

Cutting List

Description	Quantity/Size	Material
Foundation		
Drainage material	1 cu. yard	Compactable gravel
Skids	3 @ 10'	6 × 6 treated timbers
Floor Framing		
Rim joists	2 @ 10'	2 × 6 pressure-treated
Joists	9 @ 8'	2 × 6 pressure-treated
Joist clip angles	18	3 × 3 × 3" × 18-gauge galvanized
Floor sheathing	3 sheets @ 4 × 8'	¾" tongue-&-groove ext.-grade plywood
Wall Framing		
Posts	6 @ 8'	4 × 4 rough-sawn cedar
Window posts	2 @ 4'	4 × 4 rough-sawn cedar
Girts	2 @ 10', 2 @ 8'	4 × 4 rough-sawn cedar
Beams	2 @ 10', 2 @ 8'	4 × 6 rough-sawn cedar
Braces	8 @ 2'	4 × 4 rough-sawn cedar
Post bases	6, with nails	Simpson BC40
Post-beam connectors	8 pieces, with nails	Simpson LCE
L-connectors	4, with nails	Simpson A34
Additional posts	6 @ 8'	4 × 4 rough-sawn cedar
Roof Framing		
Rafters	12 @ 7'	2 × 4
Collar ties	2 @ 10'	2 × 4
Ridge board	1 @ 10'	2 × 6
Metal anchors—rafters	8, with nails	Simpson H1
Gable-end blocking	4 @ 7'	2 × 2
Exterior Finishes		
Siding	2 @ 14' 8 @ 12' 10 @ 10' 29 @ 9'	1 × 8 V-joint rough-sawn cedar
Corner trim	8 @ 9'	1 × 4 rough-sawn cedar
Fascia	4 @ 7', 2 @ 12'	1 × 6 rough-sawn cedar
Fascia trim	4 @ 7', 2 @ 12'	1 × 2 rough-sawn cedar
Subfascia	2 @ 12'	1 × 4 pine
Plywood soffits	1 sheet 4 × 8'	⅜" cedar or fir plywood
Soffit vents (optional)	4 @ 4 × 12"	Louver with bug screen
Flashing (door)	4 linear ft.	Galvanized—18 gauge

Description	Quantity/Size	Material
Roofing		
Roof sheathing	6 sheets @ 4 × 8'	½" ext.-grade plywood
Asphalt shingles	1.7 squares	
15# building paper	140 sq. ft.	
Metal drip edge	2 @ 12', 4 @ 7'	Galvanized metal
Roof vents (optional)	2 units	
Door		
Frame	2 @ 7' 1 @ 4'	¾ × 4¼" (actual) S4S cedar
Stops	2 @ 7', 1 @ 4'	1 × 2 S4S cedar
Panel material	7 @ 7'	1 × 6 T&G V-joint rough-sawn cedar
Z-brace	1 @ 8' to 2 @ 8'	1 × 6 rough-sawn cedar
Strap hinges	3	
Trim	5 @ 7'	1 × 3 rough-sawn cedar
Flashing	42" metal flashing	
Fasteners		
60d common nails	16 nails	
20d common nails	32 nails	
16d galvanized common nails	3½ lbs.	
10d common nails	1 lb.	
10d galvanized casing nails	½ lb.	
8d galvanized box nails	1½ lbs.	
8d galvanized finish nails	7 lbs.	
8d box nails	¼ lb.	
6d galvanized finish nails	40 nails	
3d galvanized finish nails	50 nails	
1½" joist hanger nails	72 nails	
2½" deck screws	25 screws	
1½" wood screws	50 screws	
⅞" galvanized roofing nails	2 lbs.	
⅜" × 6" lag screws, w/washers	16 screws	
¼" × 6" lag screws, w/washers		
Construction adhesive	4 tubes	

Note: Additional posts may be added as a safety precaution to prevent eave beam deflection.

Front Framing Elevation

Left Side Framing Elevation

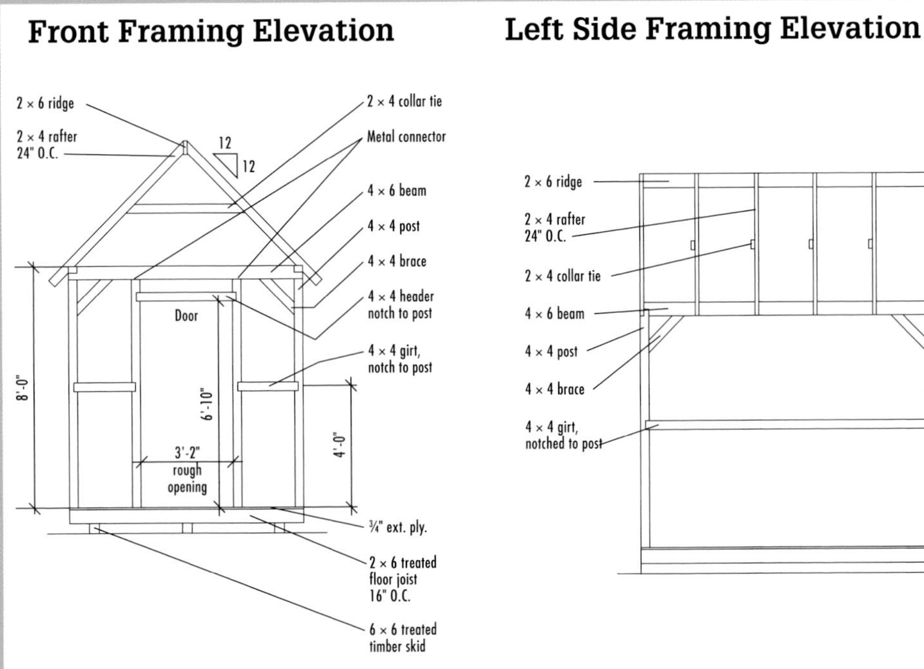

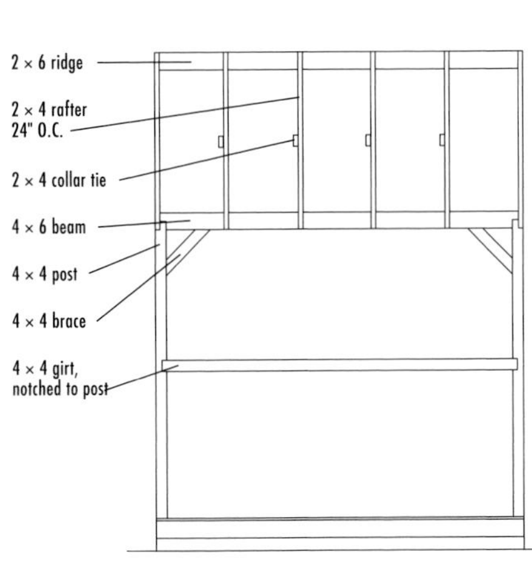

Rear Framing Elevation

Right Side Framing Elevation

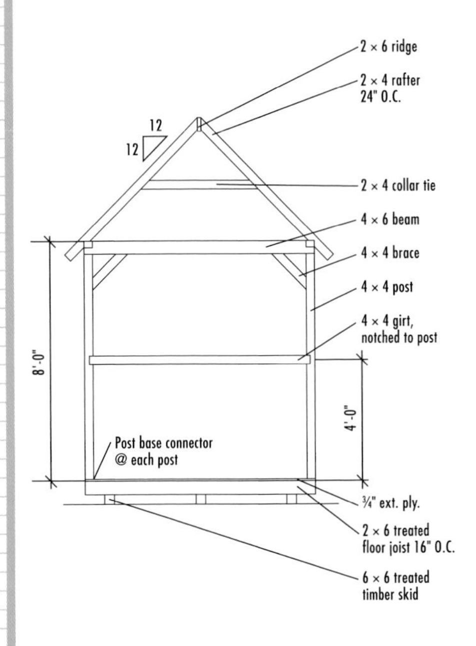

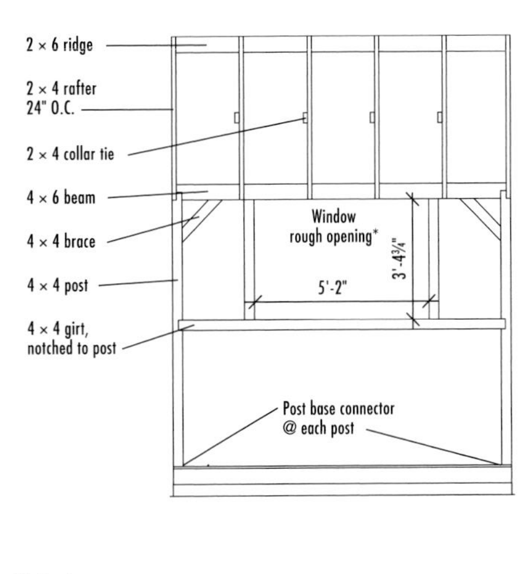

*Optional

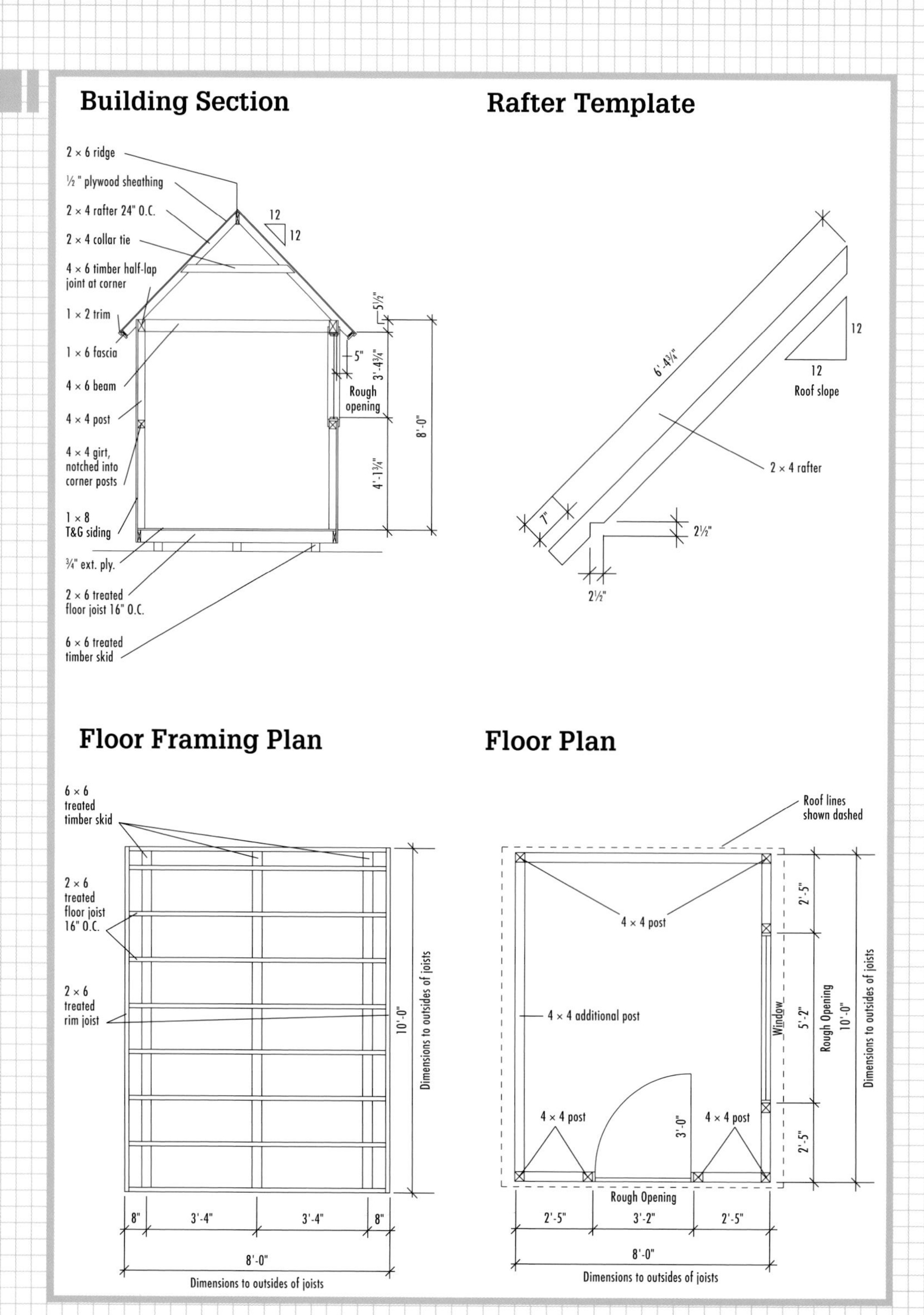

Building Section
2 × 6 ridge
½" plywood sheathing
2 × 4 rafter 24" O.C.
2 × 4 collar tie
4 × 6 timber half-lap joint at corner
1 × 2 trim
1 × 6 fascia
4 × 6 beam
4 × 4 post
4 × 4 girt, notched into corner posts
1 × 8 T&G siding
¾" ext. ply.
2 × 6 treated floor joist 16" O.C.
6 × 6 treated timber skid
12
12
5½"
5"
3'-4¾"
Rough opening
4'-1¾"
8'-0"
Rafter Template
6'-4¾"
12
12
Roof slope
2 × 4 rafter
7"
2½"
2½"
Floor Framing Plan
6 × 6 treated timber skid
2 × 6 treated floor joist 16" O.C.
2 × 6 treated rim joist
10'-0"
Dimensions to outsides of joists
8"
3'-4"
3'-4"
8"
8'-0"
Dimensions to outsides of joists
Floor Plan
Roof lines shown dashed
4 × 4 post
4 × 4 additional post
Window
2'-5"
5'-2"
2'-5"
Rough Opening
10'-0"
Dimensions to outsides of joists
4 × 4 post
3'-0"
4 × 4 post
Rough Opening
2'-5"
3'-2"
2'-5"
8'-0"
Dimensions to outsides of joists

Front Elevation

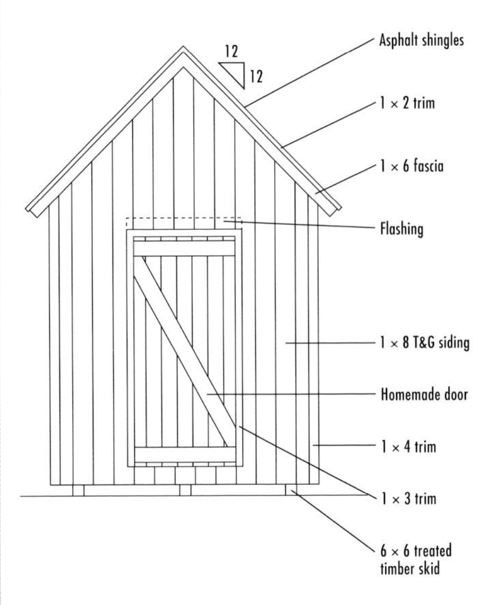

Left Side Elevation

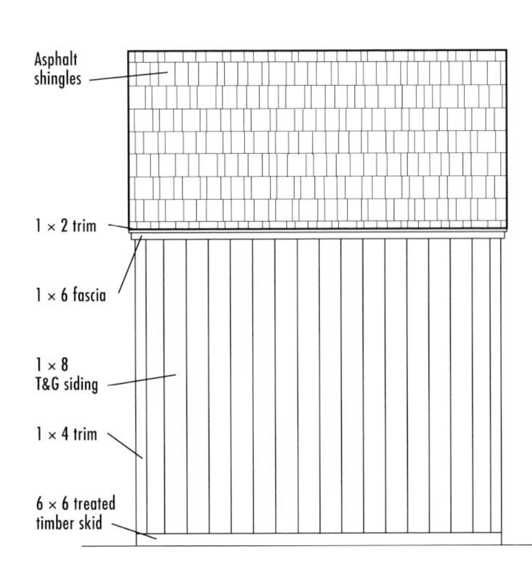

Rear Elevation

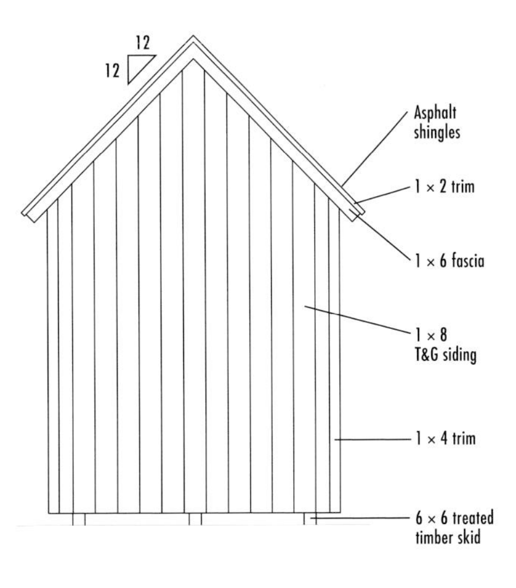

Right Side Elevation

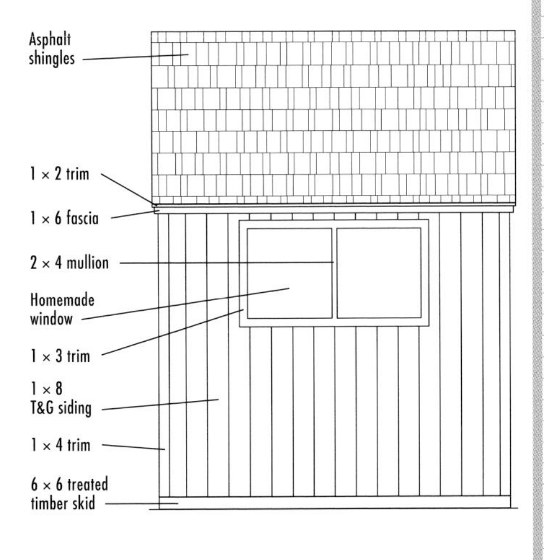

Gable Overhang Detail

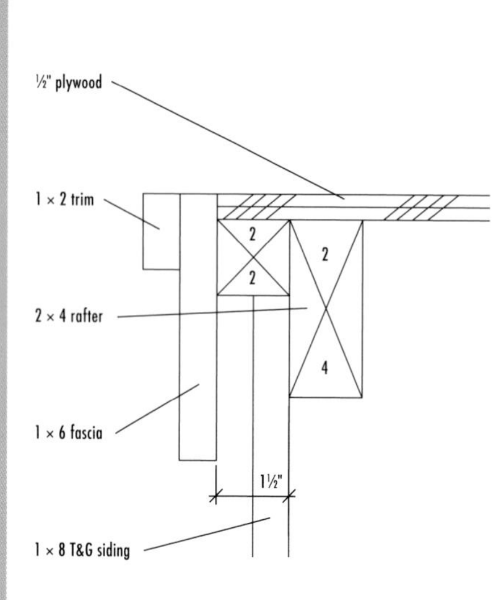

Eave Detail

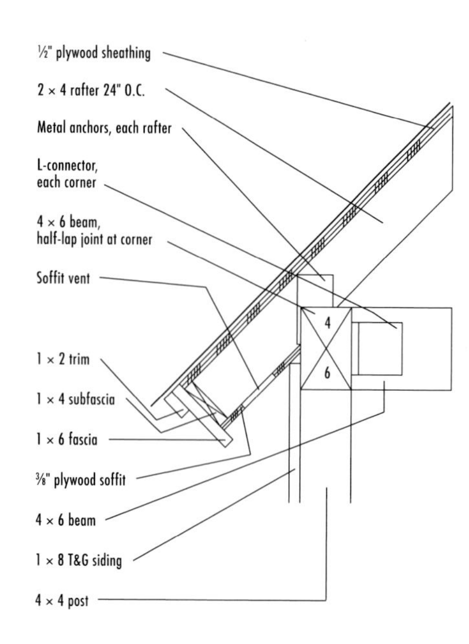

Door Jamb Detail

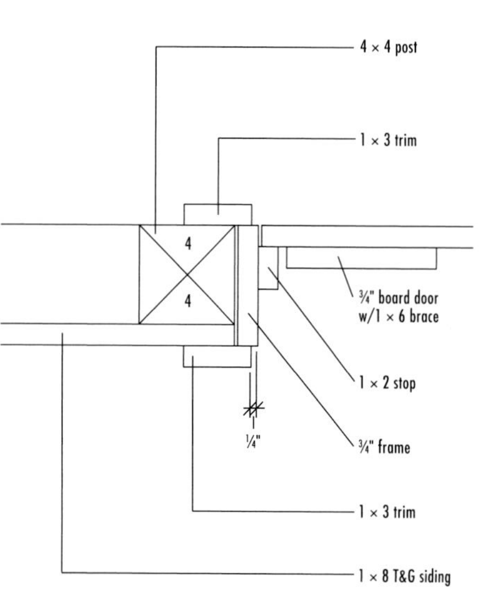

Door Detail

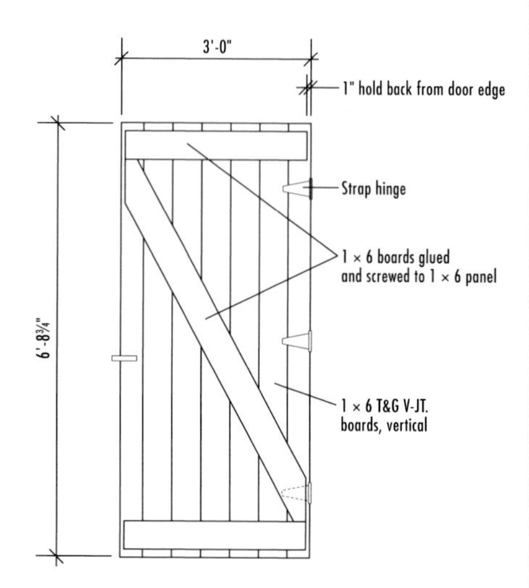

How to Build a Timber-frame Shed

Prepare the foundation site with a 4"-deep layer of compacted and leveled gravel. Cut three 6 × 6 treated timber skids (120"). Place the skids following the FLOOR FRAMING PLAN (page 178). Lay a straight 2 × 4 across the skids to make sure they are level.

Cut two 2 × 6 rim joists (120") and nine joists (93"). Assemble the floor frame with galvanized nails, as shown in the FLOOR FRAMING PLAN. Check the frame to make sure it is square by measuring the diagonals.

Position the floor frame on top of the skids and measure the diagonals to make sure it's square. Install joist clip angles at each joist along the two outer skids with galvanized nails. Toenail each joist to the center skid.

Install the tongue-and-groove plywood floor sheathing, starting with a full sheet at one corner of the frame. The flooring should extend to the outside edges of the floor frame.

(continued)

To prepare the wall posts, cut six 4 × 4 posts (90½"), making sure both ends are square (see page 369). On the four corner posts, mark for 3½"-long × 1½"-deep notches (to accept the girts) on the two adjacent inside faces of each post. Start the notches 46¼" from the bottom ends of the posts.

Mark the door frame posts for notches to receive a girt at 46¼" and for the door header at 82"; see the FRONT FRAMING ELEVATION (page 177). Remove the waste from the notch areas with a circular saw and clean up with a broad wood chisel. Test-fit the notches to make sure the 4 × 4 girts will fit snugly.

Position the post bases so the posts will be flush with the outsides of the shed floor. Install the bases with 16d galvanized common nails. The insides of the door posts should be 29" from the floor sides. Brace each post so it is perfectly plumb, and then fasten it to its base using the base manufacturer's recommended fasteners.

Cut two 4 × 6 beams at 10 ft. and two at 8 ft. Notch the ends of the beams for half-lap joints: Measure the width and depth of the beams and mark notches equal to the width × ½ the depth. Orient the notches as shown in the FRAMING ELEVATIONS (page 177). Cut the notches with a handsaw, then test-fit the joints, and make fine adjustments with a chisel.

Set an 8-ft. beam onto the front wall posts and tack it in place with a 16d nail at each end. Tack the other 8-ft. beam to the back posts. Then, position the 10-ft. beams on top of the short beam ends, forming the half-lap joints. Measure the diagonals of the front wall frame to make sure it's square, and then anchor the beams with two 60d galvanized nails at each corner (drill pilot holes for the nails).

Reinforce the beam connections with a metal post-beam connector on the outside of each corner and on both sides of the door posts using the recommended fasteners. Install an L-connector on the inside of the beam-to-beam joints; see the EAVE DETAIL (page 180).

Cut eight 4 × 4 corner braces (20"), mitering the ends at 45°. Install the braces flush with the outsides of the beams and corner posts using two ⅜ × 6" lag screws (with washers) driven through counterbored pilot holes.

Measure between the posts at the notches, and cut the 4 × 4 girts to fit. To allow the girts to meet at the corner posts, make a 1½ × 1½" notch at both ends of the rear wall girts and the outside ends of the front wall girts. Install the girts with construction adhesive and two 20d nails driven through the outsides of the posts (drill pilot holes first). Cut and install the 4 × 4 door header in the same fashion.

(continued)

Frame the roof: Cut two pattern rafters using the RAFTER TEMPLATE (page 178). Test-fit the patterns, and then cut the remaining ten rafters. Cut the 2 × 6 ridge (120"). Install the rafters and ridge using 24" on-center spacing. Cut four 2 × 2s to extend from the roof peak to the rafter ends, and install them flush with the tops of the end rafters; see the GABLE OVERHANG DETAIL (page 180). Add framing connectors at the rafter-beam connections (except at the outer rafters).

Cut four 2 × 4 collar ties (58"), mitering the ends at 45°. Install the ties ½" below the tops of the rafters, as shown in the FRAMING ELEVATIONS.

Install the 1 × 8 siding on the front and rear walls so it runs from the 2 × 2s along the end rafters down to ¾" below the bottom of the floor frame. Fasten the siding with 8d corrosion-resistant finish nails or siding nails. Don't nail the siding to the door header in this step.

Cover the rafter ends along the eaves with 1 × 4 subfascia, flush with the tops of the rafters; see the EAVE DETAIL. Install the 1 × 6 fascia and 1 × 2 trim at the gable ends and along the eaves, mitering the corner joints. Keep the fascia and trim ½" above the rafters so it will be flush with the roof sheathing.

Rip the plywood soffit panels to fit between the wall framing and the fascia; install them with 3d galvanized box nails; see the EAVE DETAIL. Apply siding along the side walls, butting the top ends up to the soffits.

Deck the roof with ½" plywood sheathing, starting at the bottom corners. Cover the sheathing with building paper, overhanging the 1 × 2 fascia trim by ¾". Install the cedar shingle roofing or asphalt shingles following the steps on pages 56 to 59. Include roof vents, if desired (they're a good idea). Finish the roof at the peak with a 1× ridge cap.

Construct the door frame from ¾ × 4¼" stock. Cut the head jamb at 37¾" and the side jambs at 81". Fasten the head jamb over the ends of the side jambs with 2½" deck screws. Install the frame in the door opening using shims and 10d galvanized casing nails. Add 1 × 2 stops to the jambs, ¾" from the outside edges.

Build the door with seven pieces of 1 × 6 siding cut at 80¾". Fit the boards together, then mark and trim the outer pieces so the door is 36" wide. Install the 1 × 6 Z-bracing with adhesive and 1¼" wood screws, as shown in the DOOR DETAIL (page 180). Install flashing over the outside of the door, then add 1 × 3 trim around both sides of the door opening, as shown in the DOOR JAMB DETAIL (page 180). Hang the door with two or three strap hinges.

Gothic Playhouse

Playhouses are all about stirring the imagination. Loaded with fancy American Gothic details, this charming little house makes a special play home for kids and an attractive backyard feature for adults. In addition to its architectural character, what makes this a great playhouse design is its size—the enclosed house measures 5 × 7½ ft. and includes a 5-ft.-tall door and plenty of headroom inside. This means your kids will likely "outgrow" the playhouse before they get too big for it. And you can always give the house a second life as a storage shed.

At the front of the house is a 30"-deep porch complete with real decking boards and a nicely decorated railing. Each side wall features a window and flower box, and the "foundation" has the look of stone created by wood blocks applied to the floor framing. All of these features are optional, but each one adds to the charm of this well-appointed playhouse.

As shown here, the floor of the playhouse is anchored to four 4 × 4 posts buried in the ground. As an alternative, you can set the playhouse on 4 × 6 timber skids. Another custom variation you might consider is in the styling of the verge boards (the gingerbread gable trim). Instead of using the provided pattern, you can create a cardboard template of your own design. Architectural plan and pattern books from the Gothic period are full of inspiration for decorative ideas.

Gothic Style ▸

The architectural style known as American Gothic (also called Gothic Revival and Carpenter Gothic) dates back to the 1830s and essentially marks the beginning of the Victorian period in American home design. Adapted from a similar movement in England, Gothic style was inspired by the ornately decorated stone cathedrals found throughout Europe. The style quickly evolved in America as thrifty carpenters learned to re-create and reinterpret the original decorative motifs using wood instead of stone.

American Gothic's most characteristic feature is the steeply pitched roof with fancy scroll-cut bargeboards, or verge boards, which gave the style its popular nickname, "gingerbread." Other typical features found on Gothic homes (and the Gothic Playhouse) include board-and-batten siding, doors and windows shaped with Gothic arches, and spires or finials adorning roof peaks.

The roomy interior of this playhouse is a perfect space for imaginative play within view of your home.

Finish the gothic playhouse with bright colors to enhance its decorative appeal.

Cutting List

Description	Quantity/Size	Material
Foundation/Floor		
Drainage material	1 cu. yd.	Compactable gravel
Foundation posts	4 @ field measure	4 × 4 pressure-treated landscape timbers
Concrete	Field measure	3,000 psi concrete
Rim joists	3 @ 10', 1 @ 8'	2 × 12 pressure-treated, rated for ground contact
Floor joists	1 @ 10', 2 @ 8'	2 × 6 pressure-treated
Box sills (rim joists)	2 @ 12'	2 × 4 pressure-treated
Floor sheathing	2 sheets @ 4 × 8'	¾" ext.-grade plywood
Porch decking	5 @ 10'	1 × 6 pressure-treated decking
Foundation "stones"	7 @ 10'	5/4 × 6 treated decking w/radius edge (R.E.D.), rated for ground contact
Framing		
Wall framing & railings	29 @ 12'	2 × 4
Rafters & spacers	7 @ 12'	2 × 4
Ridge board	1 @ 8'	1 × 6
Collar ties	1 @ 10'	1 × 4
Exterior Finishes		
Siding, window boxes & door trim	26 @ 10'	1 × 8 pressure-treated or cedar
Battens & trim	30 @ 8'	1 × 2 pressure-treated or cedar
Door panel, verge boards & fascia	10 @ 10'	1 × 6 pressure-treated or cedar
Door braces, trim & railing trim	2 @ 10'	1 × 4 pressure-treated or cedar
Deck railing posts	2 @ 8'	4 × 4 pressure-treated or cedar
Railing balusters	4 @ 8'	2 × 2 pressure-treated or cedar
Deck railing	1 @ 8'	2 × 4 pressure-treated or cedar
Window stops	2 @ 8'	⅜" pressure-treated or cedar quarter-round molding
Window glazing (optional)	4 @ 20 × 9½"	¼" plastic glazing
Spire		
Post	1 @ 8'	4 × 4 pressure-treated
Trim	1 @ 4'	1 × 2 pressure-treated

Description	Quantity/Size	Material
Molding	1 @ 4'	Cap molding, pressure-treated
Balls	2 @ 3"-dia.	Wooden sphere, pressure-treated
Roofing		
Sheathing	4 sheets @ 4 × 8'	½" exterior-grade plywood roof sheathing
15# building paper	1 roll	
Drip edge	40 linear ft.	Metal drip edge
Shingles	1 square	Asphalt shingles—250# per sq. min.
Fasteners & Hardware		
16d galvanized common nails	3½ lbs.	
16d common nails	5 lbs.	
10d common nails (for double top plates)	½ lb.	
10d galvanized finish/casing nails	4 lbs.	
8d galvanized common nails	1 lb.	
8d box nails	2 lbs.	
8d galvanized siding nails	8 lbs.	
1" galvanized roofing nails	3 lbs.	
2" deck screws (for porch decking)	1 lb.	
6d galvanized finish nails	2 lbs.	
3½" galvanized wood screws	24 screws	
1¼" galvanized wood screws	12 screws	
Galvanized dowel screws (for spire)	3 screws	
½" Galvanized lag screws w/washers	2 @ 6"	
Door hinges w/screws	3	Corrosion-resistant hinges
Door handle/latch	1	
Exterior wood glue		
Clear exterior caulk (for optional window panes)		
Construction adhesive		

Playhouse

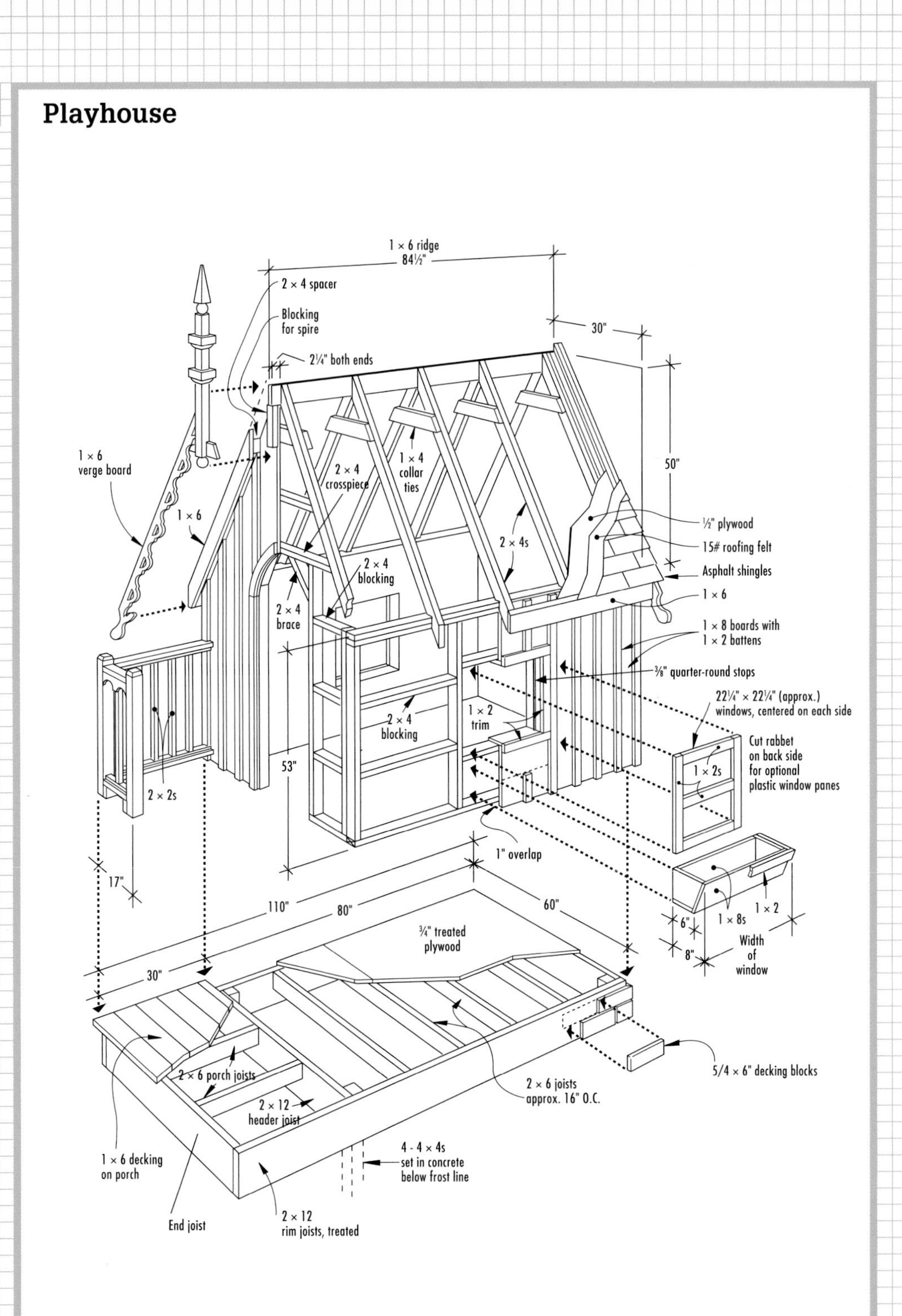

1 × 6 ridge
84½"
2 × 4 spacer
Blocking for spire
2¼" both ends
30"
50"
1 × 6 verge board
1 × 6
2 × 4 crosspiece
1 × 4 collar ties
2 × 4s
½" plywood
15# roofing felt
Asphalt shingles
2 × 4 blocking
2 × 4 brace
1 × 6
1 × 8 boards with 1 × 2 battens
⅜" quarter-round stops
22¼" × 22¼" (approx.) windows, centered on each side
2 × 4 blocking
1 × 2 trim
Cut rabbet on back side for optional plastic window panes
1 × 2s
53"
2 × 2s
1" overlap
17"
110"
80"
60"
¾" treated plywood
6"
1 × 8s
1 × 2
8"
Width of window
30"
2 × 6 porch joists
2 × 12 header joist
2 × 6 joists approx. 16" O.C.
5/4 × 6" decking blocks
1 × 6 decking on porch
4 - 4 × 4s set in concrete below frost line
End joist
2 × 12 rim joists, treated

Floor Plan

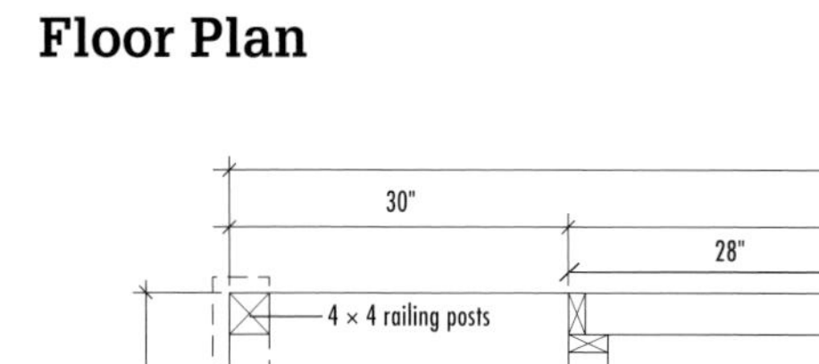
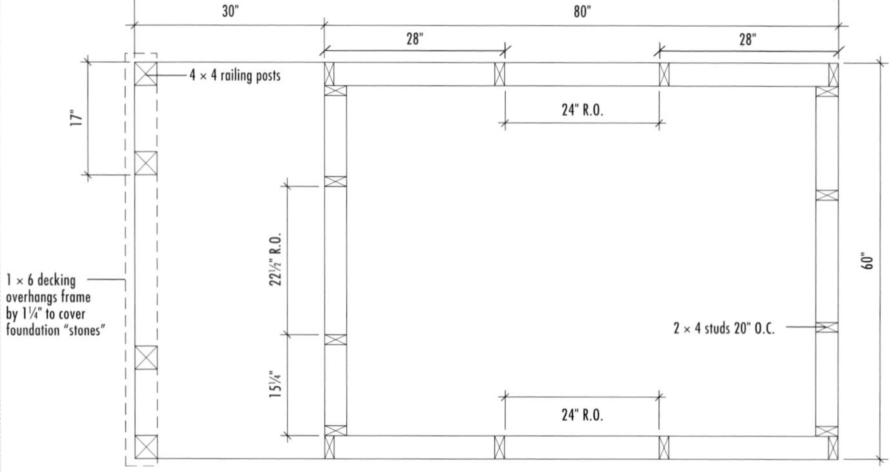

Verge Board Template

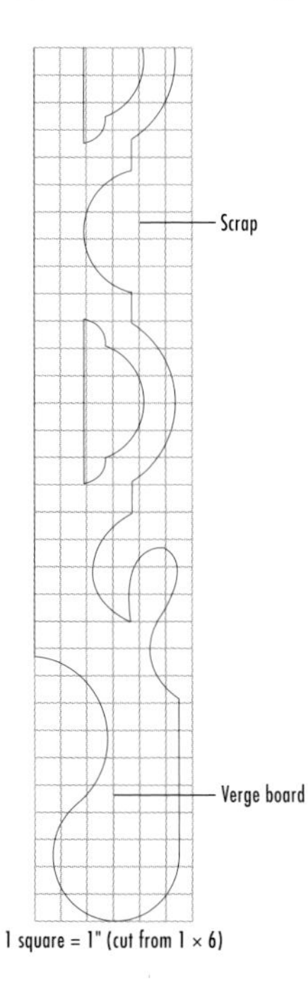

Deck Railing Detail

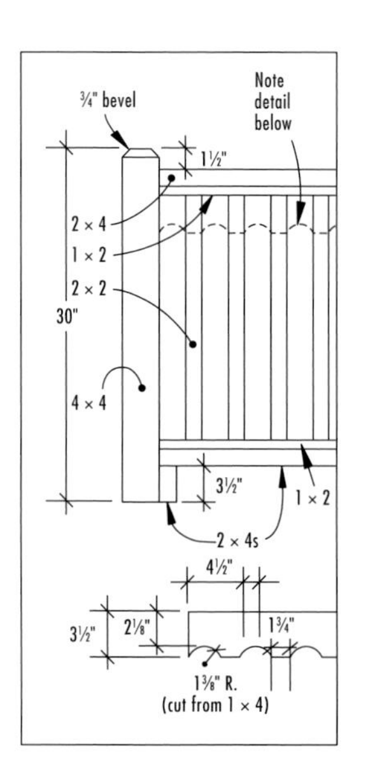

Spire Detail

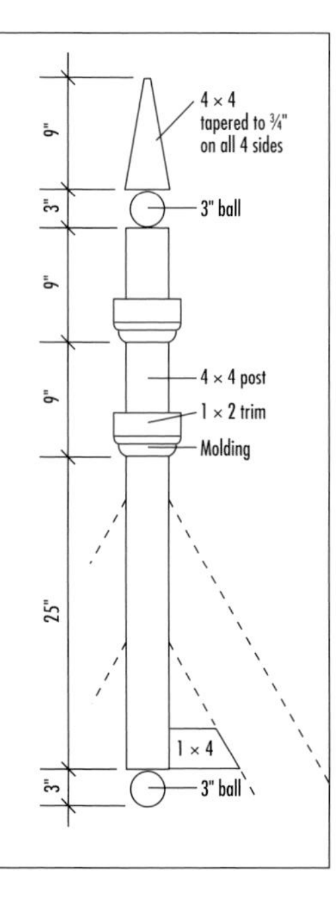

Door Detail

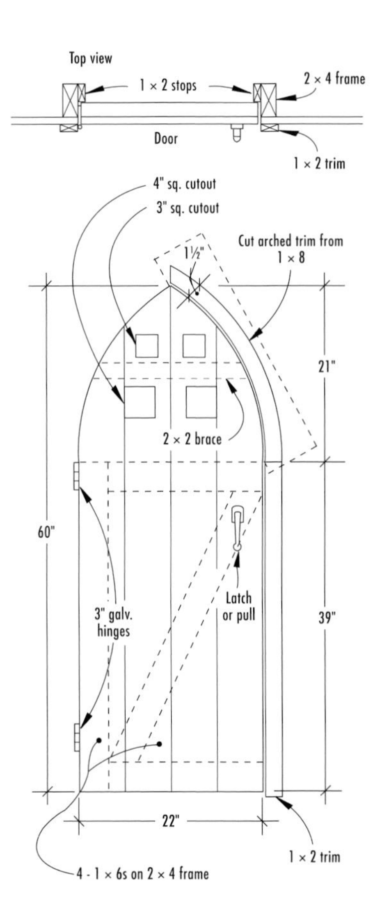

Door Arch Template

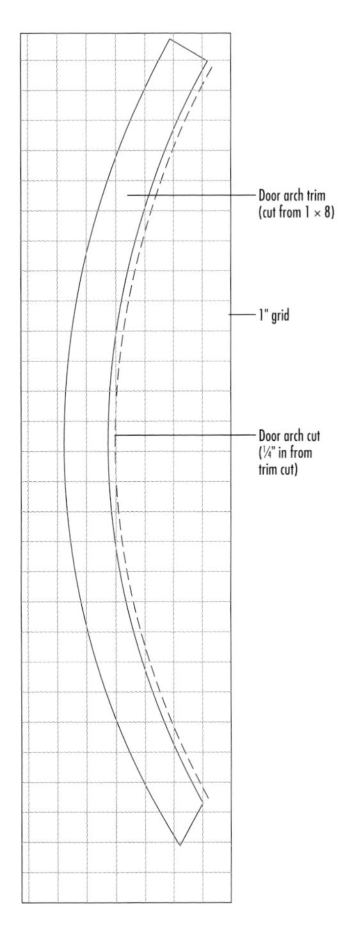

Board & Batten Detail

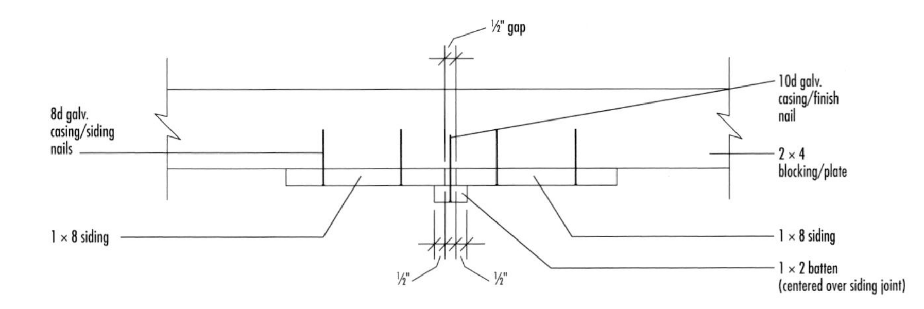

Front Framing

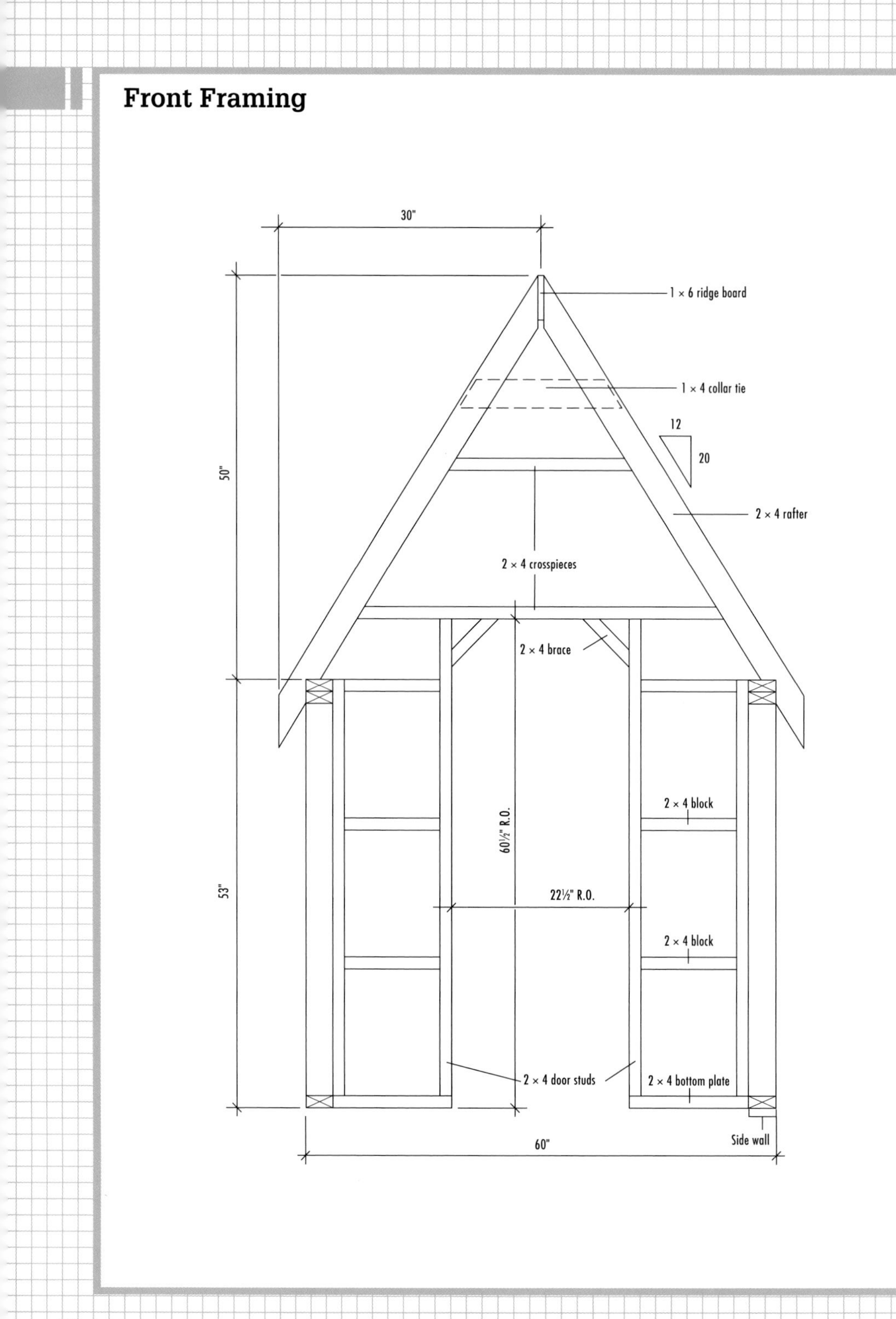

Floor Framing Plan

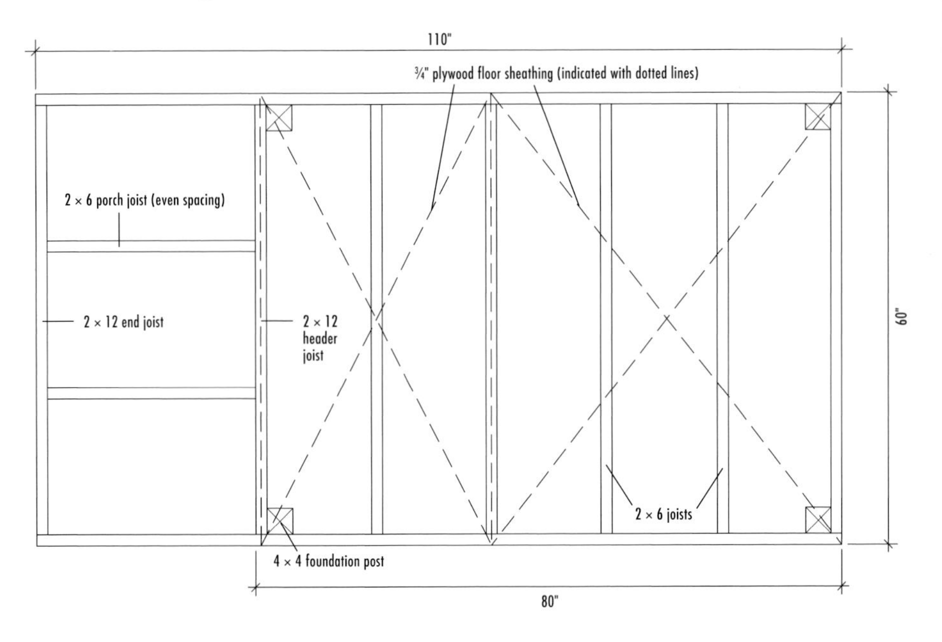

Side Framing

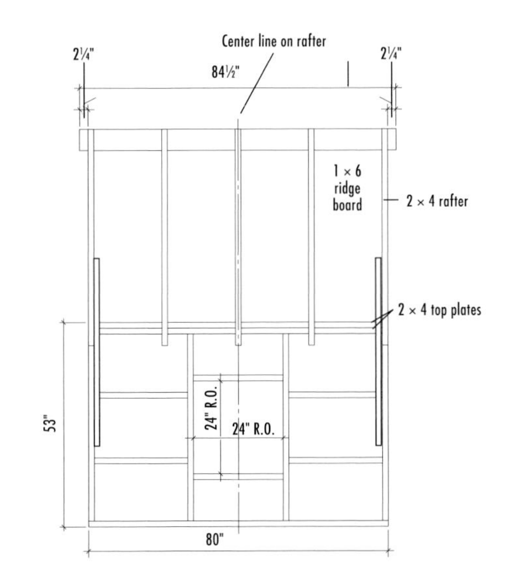

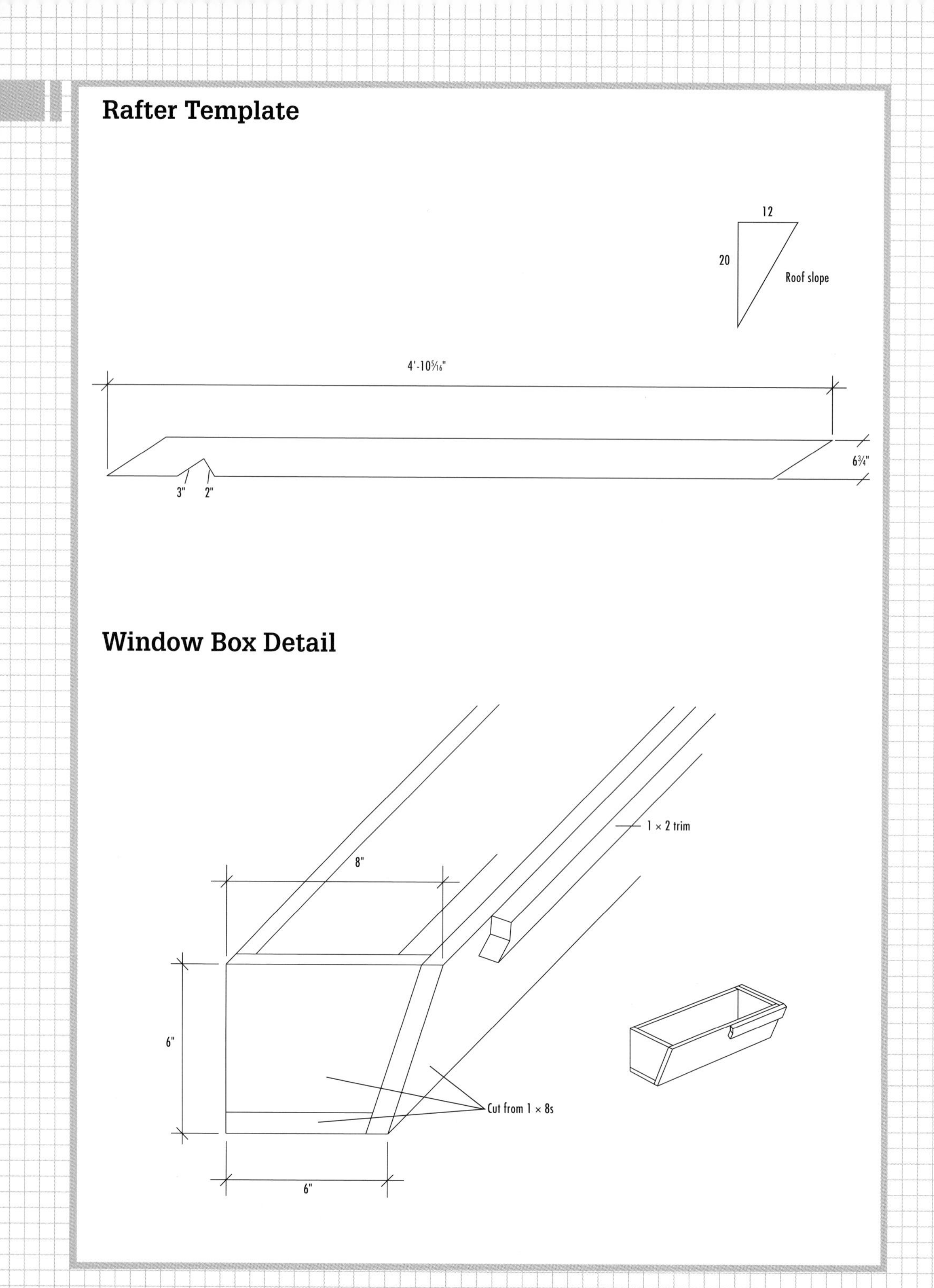
Rafter Template
12
20
Roof slope
4'-10⁵⁄₁₆"
6¾"
3"
2"
Window Box Detail
1 × 2 trim
8"
6"
Cut from 1 × 8s
6"

How to Build a Gothic Playhouse

Set up stakes and perpendicular mason's lines to plot out the excavation area and the post hole locations, as shown in FLOOR FRAMING PLAN (page 193). Excavate and grade the construction area, preparing for a 4"-thick gravel base. Dig 12"-dia. holes to a depth below the frost line, plus 4". Add 4" of gravel to each hole. Set the posts in concrete so they extend about 10" above the ground.

After the concrete dries (overnight) add compactable gravel and tamp it down so it is 4" thick and flat and level. Cut two 2 × 12 rim joists for the floor frame, two 2 × 12 end joists, and one header joist. Cut four 2 × 6 joists at 57" and two porch joists at 27¾". Assemble the floor frame with 16d galvanized common nails following FLOOR FRAMING PLAN.

Make sure the frame is square and level (prop it up temporarily), and then fasten it to the posts with 16d galvanized common nails.

Cover the interior floor with plywood, starting at the rear end. Trim the second piece so it covers ½ of the header joist. Install the 1 × 6 porch decking starting at the front edge and leaving a ⅛" gap between boards. Extend the porch decking 1¼" beyond the front and sides of the floor frame.

(continued)

Frame the side walls as shown in SIDE FRAMING (page 193) and FLOOR PLAN (page 190). Each wall has four 2 × 4 studs at 48½", a top and bottom plate at 80", and a 2 × 4 window header and sill at 24". Install the horizontal 2 × 4 blocking, spaced evenly between the plates. Install only one top plate per wall at this time.

Build the rear wall. Raise the side and rear walls, and fasten them to each other and to the floor frame. Add double top plates. Both sidewall top plates should stop flush with the end stud at the front of the wall.

To frame the front wall, cut two treated bottom plates at 15¼", two end studs at 51½", and two door studs at 59". Cut a 2 × 4 crosspiece and two braces, mitering the brace ends at 45°. Cut six 2 × 4 blocks at 12¼". Assemble the wall as shown in the FRONT FRAMING (page 192). Raise the front wall and fasten it to the floor and sidewall frames.

Cut one set of 2 × 4 pattern rafters following the RAFTER TEMPLATE (page 194). Test-fit the rafters and make any necessary adjustments. Use one of the pattern rafters to mark and cut the remaining eight rafters. Also cut four 2 × 4 spacers—these should match the rafters but have no bird's-mouth cuts.

Cut the ridge board to size and mark the rafter layout following SIDE FRAMING, and then screw the rafters to the ridge. Cut five 1 × 4 collar ties, mitering the ends at 31°. Fasten the collar ties across each set of rafters so the ends of the ties are flush with the rafter edges. Fasten the 2 × 4 crosspiece above the door to the two end rafters. Install the remaining crosspieces as shown in FRONT FRAMING.

Install the 1 × 8 siding boards so they overlap the floor frame by 1" at the bottom and extend to the tops of the side walls, and to the tops of the rafters on the front and rear walls. Gap the boards ½", and fasten them to the framing with pairs of 8d galvanized casing nails or siding nails. Install the four 2 × 4 spacers on top of the siding at the front and rear so they match the rafter placement.

Cut the arched sections of door trim from 1 × 8 lumber, following the DOOR ARCH TEMPLATE (page 191). Install the arched pieces and straight 1 × 2 side pieces flush with the inside of the door opening. Wrap the window openings with ripped 1 × 6 boards, and then frame the outsides of the openings with 1 × 2 trim. Install a 1 × 2 batten over each siding joint as shown in step 10.

Build the 1 × 2 window frames to fit the trimmed openings. Assemble the parts with exterior wood glue and galvanized finish nails. If desired, cut a ¼" rabbet in the back side and install plastic windowpanes with silicone caulk. Secure the window frames in the openings with ⅜" quarter-round molding. Construct the window boxes as shown in the WINDOW BOX DETAIL (page 194). Install the boxes below the windows with 1¼" screws.

(continued)

To build the spire, start by drawing a line around a 4 × 4 post, 9" from one end. Draw cutting lines to taper each side down to ¾", as shown in the SPIRE DETAIL (page 190). Taper the end with a circular saw or handsaw, and then cut off the point at the 9" mark. Cut the post at 43". Add 1 × 2 trim and cap molding as shown in the detail, mitering the ends at the corners. Drill centered pilot holes into the post, balls, and point, and join the parts with dowel screws.

To cut the verge boards, enlarge the VERGE BOARD TEMPLATE (page 190) on a photocopier so the squares measure 1". Draw the pattern onto a 1 × 6. Cut the pattern shape with a jigsaw. Test-fit the board and adjust as needed. Use the cut board as a pattern to mark and cut the remaining verge boards.

Add a 1 × 2 block under the front end of the ridge board. Center the spire at the roof peak, drill pilot holes, and anchor the post with 6" lag screws. Cut and install the 1 × 6 front fascia to run from the spire to the rafter ends, keeping the fascia ½" above the tops of the rafters. Install the rear fascia so it covers the ridge board. Install the verge boards over the front end near fascia, flush with the tips of the fascia. Cut and install two 1 × 4 brackets to fit between the spire post and front fascia, as shown in the SPIRE DETAIL.

Cut the 1 × 6 eave fascia to fit between the verge boards, and install it so it will be flush with the top of the roof sheathing. Cut and install the roof sheathing. Add building paper, metal drip edge, and asphalt shingles, following the steps on page 60 to 67.

Mark the deck post locations 1¼" in from the ends and front edge of the porch decking, as shown in the FLOOR PLAN. Cut four 4 × 4 railing posts at 30". Bevel the top edges of the posts at 45°, as shown in DECK RAILING DETAIL (page 190). Fasten the posts to the decking and floor frame with 3½" screws. Cut six 2 × 4 treated blocks at 3½". Fasten these to the bottoms of the posts, on the sides that will receive the railings.

Assemble the railing sections following the DECK RAILING DETAIL. Each section has a 2 × 4 top and bottom rail, two 1 × 2 nailers, and 2 × 2 balusters spaced so the edges of the balusters are no more than 4" apart. You can build the sections completely and then fasten them to the posts and front wall, or you can construct them in place starting with the rails. Cut the shaped trim boards from 1 × 4 lumber using a jigsaw. Notch the rails to fit around the house battens as needed.

Construct the door with 1 × 6 boards fastened to 2 × 4 Z-bracing, as shown in the DOOR DETAIL. Fasten the boards to the bracing with glue and 6d finish nails. Cut the square notches and the top of the door with a jigsaw. Add the 2 × 2 brace as shown. Install the door with two hinges, leaving a ¼" gap all around. Add a knob or latch as desired.

Make the foundation "stones" by cutting 116 6"-lengths of 5/4 × 6 deck boards (the pieces in the top row must be ripped down 1"). Round over the cut edges of all pieces with a router. Attach the top row of stones using construction adhesive and 6d galvanized finish nails. Install the bottom row, starting with a half-piece to create a staggered joint pattern. If desired, finish the playhouse interior with plywood or tongue-and-groove siding.

Lean-to Tool Bin

The lean-to is a classic outbuilding intended as a supplementary structure for a larger building. Its simple shed-style roof helps it blend with the neighboring structure and directs water away and keeps leaves and debris from getting trapped between the two buildings. When built to a small shed scale, the lean-to (sometimes called a closet shed) is most useful as an easy-access storage locker that saves you extra trips into the garage for often-used lawn and garden tools and supplies.

This lean-to tool bin is not actually attached to the house, though it appears to be. It is designed as a freestanding building with a wooden skid foundation that makes it easy to move. With all four sides finished, the bin can be placed anywhere, but it works best when set next to a house, or garage wall, or a tall fence. If you locate the bin out in the open—where it won't be protected against wind and extreme weather—be sure to anchor it securely to the ground to prevent it from blowing over.

As shown here, the bin is finished with asphalt shingle roofing, T1-11 plywood siding, and 1× cedar trim, but you can substitute any type of finish to match or complement a neighboring structure. Its 65"-tall double doors provide easy access to its 18 square feet of floor space. The 8-ft.-tall rear wall can accommodate a set of shelves while leaving enough room below for long-handled tools.

Because the tool bin sits on the ground, in cold climates it will be subject to shifting with seasonal freeze-thaw cycles. Therefore, do not attach the tool bin to your house or any other building set on a frost-proof foundation.

The 8-ft. wall at the back of the tool bin is perfect for storing tall garden tools, or shelving

Locate the tool bin against a house or garage wall for greatest stability. Do not, however, attach the temporary structure to the permanent structure.

Cutting List

Description	Quantity/Size	Material
Foundation		
Drainage material	0.5 cu. yd.	Compactable gravel
Skids	2 @ 6'	4 × 4 treated timbers
Floor framing		
Rim joists*	2 @ 6'	2 × 6 pressure-treated
Joists	3 @ 8'	2 × 6 pressure-treated
Floor sheathing	1 sheet @ 4 × 8	¾" tongue-&-groove ext.-grade plywood
Joist clip angles	4	3 × 3 × 3" × 16-gauge galvanized
Wall Framing		
Bottom plates	1 @ 8', 2 @ 6'	2 × 4
Top plates	1 @ 8', 3 @ 6'	2 × 4
Studs	14 @ 8', 8 @ 6'	2 × 4
Header	2 @ 6'	2 × 6
Header spacer	1 piece @ 6'	½" plywood — 5" wide
Roof Framing		
Rafters	6 @ 6'	2 × 6
Ledger	1 @ 6'	2 × 6
Roofing		
Roof sheathing	2 sheets @ 4 × 8'	½" ext.-grade plywood
Shingles	30 sq. ft.	250# per square min.
Roofing starter strip	7 linear ft.	
15# building paper	30 sq. ft.	
Metal drip edge	24 linear ft.	Galvanized metal
Roofing cement	1 tube	
Exterior Finishes		
Plywood siding	4 sheets @ 4 × 8'	⅝" texture 1-11 plywood siding, grooves 8" O.C.
Door trim	2 @ 8' 2 @ 6'	1 × 10 S4S cedar 1 × 8 S4S cedar
Corner trim	6 @ 8'	1 × 4 S4S cedar
Fascia	3 @ 6' 1 @ 6'	1 × 8 S4S cedar 1 × 4 S4S cedar
Bug screen	8" × 6'	Fiberglass
Doors		
Frame	3 @ 6'	¾" × 3½" (actual) cedar
Stops	3 @ 6'	1 × 2 S4S cedar
Panel material	12 @ 6'	1 × 6 T&G V-joint S4S cedar
Z-braces	2 @ 10'	1 × 6 S4S cedar
Construction adhesive	1 tube	
Interior trim (optional)	3 @ 6'	1 × 3 S4S cedar
Strap hinges	6, with screws	
Fasteners		
16d galvanized common nails	3½ lbs.	
16d common nails	3½ lbs.	
10d common nails	12 nails	
10d galvanized casing nails	20 nails	
8d galvanized box nails	½ lb.	
8d galvanized finish nails	2 lbs.	
8d common nails	24 nails	
8d box nails	½ lb.	
1½" joist hanger nails	16 nails	
⅞" galvanized roofing nails	¼ lb.	
2½" deck screws	6 screws	
1¼" wood screws	60 screws	

**Note: 6-foot material is often unavailable at local lumber stores, so buy half as much of 12-foot material.*

Floor Framing Plan

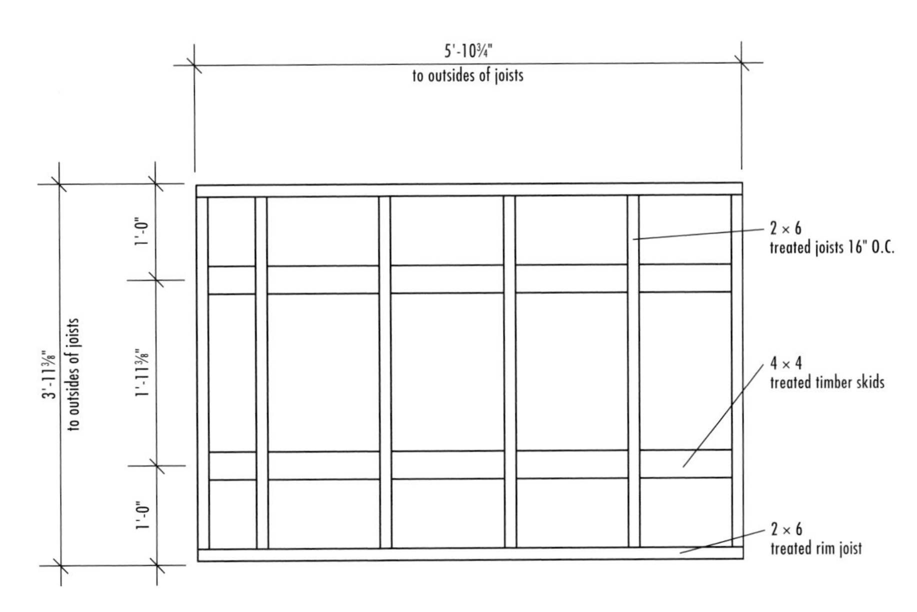

Roof Framing Plan

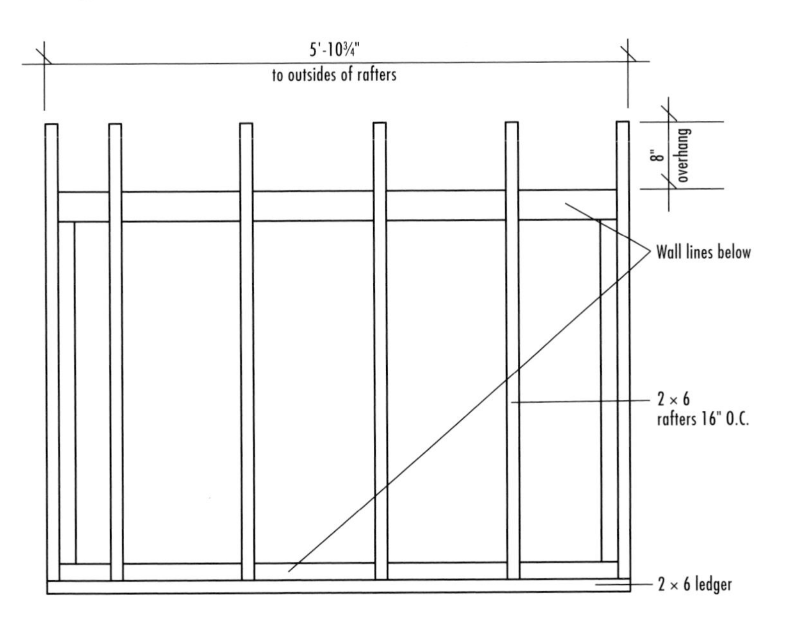

Front Framing Elevation

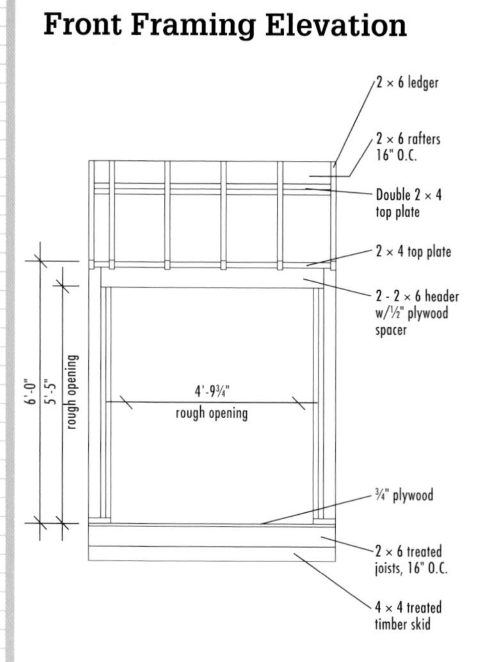

Left Framing Elevation

Rear Framing Elevation

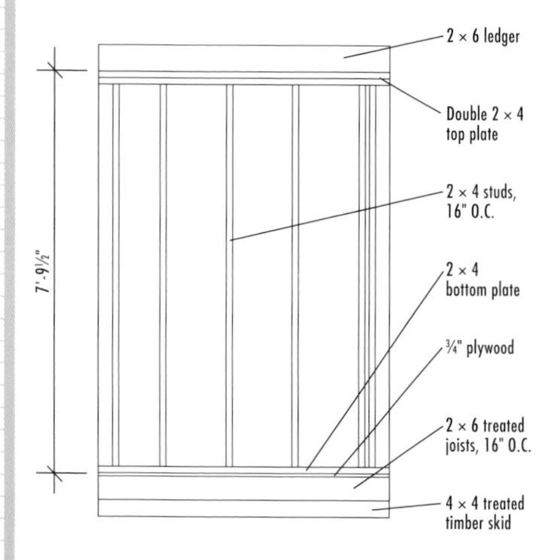

Right Side Framing Elevation

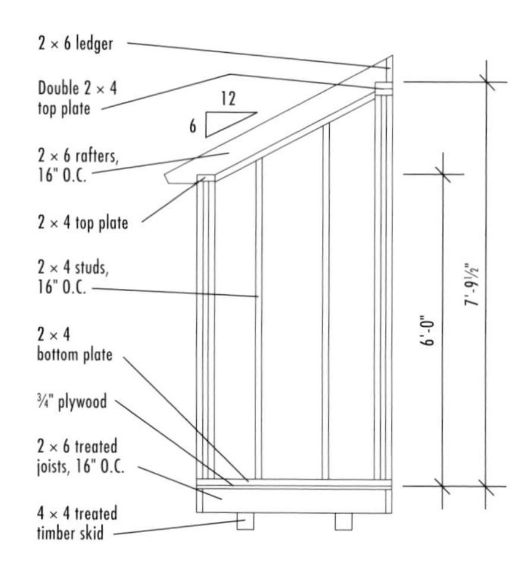

Building Section

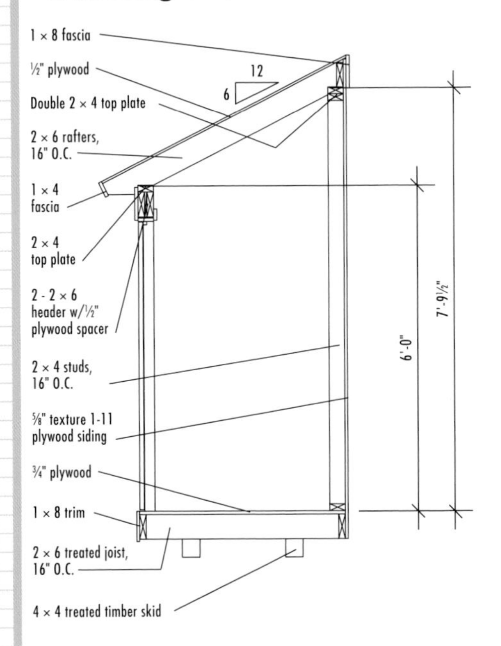

Side Elevation

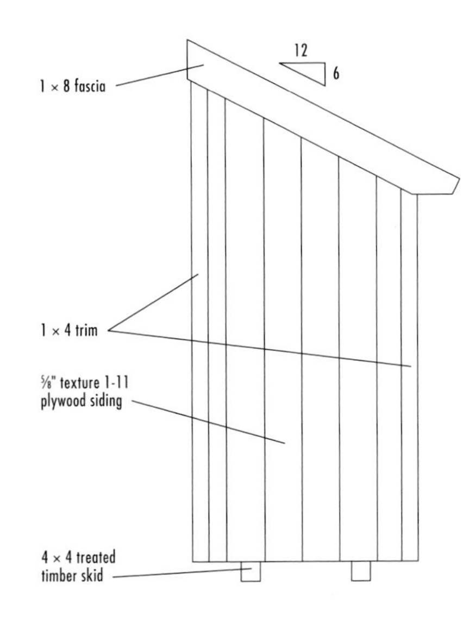

Front Elevation

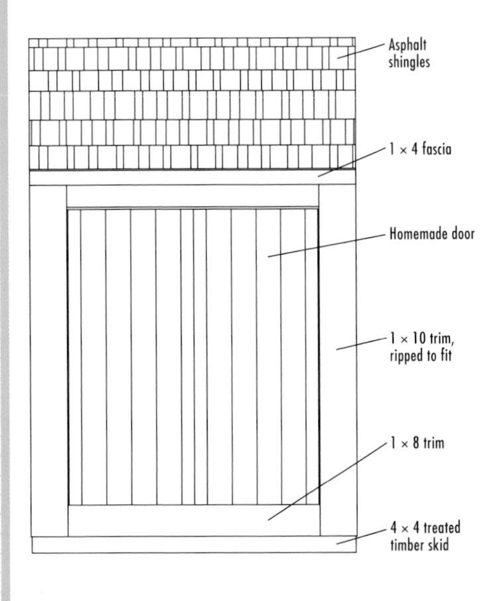

Rear Elevation

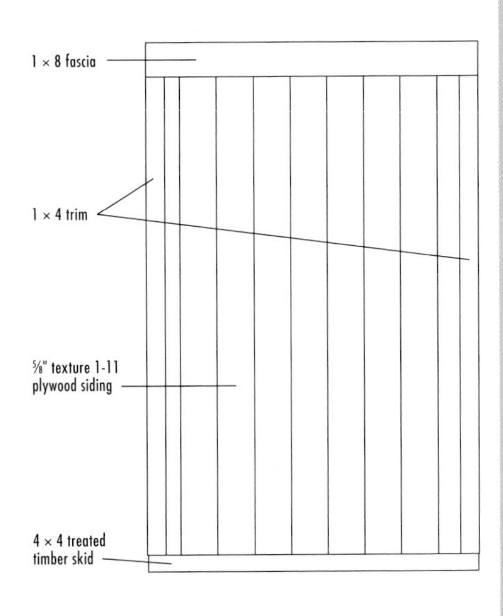

Floor Plan

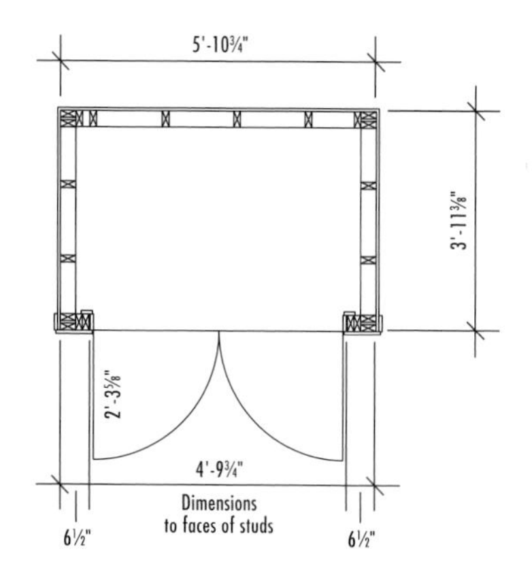

Rafter Template

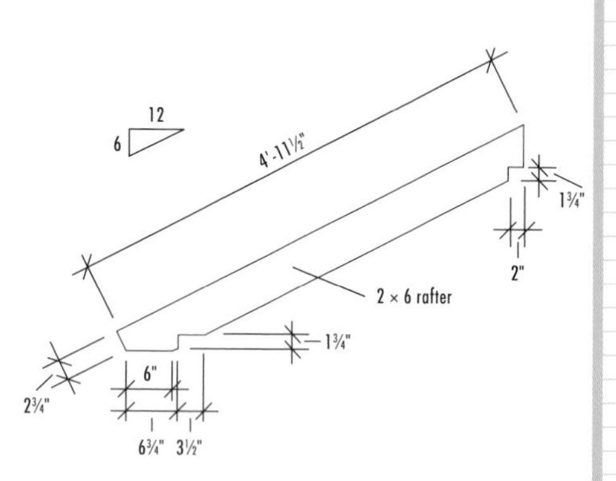

Side Roof Edge Detail

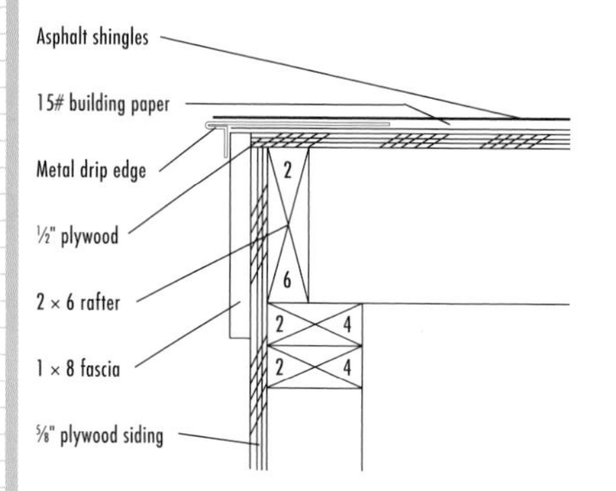

Overhang Detail

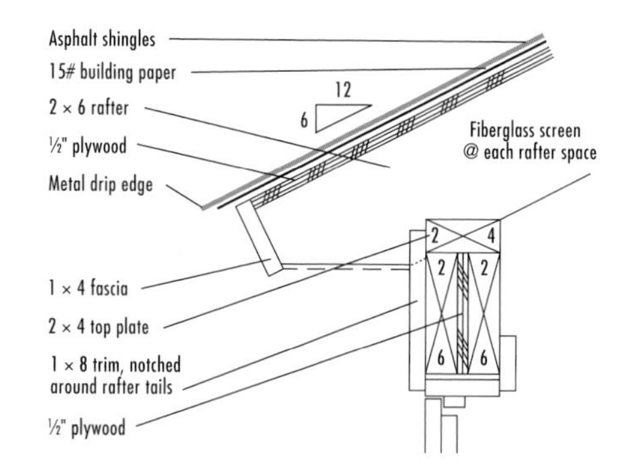

Door Jamb Detail

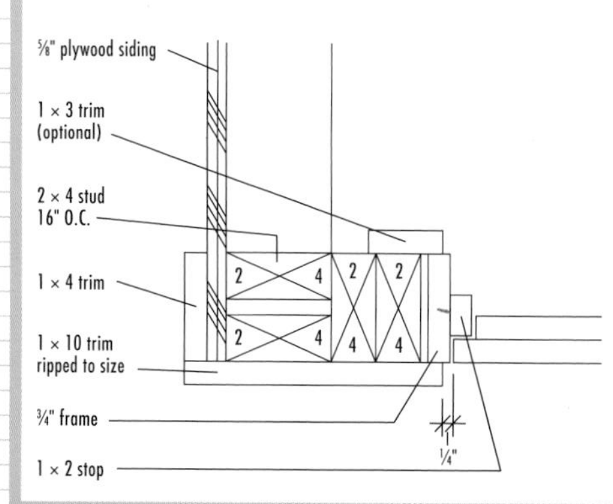

Door Elevation

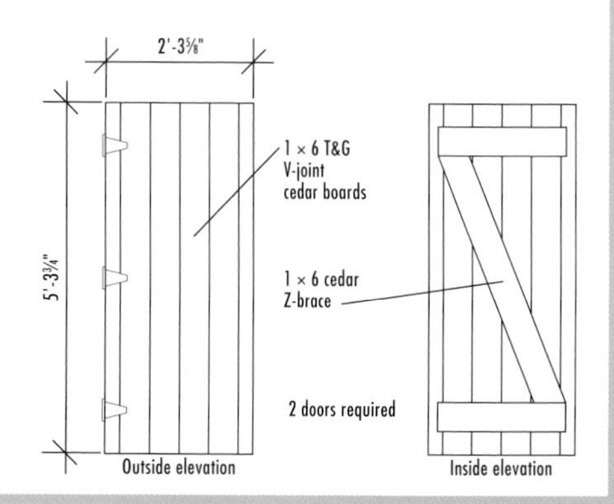

How to Build a Lean-to Tool Bin

Prepare the site with a 4" layer of compacted gravel. Cut the two 4 × 4 skids at 70¾". Set and level the skids following FLOOR FRAMING PLAN (page 203). Cut two 2 × 6 rim joists at 70¾" and six joists at 44⅜". Assemble the floor and set it on the skids as shown in the FLOOR FRAMING PLAN. Check for square, and then anchor the frame to the skids with four joist clip angles (inset photo). Sheath the floor frame with ¾" plywood.

Cut plates and studs for the walls: Side walls—two bottom plates at 47⅜", four studs at 89", and four studs at 69"; Front wall—one bottom plate at 63¾", one top plate at 70¾", and four jacks studs at 63½". Rear wall—one bottom plate at 63¾", two top plates at 70¾", and six studs at 89". Mark the stud layouts onto the plates.

Fasten the four end studs of each side wall to the bottom plate. Install these assemblies. Construct the built-up 2 × 6 door header at 63¾". Frame and install the front and rear walls, leaving the top plates off at this time. Nail together the corner studs, making sure they are plumb. Install the rear top plates flush to the outsides of the side wall studs. Install the front top plate in the same fashion.

Cut the six 2 × 6 rafters following the RAFTER TEMPLATE (page 206). Cut the 2 × 6 ledger at 70¾" and bevel the top edge at 26.5° so the overall width is 4⁵⁄₁₆". Mark the rafter layout onto the wall plates and ledger, as shown in the ROOF FRAMING PLAN (page 203), then install the ledger flush with the back side of the rear wall. Install the rafters.

(continued)

Complete the side wall framing: Cut a top plate for each side to fit between the front and rear walls, mitering the ends at 26.5°. Install the plates flush with the outsides of the end rafters. Mark the stud layouts onto the side wall bottom plates, then use a plumb bob to transfer the marks to the top plate. Cut the two studs in each wall to fit, mitering the top ends at 26.5°. Install the studs.

Sheath the side walls and rear walls with plywood siding, keeping the bottom edges ½" below the floor frame and the top edges flush with the tops of the rafters. Overlap the siding at the rear corners, and stop it flush with the face of the front wall.

Add the 1 × 4 fascia over the bottom rafter ends as shown in the OVERHANG DETAIL (page 206). Install 1 × 8 fascia over the top rafter ends. Position all fascia ½" above the rafters so it will be flush with the roof deck. Overhang the front and rear fascia to cover the ends of the side fascia, or plan to miter all fascia joints. Cut the 1 × 8 side fascia to length, and then clip the bottom front corners to meet the front fascia. Install the side fascia.

Install the ½" roof sheathing, starting with a full-width sheet at the bottom edge of the roof. Fasten metal drip edge along the front edge of the roof. Cover the roof with building paper, then add the drip edge along the sides and top of the roof. Shingle the roof (see pages 60 to 67), and finish the top edge with cut shingles or a solid starter strip.

Cut and remove the bottom plate inside the door opening. Cut the 1 × 4 head jamb for the door frame at 57⅛" and cut the side jambs at 64". Fasten the head jamb over the sides with 2½" deck screws. Install 1 × 2 doorstops ¾" from the front edges of jambs, as shown in the DOOR JAMB DETAIL (page 206). Install the frame in the door opening using shims and 10d casing nails.

For each door, cut six 1 × 6 tongue-and-groove boards at 63¾". Fit them together, then mark and trim the two end boards so the total width is 27⅝". Cut the 1 × 6 Z-brace boards following the DOOR ELEVATION (page 206). The ends of the horizontal braces should be 1" from the door edges. Attach the braces with construction adhesive and 1¼" screws. Install each door with three hinges.

Staple fiberglass insect screen along the underside of the roof, securing it to each rafter. Cut and install the 1 × 8 trim above the door, overlapping the side door jambs about ¼" on each side (see the OVERHANG DETAIL, page 206).

Rip vertical and horizontal trim boards to width. Notch the ripped 1 × 8 to fit around the rafters, as shown in the DOOR OVERHANG DETAIL (page 206). Notch the top ends of the ripped 1 × 10s to fit between the rafters and install them. Add the notched 1 × 8 trim horizontally below the door, between the 1 × 10s. Install the 1 × 4 corner trim, overlapping the pieces at the rear corners.

Additional Barn & Shed Plans

Post & Beam Barn

- 14 × 20-ft, 280 square feet
- Wood board flooring on skid foundation
- Height, floor to peak: 10'-6"
- 6-ft. double doors with 3 × 6-ft. ramp
- 1 × 12 rough-sawn pine board and batten siding
- Corrugated metal roofing
- Available as a kit, or purchase the complete plans and instructions

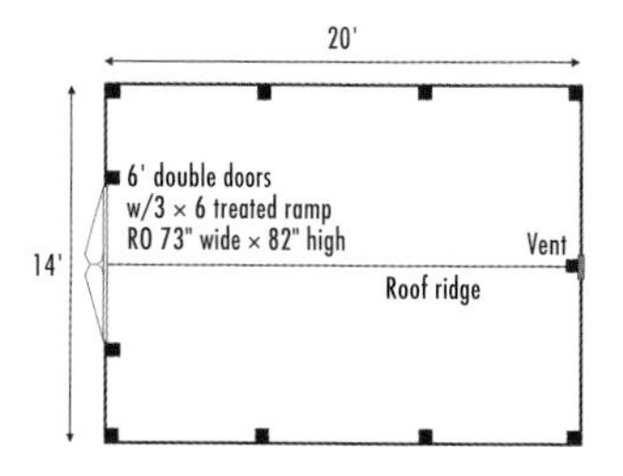

Post & Beam Shed

- 10 × 14-ft., 140 square feet
- Wood board flooring on skid foundation
- Height, floor to peak: 8'-6"
- 5-ft. double doors with 3 × 5-ft. ramp (optional)
- 1 × 12 rough-sawn pine board siding
- Corrugated metal roofing
- Available as a kit, or purchase the complete plans and instructions

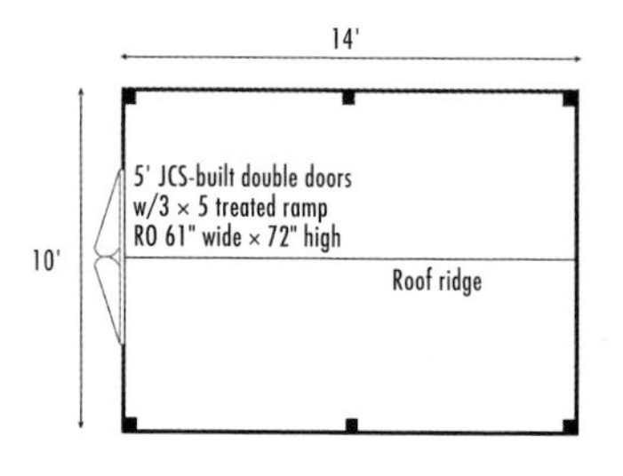

Visit www.jamaicacottageshop.com to order and view additional projects.

Combined Firewood & Storage Shed

- 8 × 16-ft., 128 square feet
- 8 × 10-ft. open firewood storage, 6 × 8-ft. closed storage
- Hemlock or plywood floor on skid foundation
- Height, floor to peak: 10'-6"
- 5-ft. double doors
- 1 × 12 rough-sawn pine board and batten siding
- Corrugated metal roofing
- Available as a kit, or purchase the complete plans and instructions

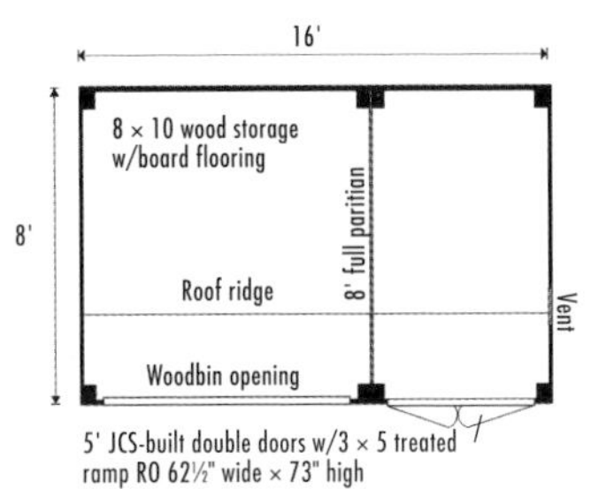

Run-in Shed

- 12 × 20-ft., 240 square feet
- Crushed gravel foundation
- Height, floor to peak: 10'-6"
- 36"-high solid kick plate around interior
- 8-ft.-wide doorway openings
- 1"-thick pine board-and-batten siding
- Corrugated metal roofing
- Available as a kit, or purchase the complete plans and instructions

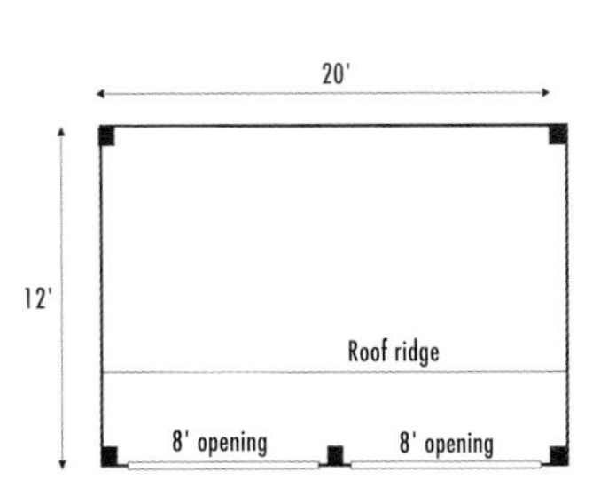

Visit www.jamaicacottageshop.com to order and view additional projects.

Barn Storage Shed with Loft

- Three popular sizes:
  12' × 12'
  12' × 16'
  12' × 20'
- Wood floor on concrete pier foundation or concrete floor
- Height, floor to peak: 12'-10"
- Ceiling height: 7'-4"
- 4'-0" × 6'-8" double door for easy access
- Complete list of materials
- Step-by-step instructions

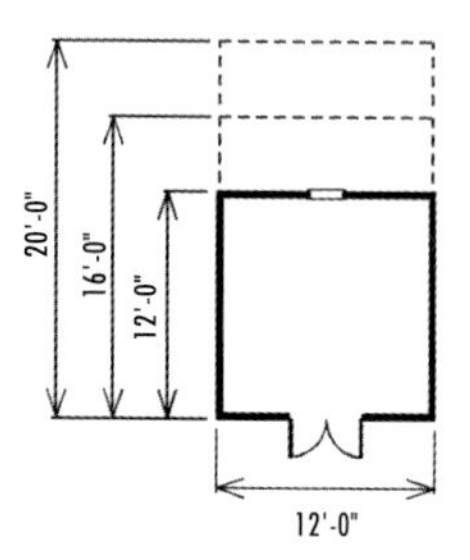

Barn Storage Sheds

- Three popular sizes:
  12' × 8'
  12' × 12'
  12' × 16'
- Wood floor on concrete pier foundation or concrete floor
- Height, floor to peak: 9'-10"
- Ceiling height: 7'-10"
- 5'-6" × 6'-8" double door for easy access
- Gambrel roof design
- Complete list of materials
- Step-by-step instructions

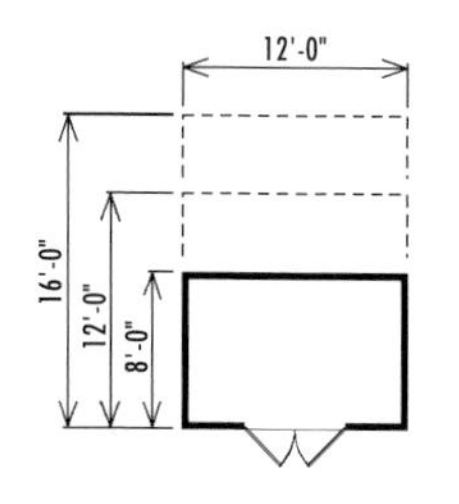

Visit www.projectplans.com to order and view additional projects.

Gable Storage Shed with Cupola

- Size: 12' × 10'
- Wood floor on concrete piers or concrete floor
- Height, floor to peak: 9'-8"
- Ceiling height: 7'-4"
- 3'-0" × 6'-8" door
- Made of cedar plywood with battens
- Complete list of materials
- Step-by-step instructions

12'-0"
10'-0"

Large Gable Storage Sheds

- Three popular sizes:
  10' × 12'
  10' × 16'
  10' × 20'
- Wood floor on 4 × 4 runners
- Height, floor to peak: 8'-8½"
- Ceiling height: 7'
- 4'-0" × 6'-4" double door for easy access
- Complete list of materials
- Step-by-step instructions

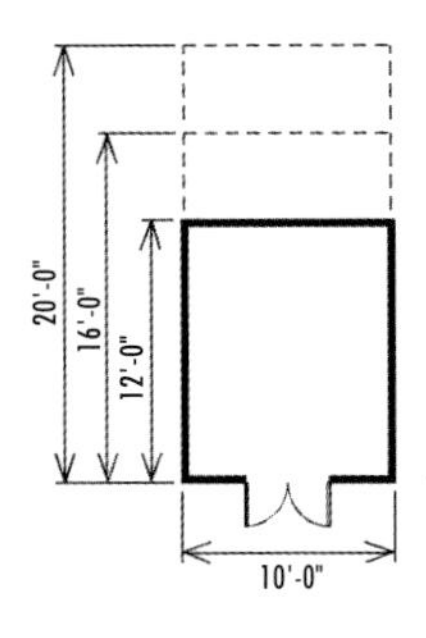

Visit www.projectplans.com to order and view additional projects.

Greenhouses & Garden Buildings

Gardening structures are available in a wide variety of styles and sizes to match just about any budget. New materials and methods have reduced the costs of building and maintaining greenhouses and other garden buildings, and kits are making them ever easier to build. But a wealth of choices leads to a host of questions, such as: What style should I choose? What design is best suited to my property? How big should it be? Which will better suit my needs, a shed or a greenhouse?

The structures in this chapter will not only help you answer these questions and get your creative juices flowing, they each clearly elucidate how possible it is for you to build a new garden shed or greenhouse in your yard. Create a space for your living things to thrive with one of the greenhouse projects, or add style and storage space at once with one of the attractive garden sheds. Go ahead—start spring a little earlier this year, or keep your tomatoes growing year-round. Surround yourself with living things and spruce up your home's landscape with a garden structure today.

In this chapter:

- Types of Greenhouses
- Hoophouse
- A-frame Greenhouse
- Hard-sided Greenhouse Kit
- Mini Garden Shed
- Sunlight Garden Shed

Types of Greenhouses

Greenhouses can take many forms, from simple, three-season A-frame structures to elaborate buildings the size of a small backyard. They can be custom-designed or built from a kit, freestanding or attached to a building, framed in metal or wood, glazed with plastic or glass. Although it's important to choose a design that appeals to you and complements your house and yard, you'll need to consider many other factors when making a decision. Ask yourself the following questions to help determine which type of greenhouse will best suit your needs:

How will it be used? The type of plants you intend to grow will determine, primarily, whether or not your greenhouse will need to be heated—as well as its size, type, and location on your property.

Should it be freestanding or a lean-to? Lean-to structures may be located conveniently close to a building's utilities but are also subject to the building's design and may be overly shaded by its mass. A freestanding greenhouse can be sited anywhere, but it may be more expensive to build and involve more complicated to install utilities.

How large should it be?

How much can I afford to spend?

How much time am I prepared to invest?

Remember, the ideal greenhouse has flexibility, combining some built-in features with ample open spaces so you can adjust the way the structure is used as your needs and interests change.

Traditional Span

This type of greenhouse has vertical side walls and an even-span roof, with plenty of headroom in the center. Side walls are typically about five feet high; the roof's central ridge stands seven to eight feet above the floor. This model shows a low base wall, known as a kneewall, but glass-to-ground models are also widely available. Kneewalls help to conserve heat, block wind, and provide impact protection, but they do limit light below the bench area.

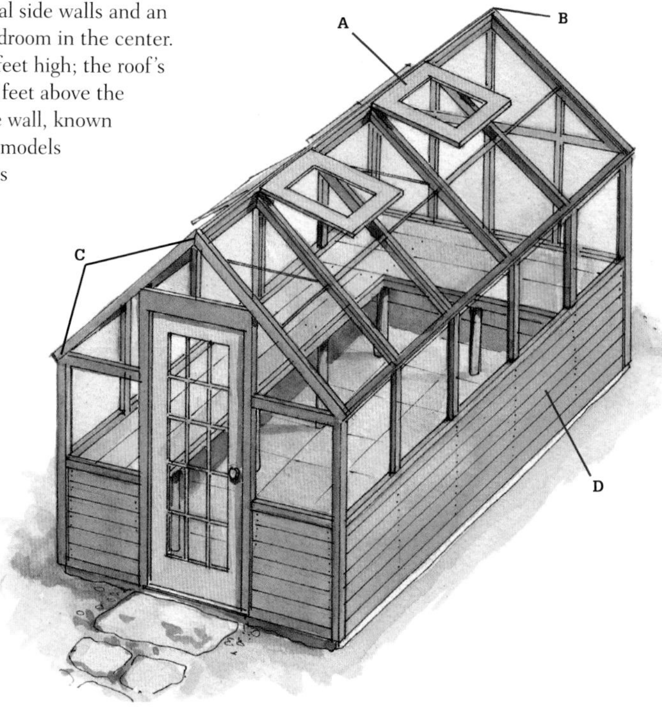

Features of the Traditional Span greenhouse include ventilating roof windows (A), a high gable peak that provides headroom (B), a 45° roof angle that encourages runoff (C), and solid kneewalls (D).

Three-Quarter Span

Attached to a building, this type of greenhouse offers the benefits of a lean-to structure—easy access to utilities and absorption of heat from the adjoining structure—without sacrificing headroom. This model also offers excellent light transmission, though it does have less light than a freestanding model. Because of the additional framing and glazing, this style is more expensive to build than standard lean-to structures.

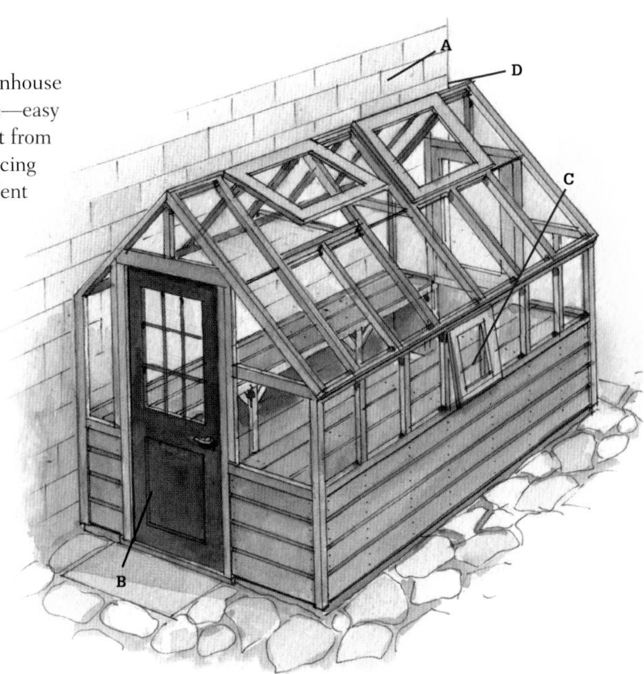

Features of the Three-Quarter Span greenhouse include shelter provided by the adjoining building (A), a half-lite door that insulates, but allows some light in (B), an operating side vent (C), and a gable roof that creates headroom (D).

Mini Greenhouse

A relatively inexpensive option that requires little space, this greenhouse is typically made of aluminum framing and can be placed against a house, garage, or even a fence, preferably facing southeast or southwest to receive maximum light exposure. Space and access are limited, however. And without excellent ventilation, a mini-greenhouse can become dangerously overheated. Because the temperature inside is difficult to control, this type of greenhouse is not recommended for winter use.

Features of the Mini Greenhouse include heat retained due to its adjoining wall (A), a convenient upper shelf that does not block airflow (B), a full-depth lower shelf that creates a hot spot below (C), and a full-lite storm door (D).

Mansard

The slanting sides and roof panels that characterize the Mansard are designed to allow maximum light transmission any time of the day or year. This style is excellent for plants that need a lot of light during the winter. However, the structure itself provides minimal insulation, and heating costs may be expensive.

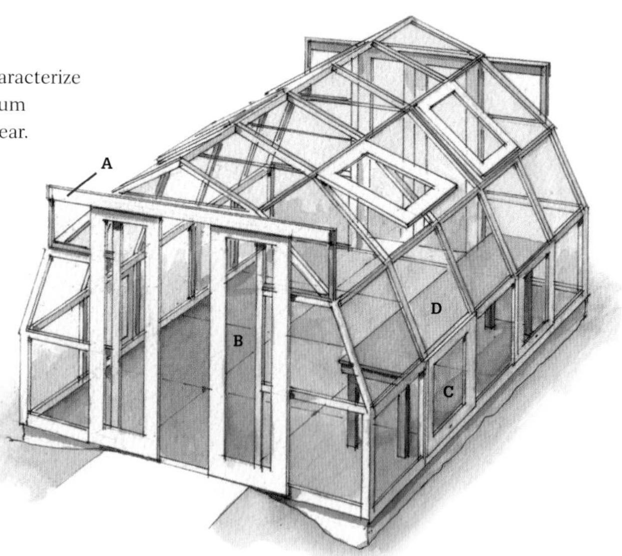

Features of the Mansard greenhouse include a full-width door frame (A), sliding doors that can be adjusted for ventilation (B), low side vents that encourage airflow (C), and stepped angles (D).

Tip: Free Greenhouse Design Software ▸

The United States Department of Agriculture (USDA) has developed a computer software program, called *Virtual Grower*, that you can use to create your own custom greenhouse design. It helps you make decisions about roof and sidewall materials, operating temperatures, and other variables. It even has a calculator for estimating heating costs. The software can be downloaded free of charge: www.ars.usda.gov/services/software/download.htm?softwareid=108

Siting a Greenhouse ▸

When the first orangeries were built, heat was thought to be the most important element for successfully growing plants indoors. Most orangeries had solid roofs and walls with large windows. Once designers realized that light was more important than heat for plant growth, they began to build greenhouses from glass.

All plants need at least six (and preferably 12) hours of light a day year-round, so when choosing a site for a greenhouse, you need to consider a number of variables. Be sure that it is clear of shadows cast by trees, hedges, fences, your house, and other buildings. Don't forget that the shade cast by obstacles changes throughout the year. Take note of the sun's position at various times of the year: A site that receives full sun in the spring and summer can be shaded by nearby trees when the sun is low in winter. Winter shadows are longer than those cast by the high summer sun, and during winter, sunlight is particularly important for keeping the greenhouse warm. If you are not familiar with the year-round sunlight patterns on your property, you may have to do a little geometry to figure out where shadows will fall. Your latitude will also have a bearing on the amount of sunlight available; greenhouses at northern latitudes receive fewer hours of winter sunlight than those located farther south. You may have to supplement natural light with interior lighting.

To gain the most sun exposure, the greenhouse should be oriented so that its ridge runs east to west, with the long sides facing north and south. A slightly southwest or southeast exposure is also acceptable, but avoid a northern exposure if you're planning an attached greenhouse; only shade-lovers will grow there.

Hoophouse

Made of PVC or metal framing and plastic glazing, this lightweight, inexpensive greenhouse is used for low-growing crops that require minimal protection from the elements. Because it does not provide the warm conditions of a traditional greenhouse, it is designed mainly for extending the growing season, not for overwintering plants. Ventilation in a hoophouse can be a problem, so some models have sides that roll up. This model's lightweight base makes the structure easy to move as well. See pages 220 to 225 for instructions on how to build your own hoophouse.

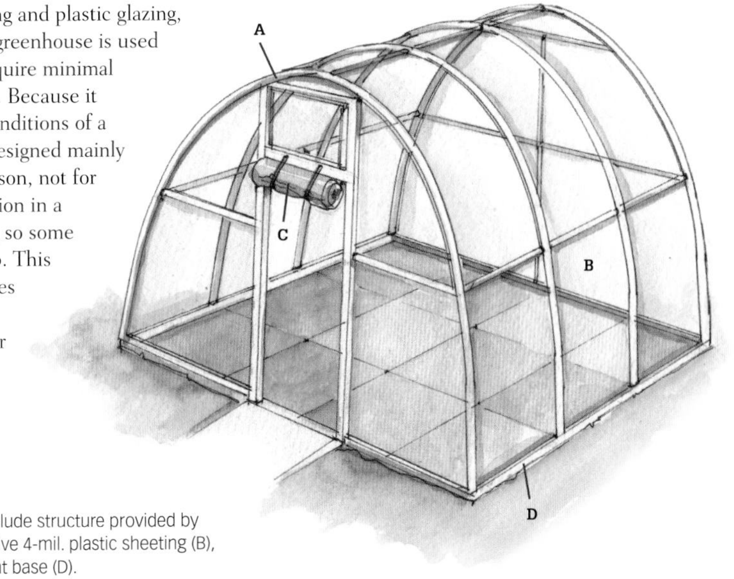

A
C
B
D

Features of the hoophouse include structure provided by bendable PVC tubes (A), inexpensive 4-mil. plastic sheeting (B), a roll-up door (C), and a lightweight base (D).

Conservation Greenhouse

With its angled roof panels, double-glazing, and insulation, the conservation greenhouse is designed to save energy. It is oriented east-to-west so that one long wall faces south, and the angled roof panels capture maximum light (and therefore heat) during the winter. To gain maximum heat absorption for the growing space, the house should be twice as long as it is wide. Placing the greenhouse against a dark-colored back wall helps to conserve heat as well—the walls radiate heat back into the greenhouse at night. The high peak offers ample headroom.

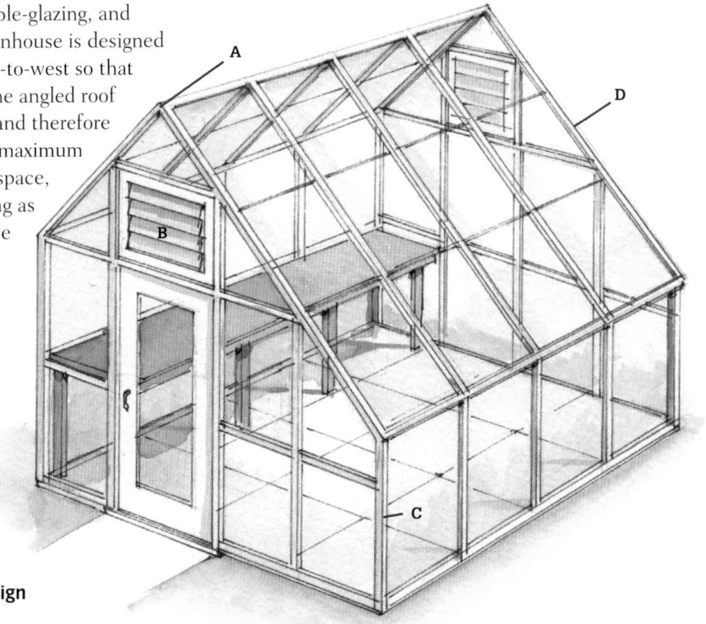

A
D
B
C

Features of the conservation design include a high peak (A), louvered wall vents (B), sturdy aluminum framing (C), and a broad roof surface (D).

Hoophouse

The hoophouse is a popular garden structure for two main reasons: it is cheap to build and easy to build. In many agricultural areas you will see hoophouses snaking across vast fields of seedlings, protecting the delicate plants at their most vulnerable stages. Because they are portable and easy to disassemble, they can be removed when the plants are established and less vulnerable.

While hoophouses are not intended as inexpensive substitutes for real greenhouses, they do serve an important agricultural purpose. And building your own is a fun project that the whole family can enjoy.

The hoophouse shown here is essentially a Quonset-style frame of bent ¾" PVC tubing draped with sheet plastic. Each semicircular frame is actually made from two 10-ft. lengths of tubing that fit into a plastic fitting at the apex of the curve. PVC tubes tend to stay together simply by friction-fitting into the fittings, so you don't normally need to solvent-glue the connections (this is important for easy disassembly and storage).

Tools & Materials ▸

For 10-ft. wide by 15-ft. long project seen here:
(11) ¾" × 10 ft. PVC pipes
(3) ½" × 10 ft. CPVC pipes
(1) 1" × 10 ft. PE pipe (black)
(3) ¾" PVC cross fittings
(2) ¾" PVC T-fittings
(4) 16-ft. pressure-treated 2 × 4
20 ft. × 16 ft. clear or translucent plastic sheeting
Stakes
Mason's string
Maul
High-visibility tape
Hacksaw
Mallet
Heavy-duty stapler
Circular saw
Drill
3" deck screws
Stapler

A hoophouse is a temporary agricultural structure designed to be low-cost and portable. Also called Quonset houses and tunnel houses, hoophouses provide shelter and shade (depending on the film you use) and protection from wind and the elements. They will boost heat during the day but are less efficient than paneled greenhouses for extending the growing season.

Anatomy of a Hoophouse

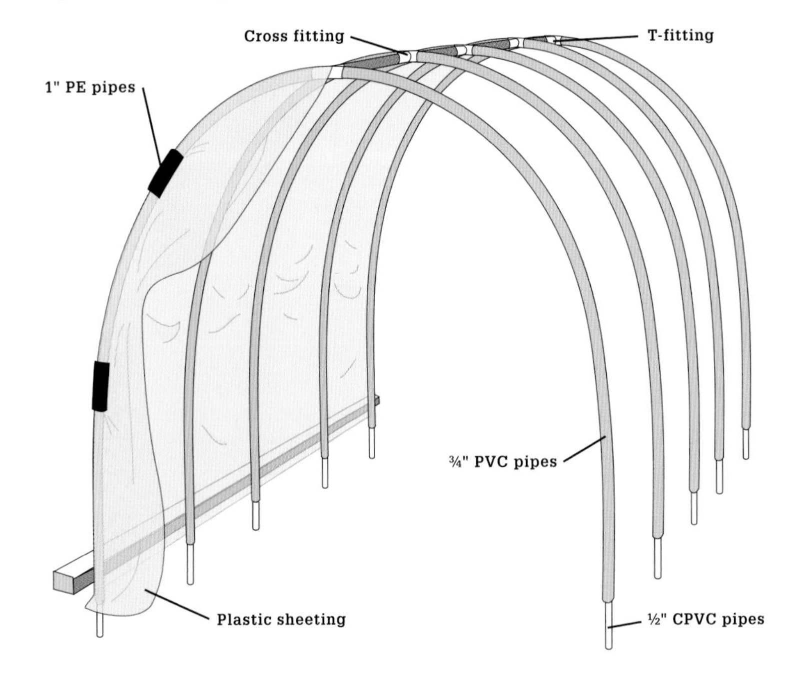

Specialty Materials

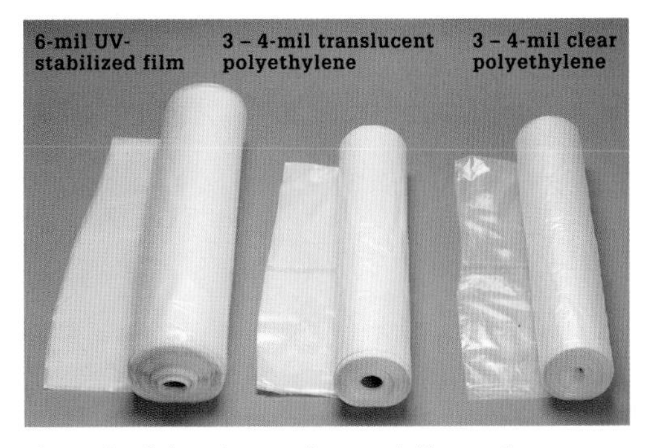

Sheet plastic is an inexpensive material for creating a greenhouse. Obviously, it is less durable than polycarbonate, fiberglass, or glass panels. But UV-stabilized films, at least 6-mil thick, can be rated to withstand four years or more of exposure. Inexpensive polyethylene sheeting (the kind you find at hardware stores) will hold up for a year or two, but it becomes brittle when exposed to sunlight. Some greenhouse builders prefer to use clear plastic sheeting to maximize the sunlight penetration, while the cloudiness of translucent poly makes it effective for diffusing light and preventing overheating. For the highest quality film coverings, look for film rated for greenhouse and agricultural use.

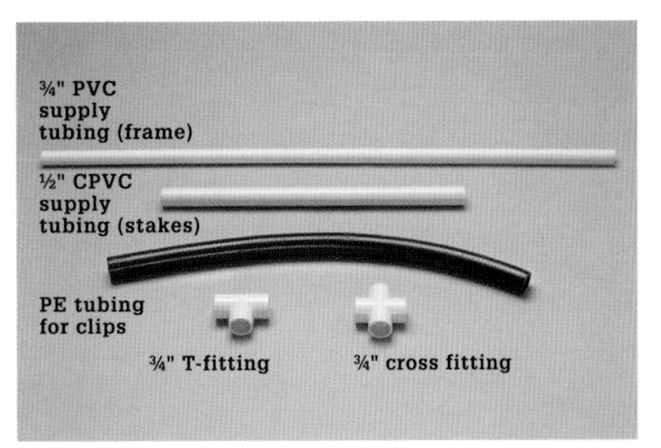

Plastic tubing and fittings used to build this hoophouse include: Light-duty ¾" PVC tubing for the frame (do not use CPVC—it is too rigid and won't bend properly); ½" CPVC supply tubing for the frame stakes (rigidity is good here); Polyethylene (PE) tubing for the cover clips; PVC T-fittings and cross fittings to join the frame members.

Successful Hoophouse Construction ▸

The fact that a hoophouse is a temporary structure doesn't give you license to skimp on the construction. When you consider how light the parts are and how many properties sheet plastic shares with boat sails, the importance of securely anchoring your hoophouse becomes obvious. Use long stakes (at least 24") to anchor the tubular frames, and make sure you have plenty of excess sheeting at the sides of the hoophouse so the cover can be held down with ballast. Creating pockets at the ends of the sheeting and inserting scrap lumber is the ballasting technique shown here, but it is also common (especially when building in a field) to weight down the sheeting by burying the ends in dirt. Attach the sheeting only at the ends of the tubular frame, and where possible, orient the structure so the prevailing winds will blow through the tunnel.

Some other tips for successful construction include:

- Space frame hoops about 3 ft. apart.
- Cut ridge members a fraction of an inch (not more than ¼") shorter than the span, which will cause the structure to be slightly shorter on top than at the base. This helps stabilize the structure.
- If you are using long-lasting greenhouse fabric for the cover, protect the investment by spray-painting the frame hoops with primer so there is no plastic-to-plastic contact.
- Because hoophouses are temporary structures that are designed to be disassembled or moved regularly, you do not need to include a base.
- The ¾" PVC pipes used to make the hoop frames are sold in 10 ft. lengths. Two pipes fitted into a Tee or cross fitting at the top will result in legs that are 10 ft. apart at the base and a ridge that is roughly 7 ft. tall.
- Clip the hoophouse covers only to the end frames. Clips fastened at the intermediate hoops will either fly off or tear the plastic cover in windy conditions.

How to Build a Hoophouse

Lay out the installation area using stakes and mason's string. Strive for square corners, but keep in mind that these are relatively forgiving structures, so you can miss by a little bit and probably won't be able to notice.

Cut a 30"-long stake from ½" CPVC supply tubing for each leg of each hoop frame. Measure out from the corners of the layout and attach a piece of high-visibility tape on the string at 3-ft. intervals; then drive a stake at each location. When the stake is fully driven, 10" should be above ground and 20" below.

Join the two legs for each frame hoop with a fitting. Use a Tee fitting for the end hoop frames and a cross fitting for the intermediate hoop frames. No priming or solvent gluing is necessary. (The friction-fit should be sufficient, but it helps if you tap on the end of the fitting with a mallet to seat it.)

Slip the open end of one hoop-frame leg over a corner stake so the leg pipe is flush against the ground. Then, bend the pipes so you can fit the other leg end over the stake at the opposite corner. If you experience problems with the pipes pulling out of the top fitting, simply tape the joints temporarily until the structure frame is completed.

Continue adding hoop frames until you reach the other end of the structure. Wait until all the hoop frames are in place before you begin installing the ridge poles. Make sure the cross fittings on the intermediate hoop frames are aligned correctly to accept the ridge poles.

Add the ridge-pole sections between the hoop frames. Pound on the end of each new section as you install it to seat it fully into the fitting. Install all of the ridge poles.

(continued)

Cut four pieces of pressure-treated 2 × 4 to the length of the hoophouse (15 ft. as shown). Cut the roof cover material to size. (We used 6-mil polyethylene sheeting.) It should be several inches longer than is necessary in each direction. Staple the cover material at one end of the 2 × 4 and then continue stapling it as you work your way toward the end. Make sure the material stays taut and crease-free as you go.

Lay a second 2 × 4 the same length as the first over the stapled plastic so the ends and edges of the 2 × 4s are flush. Drive a 3" deck screw through the top 2 × 4 and into the lower one every 24" or so, sandwiching the cover material between the boards. Lay the assembly next to one edge of the hoophouse and pull the free end of the material over the tops of the frames.

On the other side of the structure, extend the cover material all the way down so it is taut, and then position another 2 × 4 underneath the fabric where it meets the ground. Staple the plastic, and then sandwich it with a final 2 × 4.

Make clips to secure the roof cover material from a 12"-long section of hose or soft tubing. Here, 1"-dia., thin-walled PE supply tubing is slit longitudinally and then slipped over the material to clip it to the end frames. Use at least six clips per end. Do not clip at the intermediate hoop frames.

Hoophouse Variations

Make doors by clipping a piece of cover material to each end. (It's best to do this before attaching the main cover.) Then cut a slit down the center of the end material. You can tie or tape the door material to the sides when you want it open and weight down the pieces with a board or brick to keep the door shut. This solution is low-tech but effective.

More permanent hoophouse frames can be made from wood instead of PVC plastic. Wood allows you to attach plastic sheeting with staples and retainer strips.

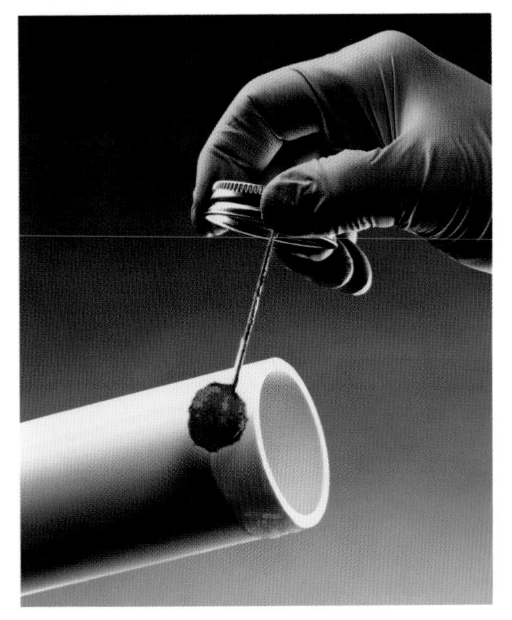

Make your hoophouse a permanent fixture by solvent-welding the pipes from the frame to the T-fitting and cross fittings. If you may be disassembling and moving the structure, it's fine to rely on tension fittings with PVC pipe.

A-frame Greenhouse

A greenhouse can be a decorative and functional building that adds beauty to your property. A greenhouse also can be a quick-and-easy, temporary structure that serves a purpose and then disappears. The wood-framed greenhouse seen here fits somewhere in between these two types. The sturdy wood construction will hold up for many seasons. The plastic sheeting covering will last one to four or five seasons, depending on the materials you choose (see page 221), and it is easy to replace when it starts to degrade.

The 5-ft.-high kneewalls in this design provide ample space for installing and working on a conventional-height potting table. The walls also provide some space for plants to grow. For a door, this plan simply employs a sheet of weighted plastic that can be tied out of the way for entry and exit. If you plan to go in and out of the greenhouse frequently, you can purchase a prefabricated greenhouse door from a greenhouse materials supplier. To allow for ventilation in hot weather, we built a wood-frame vent cover that fits over one rafter bay and can be propped open easily.

You can use hand-driven nails or pneumatic framing nails to assemble the frame if you wish, although deck screws make more sense for a small structure like this. Screws also make the greenhouse much easier to disassemble, if necessary.

A wood-frame greenhouse with sheet-plastic cover is an inexpensive, semipermanent gardening structure that can be used as a potting area as well as a protective greenhouse.

Tools, Materials & Cutting List

(1) 20 × 50-ft. roll 4- or 6-mil polyethylene sheeting
(12) 24"-long pieces of No. 3 rebar
(8) 8" timber screws
Compactable gravel (or drainage gravel)
Excavation tools
Level
Circular saw
Drill
Reciprocating saw
Maul
3" deck screws
Jigsaw
Wire brads
Brad nailer (optional)
Scissors

Key	No.	Part	Dimension	Material
A	2	Base ends	3½" × 3½" × 96"	4 × 4 landscape timber
B	2	Base sides	3½" × 3½" × 113"	4 × 4 landscape timber
C	2	Sole plates end	1½" × 3½" × 89"	2 × 4 pressure-treated
D	2	Sole plates side	1½" × 3½" × 120"	2 × 4 pressure-treated
E	12	Wall studs side	1½" × 3½" × 57"	2 × 4
F	1	Ridge support	1½" × 3½" × 91"	2 × 4
G	2	Back studs	1½" × 3½" × 76" *	2 × 4
H	2	Door frame sides	1½" × 3½" × 81" *	2 × 4
I	1	Cripple stud	1½" × 3½" × 16"	2 × 4
J	1	Door header	1½" × 3½" × 32"	2 × 4
K	2	Kneewall caps	1½" × 3½" × 120"	2 × 4
L	1	Ridge pole	1½" × 3½" × 120"	2 × 4
M	12	Rafters	1½" × 3½" × 60" *	2 × 4

**Approximate dimension; take actual length and angle measurements on structure before cutting.*

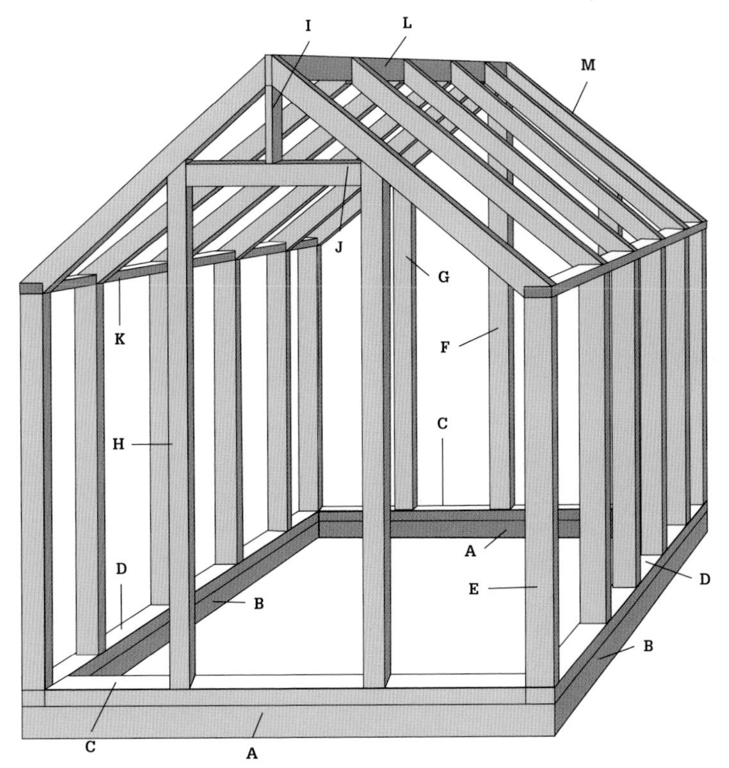

How to Build an A-frame Greenhouse

Prepare the installation area so it is flat and well drained (see pages 432 to 433); then cut the base timbers (4 × 4 landscape timbers) to length. Arrange the timbers so they are flat and level and create a rectangle with square corners. Drive a pair of 8" timber screws at each corner to assemble the frame using a drill/driver with a nut-driver bit.

Cut 12 pieces of No. 3 rebar (find it in the concrete supplies section of any building center) to 24" to use as spikes for securing the timbers to the ground. Use a metal cutoff saw or a reciprocating saw with a bimetal blade to make the cuts. Drill a ⅜" guide hole through each timber near each end and in the middle. Drive a rebar spike at each hole with a maul or sledgehammer until the top is flush with the wood.

Cut the plates and studs for the two side walls (called kneewalls). Arrange the parts on a flat surface and assemble the walls by driving three 3" deck screws through the cap and sole plates and into the ends of the studs. Make both kneewalls.

Set the base plate of each kneewall on the timber base and attach the walls by driving 3" deck screws down through the sole plates and into the timbers. For extra holding power you can apply construction adhesive to the undersides of the plates, but only if you don't plan to relocate the structure later.

Cut the end sole plates and the ridge support post to length and attach it to the center of one end sole plate. Cut another post the same length for the front (this will be a temporary post) and attach it to the plate. Fasten both plates to front and back end timbers.

Set the ridge pole on top of the posts and check that it is level. Also check that the support posts are level and plumb. Attach a 2 × 4 brace to the outer studs of the kneewalls and to the posts to hold them in square relationship. Double-check the pole and posts with the level.

Cut a 2 × 4 to about 66" to use as a rafter template. Hold the 2 × 4 against the end of the ridge pole and the top outside corner of a kneewall. Trace along the face of the ridge and the top plate of the wall to mark cutting lines. Cut the rafter and use it as a template for the other rafters on that side of the roof. Create a separate template for the other side of the roof.

Mark cutting lines for the rafters using the templates, and cut them all. Use a jigsaw or handsaw to make the birdsmouth cuts on the rafter ends that rest on the kneewalls.

(continued)

Attach the rafters to the ridge pole and the kneewalls with deck screws driven through pilot holes. Plan the rafter layout so they are aligned with the kneewall studs and are perpendicular to the ridge.

10

Mark the positions for the end wall studs onto the sole plate. At each location, hold a 2 × 4 on end on the sole plate and make it plumb. Trace a cutting line at the top of the stud where it meets the rafter. Cut the studs and install them by driving screws toenail-style.

Measure up 78" (or less if you want a shorter door) from the sole plate in the door opening and mark a cutting line on the temporary ridge post. Make a square cut along the line with a circular saw or cordless trim saw (inset). Then, cut the door header to fit between the vertical door frame members. Screw the header to the cut end of the ridge post and drive screws through the frame members and into the header.

Begin covering the greenhouse with your choice of cover material (we used 6-mil polyethylene sheeting). Start at the ends: cut the sheeting a little oversized, and fasten it by attaching screen retainer strips to the framing members at the edges of the area being covered. Tack the sheeting first at the top, then at the sides, and finally at the bottom using wire brads to secure the strips. After the strips are installed, trim the sheeting along the edges of the strips with a utility knife.

Attach the sheeting to the outside edge of the sole plate on one side. Roll sheeting over the roof and down the other side. Draw it taut and cut it slightly long with scissors. Attach retainer strips to the other sole plate and then to the outside edges of the corner studs.

Make and hang a door. We simply cut a piece of sheet plastic a little bigger than the opening (32") and hung it with retainer strips from the header. Attach a piece of 2 × 4 to the bottom of the door for weight.

Option: Make a vent window. First, cut a hole in the roof in one rafter bay and tack the cut edges of the plastic to the faces (not the edges) of the rafters, ridge pole and wall cap. Then, build a frame from 1 × 2 stock that will span from the ridge to the top of the kneewall and extend a couple of inches past the rafters at the side of the opening. Clad the frame with plastic sheeting and attach it to the ridge pole with butt hinges. Install a screw-eye latch to secure it at the bottom. Make and attach props if you wish.

Hard-Sided Greenhouse Kit

Building a greenhouse from a prefabricated kit offers many advantages. Kits are usually very easy to assemble because all parts are prefabricated and the lightweight materials are easy to handle. The quality of kit greenhouses varies widely, though, and buying from a reputable manufacturer will help ensure that you get many years of service from your greenhouse.

If you live in a snowy climate, you may need to either provide extra support within the greenhouse or be ready to remove snow whenever there is a significant snowfall because the lightweight aluminum frame members can easily bend under a heavy load. Before buying a kit, make sure to check on how snowfall may affect it.

Kit greenhouses are offered by many different manufacturers, and the assembly technique you use will depend on the specifics of your kit. Make sure you read the printed instructions carefully, as they may vary slightly from this project.

The kit we're demonstrating here is made from aluminum frame pieces and transparent polycarbonate panels and is designed to be installed over a base of gravel about 5" thick. Other kits may have different base requirements.

When you purchase your kit, make sure to uncrate it and examine all the parts before you begin. Make sure all of the pieces are there and that there are no damaged panels or bent frame members.

A perfectly flat and level base is crucial to any kit greenhouse, so make sure to work carefully. Try to do the work on a dry day with no wind, as the panels and frame pieces can be hard to manage on a windy day. Never try to build a kit greenhouse by yourself. At least one helper is mandatory, and you'll do even better with two or three.

Construction of a kit greenhouse consists of four basic steps: laying the base, assembling the frame, assembling the windows and doors, and attaching the panels.

Tools & Materials ▸

- Stakes
- Mason's string
- Tape measure
- 4-ft. level or laser level
- Long, straight 2 × 4
- Gravel or fill material
- Excavation tools
- Protective gloves
- Greenhouse kit
- Drill
- Eye protection
- Batterboards

Kit greenhouses come in a wide range of shapes, sizes, and quality. The best ones have tempered-glass glazing and are rather expensive. The house shown here is glazed with corrugated polyethylene and is at the low end of the cost spectrum.

Unpacking the Kit Contents

Organize and inspect the contents of your kit cartons to make sure all of the parts are present and in good condition. Most manuals will have a checklist. Staging the parts makes for a more efficient assembly. Just be sure not to leave any small parts loose, and do not store parts in high-traffic areas.

How to Build a Kit Greenhouse

Create an outline for the base of the greenhouse using stakes or batterboards and string. The excavation should be about 2" wider and longer than the overall dimensions of your greenhouse. To ensure that the excavation is perfectly square, measure the diagonals of the outline. If the diagonals are equal, the outline is perfectly square. If not, reposition the stakes until the outline is square.

Excavate the base area to a depth of 5". Use a long 2 × 4 and a 4-ft. level to periodically check the excavation and make sure it is level and flat. You can also use a laser level for this job.

(continued)

Assemble the base of the greenhouse using the provided corner connectors and end connectors, attaching them with base nuts and bolts. Lower the base into the excavation area, and check to make sure it's level. Measure the diagonals to see if they are equal; if not, reposition the base until the diagonals are equal, which ensures that the base is perfectly square. Pour a layer of gravel or other fill material into the excavation, to within about 1" of the top lip of the base frame. Smooth the fill with a long 2 × 4.

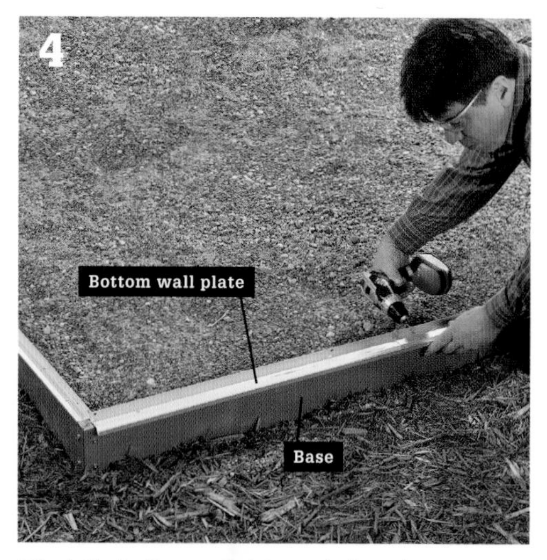

Attach the bottom wall plates to the base pieces so that the flanged edges face outside the greenhouse. In most systems, the floor plates will interlock with one another, end to end, with built-in brackets.

Fasten the four corner studs to the bottom wall plates using hold-down connectors and bolts. In this system, each corner stud is secured with two connectors.

Assemble the pieces for each side ceiling plate, then attach one assembled side plate against the inside of the two corners studs, making sure the gutter is positioned correctly. Now attach the front ceiling plate to the outside of the two corner studs that are over the front floor plate.

Attach the other side ceiling plate along the other side, flat against the inside of the corner studs. Then attach the corner brackets to the rear studs, and construct the back top plate by attaching the rear braces to the corners and joining the braces together with stud connectors.

Fasten the left and right rear studs to the outside of the rear floor plate, making sure the top ends are sloping upward, toward the peak of the greenhouse. Attach the center rear studs to the rear floor plate, fastening them to the stud connectors used to join the rear braces.

Attach the side studs on each side wall using the provided nuts and bolts. Then, attach the doorway studs to the front wall of the greenhouse.

Attach diagonal struts, as specified by the manufacturer. Periodically take diagonal measurements between the top corners of the greenhouse, adjusting as necessary so that the measurements are equal and the greenhouse square.

(continued)

Fasten the gable-end stud extensions to the front and back walls of the greenhouse. The top ends of the studs should angle upward, toward the peak of the greenhouse.

Assemble the roof frame on a flat area near the wall assembly. First assemble the crown-beam pieces, then, attach the rafters to the crown, one by one. The crown beams, or end rafters, have a different configuration, so make sure not to confuse them with the common rafters.

With at least one helper, lift the roof into place onto the wall frames. The gable-end studs should meet the outside edges of the crown beams, and the ends of the crown beams rest on the outer edges of the corner brackets. Fasten the roof in place with the provided nuts and bolts.

Attach the side braces and the roof-window support beams to the undersides of the roof rafters, as specified by the manufacturer's instructions.

Backwards and Forwards ▸

With some kits you need to go backward to go forward. Because the individual parts of your kit depend upon one another for support, you may be required to tack all the parts together with bolts first and then undo and remake individual connections as you go before you can finalize them. For example, in this kit you must undo the track/brace connections one at a time so you can insert the bolt heads for the stud connectors into the track.

Build the roof windows by first connecting the two side window frames to the top window frame. Slide the window panel into the frame, then secure it by attaching the bottom window frame. Slide the window into the slot at the top of the roof crown, then gradually lower it in place. Attach the window stop to the window support beam.

Assemble the doors, making sure the top slider/roller bar and the bottom slider bar are correctly positioned. Lift the door panels up into place onto the top and bottom wall plates.

(continued)

Install the panels one-by-one using panel clips. Begin with the large wall panels. Position each panel and secure it by snapping a clip into the frame at the intervals specified by the manufacturer's instructions.

Add the upper panels. At the gable ends, the upper panels will be supported by panel connectors that allow the top panel to be supported by the bottom panel. The lower panels should be installed already.

Install the roof panels and the roof-window panels so that the top edges fit up under the edge of the crown or window support and the bottom edges align over the gutters.

Test the operation of the doors and roof windows to make sure they operate smoothly.

Greenhouse Ventilation ▸

Whether your plants thrive depends on how well you control their environment. Adequate sunlight is a good start, but ventilation is just as important: It expels hot air, reduces humidity, and provides air circulation, which is essential year-round to move stagnant air around, keep diseases at bay, and avoid condensation problems. You have two main options for greenhouse ventilation: vents and fans.

Because hot air rises, roof vents are the most common choice. They should be staggered on both sides of the ridgeline to allow a gentle, even exchange of air and proper circulation. Roof vents are often used in conjunction with wall vents or louvers. Opening the wall vents results in a more aggressive air exchange and cools the greenhouse much faster than using roof vents alone. On hot days, you can open the greenhouse door to let more air inside. Also consider running small fans to enhance circulation.

Vents can be opened and closed manually, but this requires constant temperature monitoring, which is inconvenient and can leave plants wilting in the heat if you are away. It's far easier—and safer—to use automatic vent openers. These can be thermostat-controlled and operated by a motor, which turns on at a set temperature, or they can be solar-powered. Unlike thermostat-controlled vent openers, which require electricity, solar-powered openers use a cylinder filled with wax, which expands as the temperature rises and pushes a rod that opens the vent. When the temperature drops, the wax shrinks and the vent closes. How far the vent opens is dictated by temperature: the higher the temperature, the wider the vent opens to let in more air.

A fan ventilator is a good idea if you have a large greenhouse. The fan is installed in the back opposite the greenhouse door, and a louvered vent is set into the door wall. At a set temperature, a thermostat mounted in the middle of the greenhouse activates the fan, and the louvered vent opens. Cool air is drawn in through the vent, and the fan expels the warm air. The fan should be powerful enough to provide a complete air exchange every one to two minutes.

Greenhouse manufacturers rarely include enough vents in kits, so be sure to buy more. To determine the square footage of venting your greenhouse should have, multiply the square footage of the floor by 0.2.

Automatic openers sense heat buildup and open vents to promote airflow.

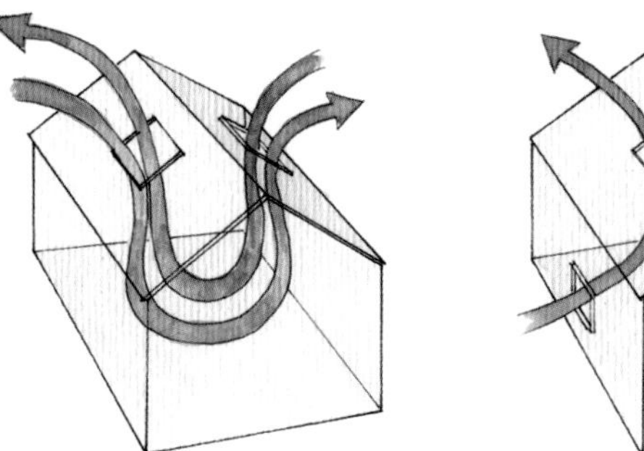

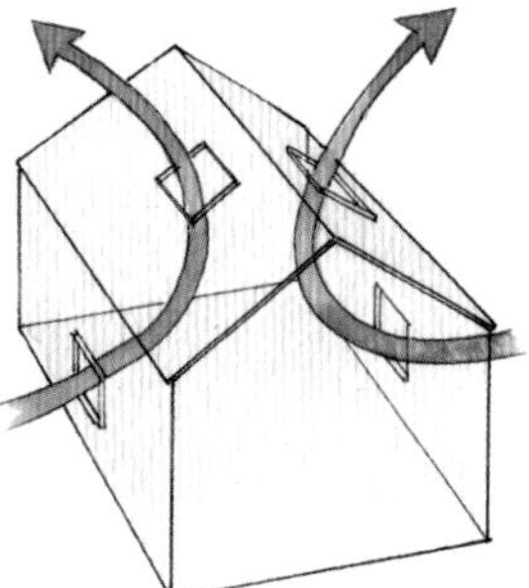

Installing at least one operable roof vent on each side of the ridgeline creates good air movement within the structure. Adding lower intake vents helps for cooling. Adding fans to the system greatly increases air movement.

Mini Garden Shed

Whether you are working in a garden or on a construction site, getting the job done is always more efficient when your tools are close at hand. Offering just the right amount of on-demand storage, this mini garden shed can handle all of your gardening hand tools but with a footprint that keeps costs and labor low.

The mini shed base is built on two 2 × 8 front and back rails that raise the shed off the ground. The rails can also act as runners, making it possible to drag the shed like a sled after it is built. The exterior is clad with vertical-board-style fiber-cement siding. This type of siding not only stands up well to the weather, it is very stable and resists rotting and warping. It also comes preprimed and ready for painting. Fiber-cement siding is not intended to be in constant contact with moisture, so the manufacturer recommends installing it at least 6" above the ground. You can paint the trim and siding any color you like. You might choose to coordinate the colors to match your house, or you might prefer a unique color scheme so that the shed stands out as a garden feature.

The roof is made with corrugated fiberglass roof panels. These panels are easy to install and are available in a variety of colors, including clear, which will let more light into the shed. An alternative to the panels is to attach plywood sheathing and then attach any roofing material you like over the sheathing. These plans show how to build the basic shed, but you can customize the interior with hanging hooks and shelves to suit your needs.

Working with Fiber-cement Siding ▸

Fiber-cement siding is sold in ¼"-thick, 4 × 8-ft. sheets at many home centers. There are specially designed shearing tools that contractors use to cut this material, but you can also cut it by scoring it with a utility knife and snapping it— just like cement tile backer board or drywall board. You can also cut cementboard with a circular saw, but you must take special precautions. Cementboard contains silica. Silica dust is a respiratory hazard. If you choose to cut it with a power saw, then minimize your dust exposure by using a manufacturer-designated saw blade designed to create less fine dust and by wearing a NIOSH/MSHA-approved respirator with a rating of N95 or better.

This scaled-down garden shed is just small enough to be transportable. Locate it near gardens or remote areas of your yard where on-demand tool storage is useful.

Tools, Materials & Cutting List

(2) 24"-wide × 8 ft. roof panels
Large utility or gate handle
Exterior-rated screws (1½", 2½")
1½" siding nails
2" galvanized finish nails
16d common nails
1" neoprene gasket screws
Excavation tools
Compactable gravel
Level
Circular saw
10d common nails
Power miter saw
Hammer
Clamps
Drill with bits
Utility knife
Straightedge
Framing square
Jigsaw with fine tooth blade
Finishing tools
Exterior-grade latex paint
(4) Door hinges

Key	Part	Dimension	Pcs.	Material
Lumber				
A	Front/back base rails	1½ × 7¼ × 55"	2	Treated pine
B	Base crosspieces	1½ × 3½ × 27"	4	Treated pine
C	Base platform	¾ × 30 × 55"	1	Ext. plywood
D	Front/back plates	1½ × 3½ × 48"	2	SPF
E	Front studs	1½ × 3½ × 81"	4	SPF
F	Door header	1½ × 3½ × 30"	1	SPF
G	Back studs	1½ × 3½ × 75"	4	SPF
H	Side bottom plate	1½ × 3½ × 30"	2	SPF
I	Top plate	1½ × 3½ × 55"	2	SPF
J	Side front stud	1½ × 3½ × 81"	2	SPF
K	Side middle stud	1½ × 3½ × 71"	2	SPF
L	Side back stud*	1½ × 3½ × 75¼"	2	SPF
M	Side crosspiece	1½ × 3½ × 27"	2	SPF
N	Door rail (narrow)	¾ × 3½ × 29¾"	1	SPF
O	Door rail (wide)	¾ × 5½ × 23"	2	SPF
P	Door stiles	¾ × 3½ × 71"	2	SPF
Q	Rafters	1½ × 3½ × 44"	4	SPF
R	Outside rafter blocking*	1½ × 3½ × 15¼"	4	SPF
S	Inside rafter blocking*	1½ × 3½ × 18¾"	2	SPF
Siding & Trim				
T	Front left panel	¼ × 20 × 85"	1	Siding
U	Front top panel	¼ × 7½ × 30"	1	Siding
V	Front right panel	¼ × 5 × 85"	1	Siding

*Not shown

Key	Part	Dimension	Pcs.	Material
W	Side panels	¼ × 30½ × 74½"	2	Siding
X	Back panel	¼ × 48 × 79"	1	Siding
Y	Door panel	¾ × 29¾ × 74"	1	Ext. plywood
Z	Front corner trim	¾ × 3½ × 85"	2	SPF
AA	Front top trim	¾ × 3½ × 50½"	1	SPF
BB	Side casing	¾ × 1½ × 81½"	2	SPF
CC	Top casing	¾ × 1½ × 30"	1	SPF
DD	Bottom casing	¾ × 2½ × 30"	1	SPF
EE	Trim rail (narrow)	¾ × 1½ × 16½"	3	SPF
FF	Trim rail (wide)	¾ × 3½ × 16½"	1	SPF
GG	Side trim	¾ × 2½ × 27"	2	SPF
HH	Side trim	¾ × 2½ × 27¾"	2	SPF
II	Side corner trim	¾ × 1¾ × 85¼"	2	SPF
JJ	Side corner trim	¾ × 1¾ × 79½"	2	SPF
KK	Side trim	¾ × 3½ × 27"	2	SPF
LL	Side trim	¾ × 1½ × 69"	2	SPF
MM	Back corner trim	¾ × 3½ × 79"	2	SPF
NN	Back trim	¾ × 3½ × 50½"	2	SPF
OO	Back trim	¾ × 1½ × 72"	2	SPF
PP	Side windows	¼ × 10 × 28"	2	Acrylic
Roof				
QQ	Purlins	1½ × 1½ × 61½"	5	
RR	Corrugated closure strips	61½" L	5	
SS	Corrugated roof panels	24 × 46"	3	

Mini Garden Shed

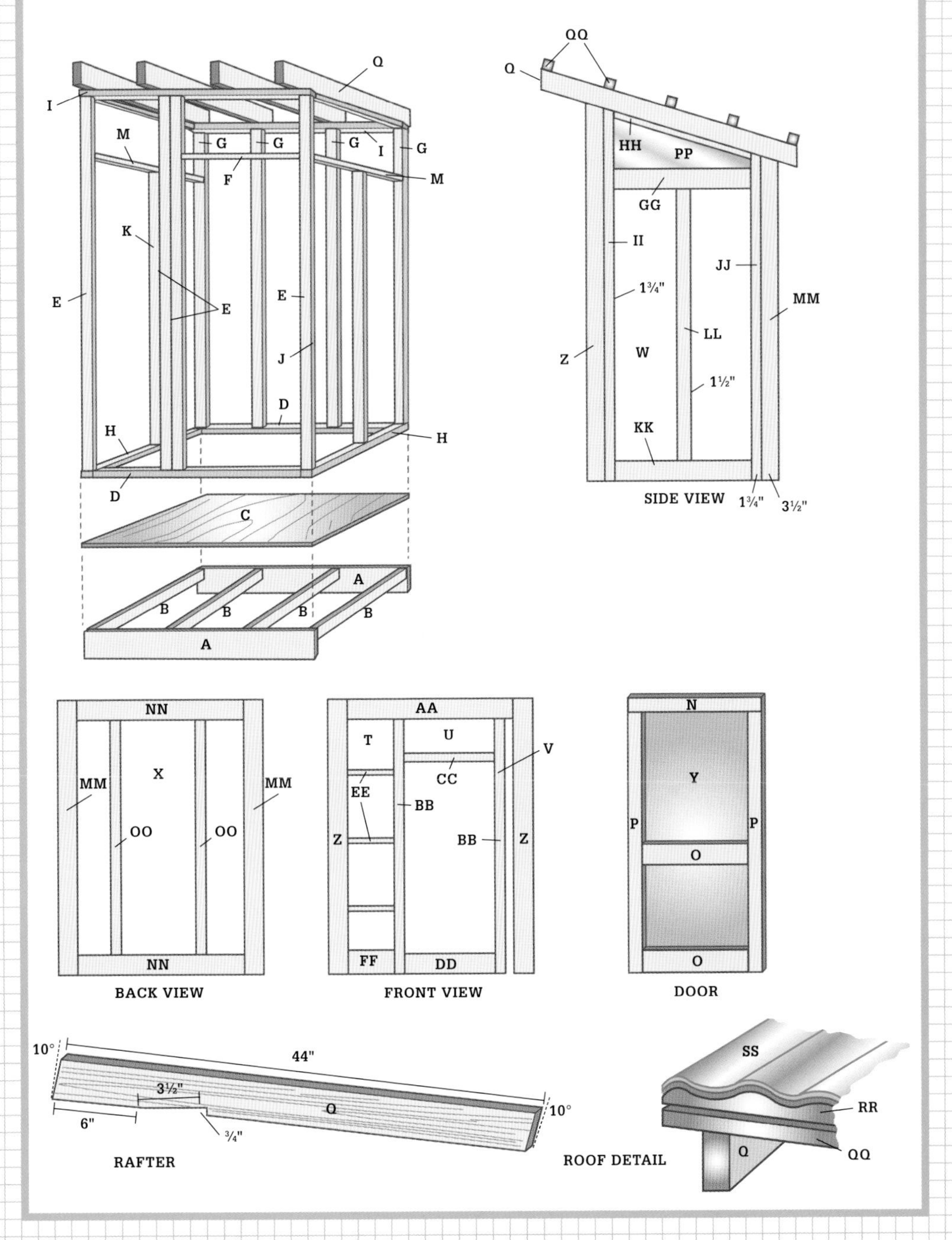

How to Build a Mini Garden Shed

BUILD THE BASE

Even though moving it is possible, this shed is rather heavy and will require several people or a vehicle to drag it if you build it in your workshop or garage. When possible, determine where you want the shed located and build it in place. Level a 3 × 5-ft area of ground. The shed base is made of treated lumber, so you can place it directly on the ground. If you desire a harder, more solid foundation, dig down 6" and fill the area with tamped compactable gravel.

Cut the front and back base rails and base crosspieces to length. Place the base parts upside-down on a flat surface and attach the crosspieces to the rails with 2½" deck screws. Working with the parts upside-down makes it easy to align the top edges flush. Cut the base platform to size. Flip the base frame over and attach the base platform (functionally, the floor) with 1½" screws. Set and level the base in position where the shed will be built.

FRAME THE SHED

Cut the front wall framing members to size, including the top and bottom plates, the front studs, and the door header. Lay out the front wall on a flat section of ground, such as a driveway or garage floor. Join the wall framing components with 16d common nails (photo 1). Then, cut the back-wall top and bottom plates and studs to length. Lay out the back wall on flat ground and assemble the back wall frame.

Cut both sidewall top and bottom plates to length, and then cut the studs and crosspiece. Miter-cut the ends of the top plate to 10°. Miter-cut the top of the front and back studs at 10° as well. Lay out and assemble the side walls on the ground. Place one of the side walls on the base platform. Align the outside edge of the wall so it is flush with the outside edge of the base platform. Get a helper to hold the wall plumb while you position the back wall. If you're working alone, attach a brace to the side of the wall and the platform to hold the wall plumb (photo 2).

Place the back wall on the platform and attach it to the side wall with 2½" deck screws (photo 3). Align the outside edge of the back wall with the edge of the platform. Place the front wall on the platform and attach it to the side wall with 2½" screws. Place the

Build the wall frames. For the front wall, attach the plates to the outside studs first and then attach the inside studs using the door header as a spacer to position the inside studs.

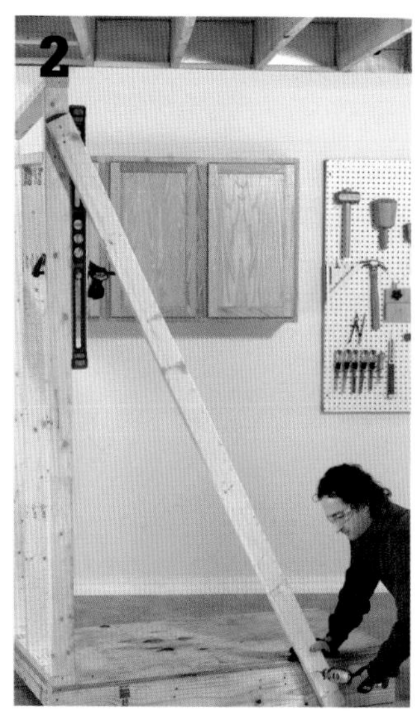

Raise the walls. Use a scrap of wood as a brace to keep the wall plumb. Attach the brace to the side-wall frame and to the base platform once you have established that the wall is plumb.

Fasten the wall frames. Attach the shed walls to one another and to the base platform with 2½" screws. Use a square and level to check that the walls are plumb and square.

Make the rafters. Cut the workpieces to length, then lay out and cut a birdsmouth notch in the bottom of the two inside rafters. These notches will keep the tops of the inside rafters in line with the outside rafters. The ends should be plumb-cut at 10°.

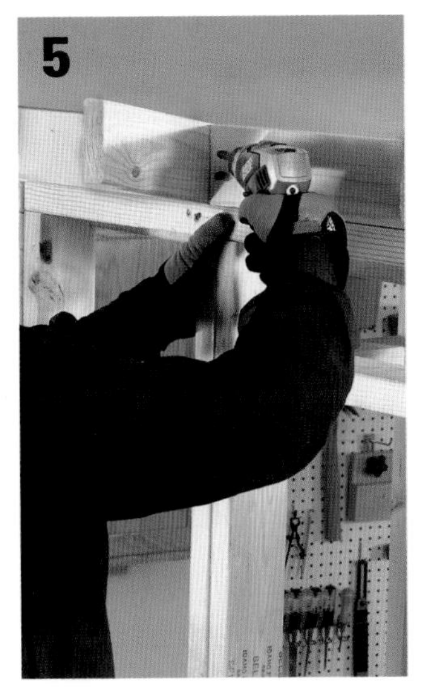

Install rafter blocking. Some of the rafter blocking must be attached to the rafters by toe-screwing (driving screws at an angle). If you own a pocket screw jig you can use it to drill angled clearance holes for the deck screw heads.

Install the roofing. Attach the corrugated roof panels with 1" neoprene gasket screws (sometimes called pole barn screws) driven through the panels at closure strip locations. Drill slightly oversized pilot holes for each screw and do not overdrive screws—it will compress the closure strips or even cause the panels to crack.

second side wall on the platform and attach it to the front and back walls with 2½" screws.

Cut the rafters to length, then miter-cut each end to 10° for making a plumb cut (this way the rafter ends will be perpendicular to the ground). A notch, referred to as a "birdsmouth," must be cut into the bottom edge of the inside rafters so the tops of these rafters align with the outside rafter tops while resting solidly on the wall top plates. Mark the birdsmouth on the inside rafters (see Diagram, page 147) and cut them out with a jigsaw (photo 4). Cut the rafter blocking to length; these parts fit between the rafters at the front and back of the shed to close off the area above the top plates. Attach the rafters to the rafter blocking and to the top plates. Use the blocking as spacers to position the rafters and then drive 2½" screws up through the top plates and into the rafters. Then, drive 2½" screws through the rafters and into the blocking (photo 5). Toe-screw any rafter blocking that you can't access to fasten through a rafter. Finally, cut the door rails and stiles to length. Attach the rails to the stiles with 2½" screws.

INSTALL THE ROOFING

This shed features 24"-wide corrugated roofing panels. The panels are installed over wood or foam closure strips that are attached to the tops of 2 × 2 purlins running perpendicular to the rafters. Position the purlins so the end ones are flush with the ends of the rafters and the inner ones are evenly spaced. The overhang beyond the rafters should be equal on the purlin ends.

Cut five 61½"-long closure strips. If the closure strips are wood, drill countersunk pilot holes through the closure strips and attach them to the purlins with 1½" screws. Some closure strips are made of foam with a self-adhesive backing. Simply peel off the paper backing and press them in place. If you are installing foam strips that do not have adhesive backing, tack them down with a few pieces of double-sided carpet tape so they don't shift around.

Cut three 44"-long pieces of corrugated roofing panel. Use a jigsaw with a fine-tooth blade or a circular saw with a fine-tooth plywood blade to cut fiberglass or plastic panels. Clamp the panels

(continued)

Cut the wall panels. Use a utility knife to score the fiber-cement panel along a straightedge. Place a board under the scored line and then press down on the panel to break the panel as you would with drywall.

Attach siding panels. Attach the fiber-cement siding with 1½" siding nails driven through pilot holes. Space the nails 8 to 12" apart. Drive the nails a minimum ⅜" away from the panel edges and 2" from the corners..

Cut the acrylic window material to size. One way to accomplish this is to sandwich the acrylic between two sheets of scrap plywood and cut all three layers at once with a circular saw (straight cuts) or jigsaw.

together between scrap boards to minimize vibration while they're being cut (but don't clamp down so hard that you damage the panels). Position the panels over the closure strips, overlapping roughly 4" of each panel and leaving a 1" overhang in the front and back.

Drill pilot holes 12" apart in the field of panels and along the overlapping panel seams. Fasten only in the valleys of the corrugation. The pilot hole diameter should be slightly larger than the diameter of the screw shanks. Fasten the panels to the closure strips and rafters with hex-head screws that are pre-fitted with neoprene gaskets (photo 6, page 245).

ATTACH THE SIDING

Cut the siding panels to size by scoring them with a utility knife blade designated for scoring concrete and then snapping them along the scored line (photo 7). Or, use a rented cementboard saw (see page 240). Drill pilot holes in the siding and attach the siding to the framing with 1½" siding nails spaced at 8 to 12" intervals (photo 8). (You can rent a cementboard coil nailer instead.) Cut the plywood door panel to size. Paint the siding and door before you install the windows and attach the wall and door trim. Apply two coats of exterior latex paint.

INSTALL THE WINDOWS

The windows are fabricated from ¼"-thick sheets of clear plastic or acrylic. To cut the individual windows to size, first mark the cut lines on the sheet. To cut acrylic with a circular saw, secure the sheet so that it can't vibrate during cutting. The best way to secure it is to sandwich it between a couple of pieces of scrap plywood and cut through all three sheets (photo 9). Drill ¼"-dia. pilot holes around the perimeter of the window pieces. Position the holes ½" from the edges and 6" apart. Attach the windows to the side wall framing on the exterior side using 1½" screws (photo 10).

ATTACH THE TRIM

Cut the wall and door trim pieces to length. Miter-cut the top end of the side front and back trim pieces to 10°. Attach the trim to the shed with 2" galvanized finish nails (photo 11). The horizontal side trim overlaps the window and the side siding panel. Be careful not to drive any nails through the plastic window panels. Attach the door trim to the door with 1¼" exterior screws.

HANG THE DOOR

Make the door and fasten a utility handle or gate handle to it. Fasten three door hinges to the door and then fasten the hinges to a stud on the edge of the door opening (photo 12). Use a scrap piece of siding as a spacer under the door to determine the proper door height in the opening. Add hooks and hangers inside the shed as needed to accommodate the items you'll be storing. If you have security concerns, install a hasp and padlock on the mini shed door.

Attach the window panels. Drill a ¼"-dia. pilot hole for each screw that fastens the window panels. These oversized holes give the plastic panel room to expand and contract. The edges of the windows (and the fasteners) will be covered by trim.

Attach the trim boards with 2" galvanized finish nails. In the areas around windows, predrill for the nails so you don't crack the acrylic.

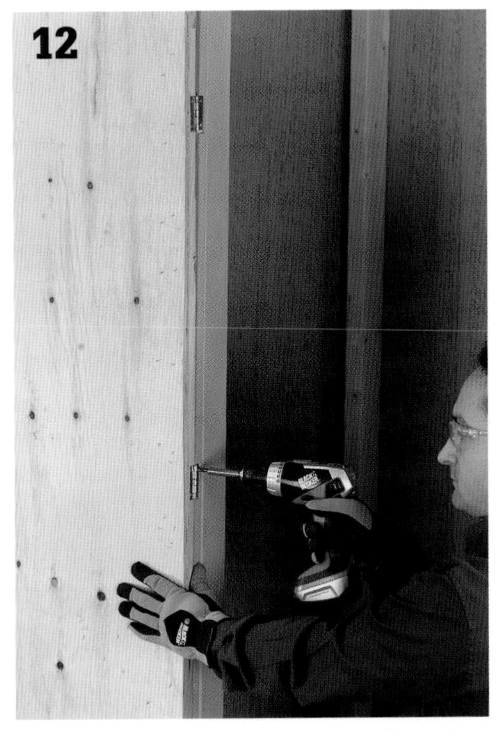

Hang the door using three exterior-rated door hinges. Slip a scrap of ¼"-thick siding underneath the door to raise it off the bottom plate while you install it.

Sunlight Garden Shed

The large windows on this garden shed's roof and front wall are perfect for providing light for seedlings.

This unique outbuilding is part greenhouse and part shed, making it perfect for a year-round garden space or backyard sunroom, or even an artist's studio. The front facade is dominated by windows—four 29 × 72" windows on the roof, plus four 29 × 18" windows on the front wall. When appointed as a greenhouse, two long planting tables inside the shed let you water and tend to plants without flooding the floor. If gardening isn't in your plans, you can omit the tables and cover the entire floor with plywood or perhaps fill in between the floor timbers with pavers or stones.

Some other details that make this 10 × 12-ft. shed stand out are the homemade Dutch door, with top and bottom halves that you can open together or independently, and its traditional saltbox shape. The roof covering shown here consists of standard asphalt shingles, but cedar shingles make for a nice upgrade.

Because sunlight plays a central role in this shed design, consider the location and orientation carefully. To avoid shadows from nearby structures, maintain a distance between the shed and the structure that's at least 2½ times the height of the obstruction. With all of that sunlight, the temperature inside the shed is another important consideration. You may want to install some roof vents to release hot air and water vapor.

Building the Sunlight Garden Shed involves a few unconventional construction steps. First, the side walls are framed in two parts: You build the square portion of the end walls first, then move onto the roof framing. After the rafters are up, you complete the "rake," or angled, sections of the side walls. This makes it easy to measure for each wall stud, rather than having to calculate the lengths beforehand. Second, the shed's 4 × 4 floor structure also serves as its foundation. The plywood floor decking goes on after the walls are installed, rather than before.

The unique, handmade Dutch door that is part of the building's design opens as one unit—or, you can open the top and bottom halves independently.

Cutting List

Description	Quantity/Size	Material
Foundation/Floor		
Foundation base & interior drainage beds	5 cu. yds.	Compactable gravel
Floor joists & blocking	7 @ 10'	4 × 4 pressure-treated landscape timbers
4 × 4 blocking	1 @ 10' 1 @ 8'	4 × 4 pressure-treated landscape timbers
Box sills (rim joists)	2 @ 12'	2 × 4 pressure-treated
Nailing cleats & 2 × 4 blocking	2 @ 8'	2 × 4 pressure-treated
Floor sheathing	2 sheets @ 4 × 8'	¾" ext.-grade plywood
Wall Framing		
Bottom plates	2 @ 12', 2 @ 10'	2 × 4 pressure-treated
Top plates	4 @ 12', 2 @ 10'	2 × 4
Studs	43 @ 8'	2 × 4
Door header & jack studs	3 @ 8'	2 × 4
Rafter header	2 @ 12'	2 × 8
Roof Framing		
Rafters—A & C, & nailers	10 @ 12'	2 × 4
Rafters—B & lookouts	10 @ 10'	2 × 4
Ridge board	1 @ 14'	2 × 6
Exterior Finishes		
Rear fascia	1 @ 14'	1 × 6 cedar
Rear soffit	1 @ 14'	1 × 8 cedar
Gable fascia (rake board) & soffit	4 @ 16'	1 × 6 cedar
Siding	10 sheets @ 4 × 8'	⅝" texture 1-11 plywood siding
Siding flashing	10 linear ft.	Metal Z-flashing
Trim*	4 @ 12' 1 @ 12'	1 × 4 cedar 1 × 2 cedar
Wall corner trim	6 @ 8'	1 × 4 cedar
Roofing		
Sheathing	5 sheets @ 4 × 8'	½" exterior-grade plywood roof sheathing
15# building paper	1 roll	
Drip edge	72 linear ft.	Metal drip edge
Shingles	2⅔ squares	Asphalt shingles—250# per sq. min.

Description	Quantity/Size	Material
Windows		
Glazing	4 pieces @ 31¼ × 76½" 4 pieces @ 31¼ × 20¾"	¼"-thick clear plastic glazing
Window stops	12 @ 10'	2 × 4
Glazing tape	60 linear ft.	
Clear exterior caulk	5 tubes	
Door		
Trim & stops	3 @ 8'	1 × 2 cedar
Surround	4 @ 8'	2 × 2 cedar
Z-flashing	3 linear ft.	
Plant Tables (optional)		
Front table, top & trim	6 @ 12'	1 × 6 cedar or pressure-treated
Front table, plates & legs	4 @ 12'	2 × 4 pressure-treated
Rear table, top & trim	6 @ 8'	1 × 6 cedar or pressure-treated
Rear table, plates & legs	4 @ 8'	2 × 4 pressure-treated
Fasteners & Hardware		
16d galvanized common nails	5 lbs.	
16d common nails	16 lbs.	
10d common nails	1½ lbs.	
8d galvanized common nails	2 lbs.	
8d galvanized box nails	3 lbs.	
10d galvanized finish nails	2½ lbs.	
8d galvanized siding nails	8 lbs.	
1" galvanized roofing nails	7 lbs.	
8d galvanized casing nails	3 lbs.	
6d galvanized casing nails	2 lbs.	
Door hinges with screws	4 @ 3½"	Corrosion-resistant hinges
Door handle	1	
Sliding bolt latch	1	
Construction adhesive	1 tube	

**Note: The 1 × 4 trim bevel at the bottom of the sloped windows can be steeper (45° or more) so the trim slopes away from the window if there is concern that the trim may capture water running down the glazing (see WINDOW DETAIL, page 256).*

Building Section

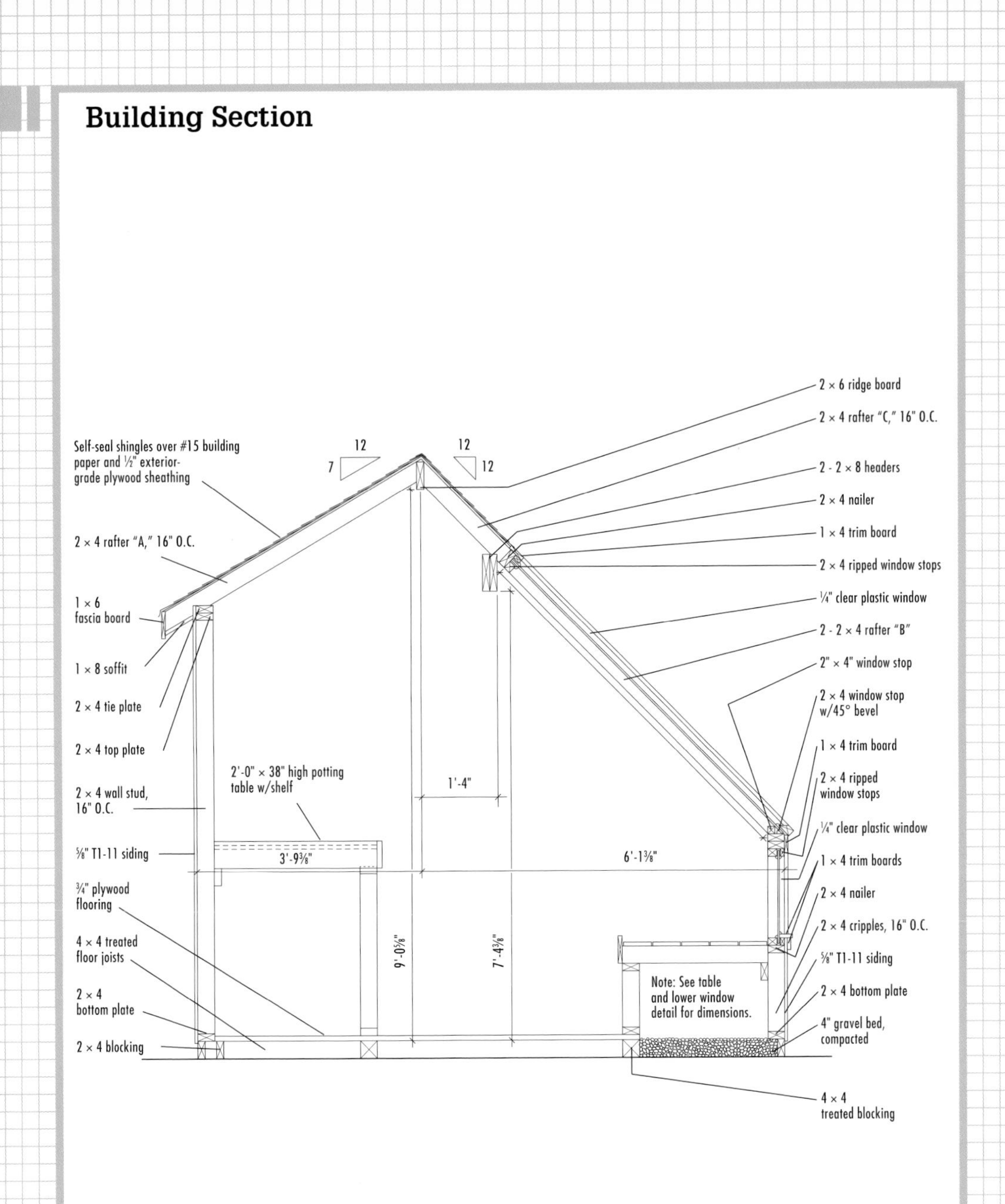

Floor Framing Plan

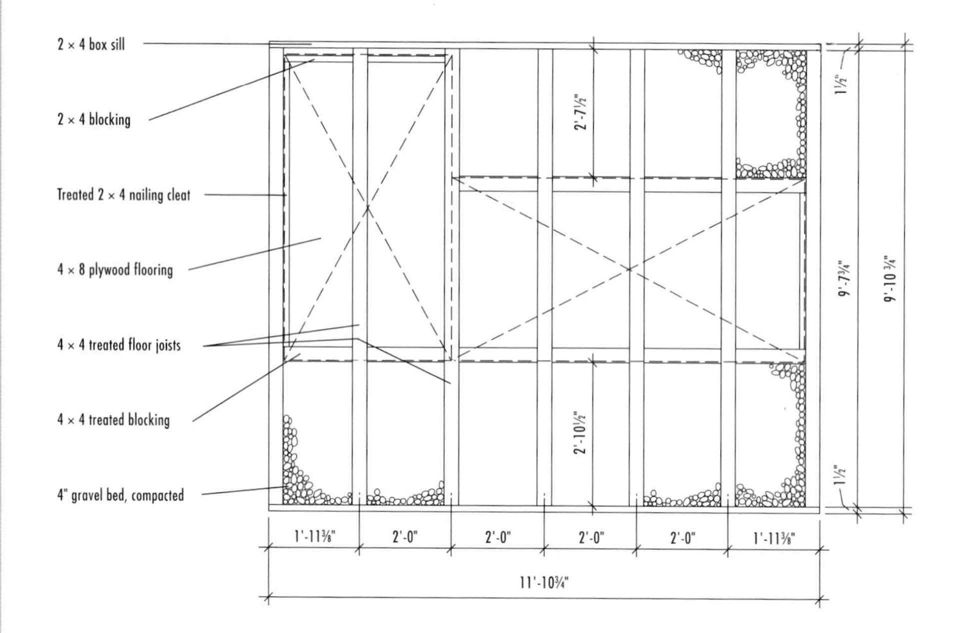

Left Side Framing

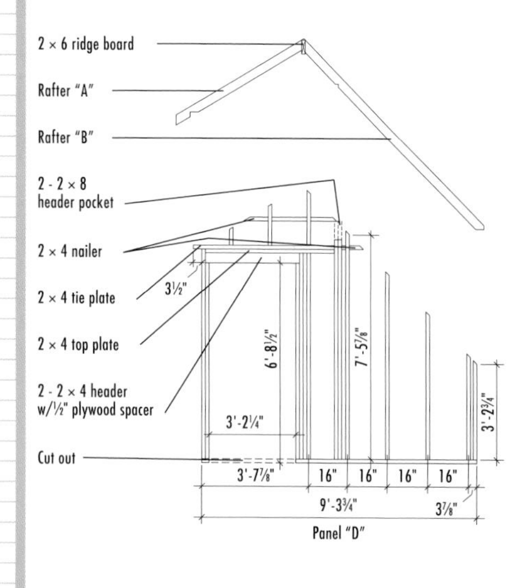

Right Side Framing

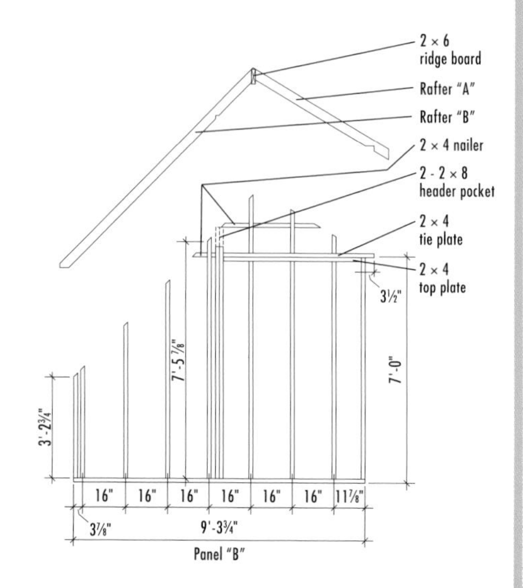

Front Framing

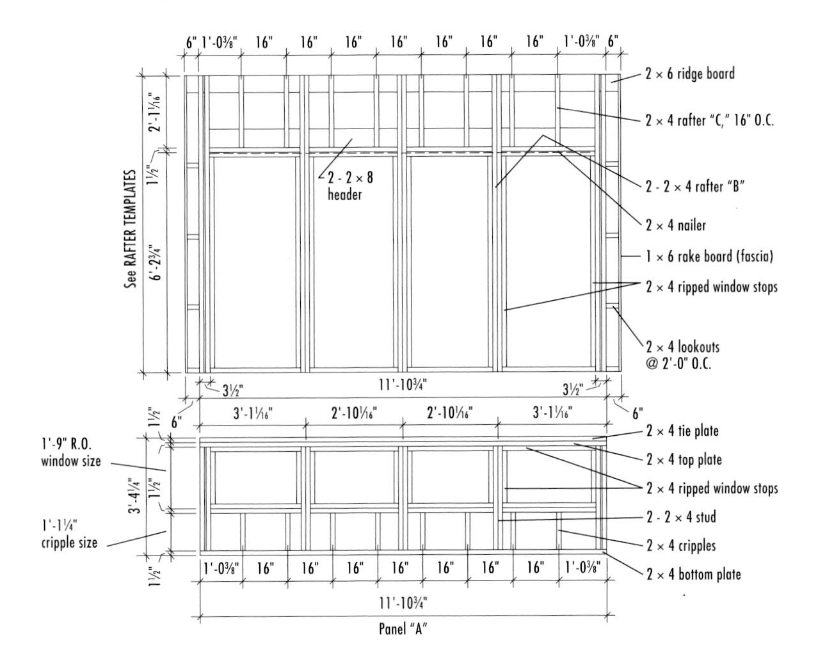

Rear Framing

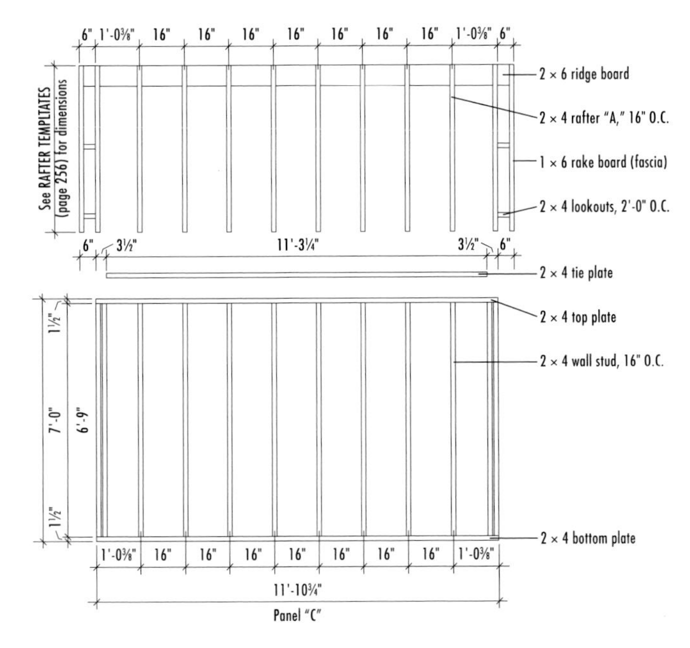

Front Elevation

Rear Elevation

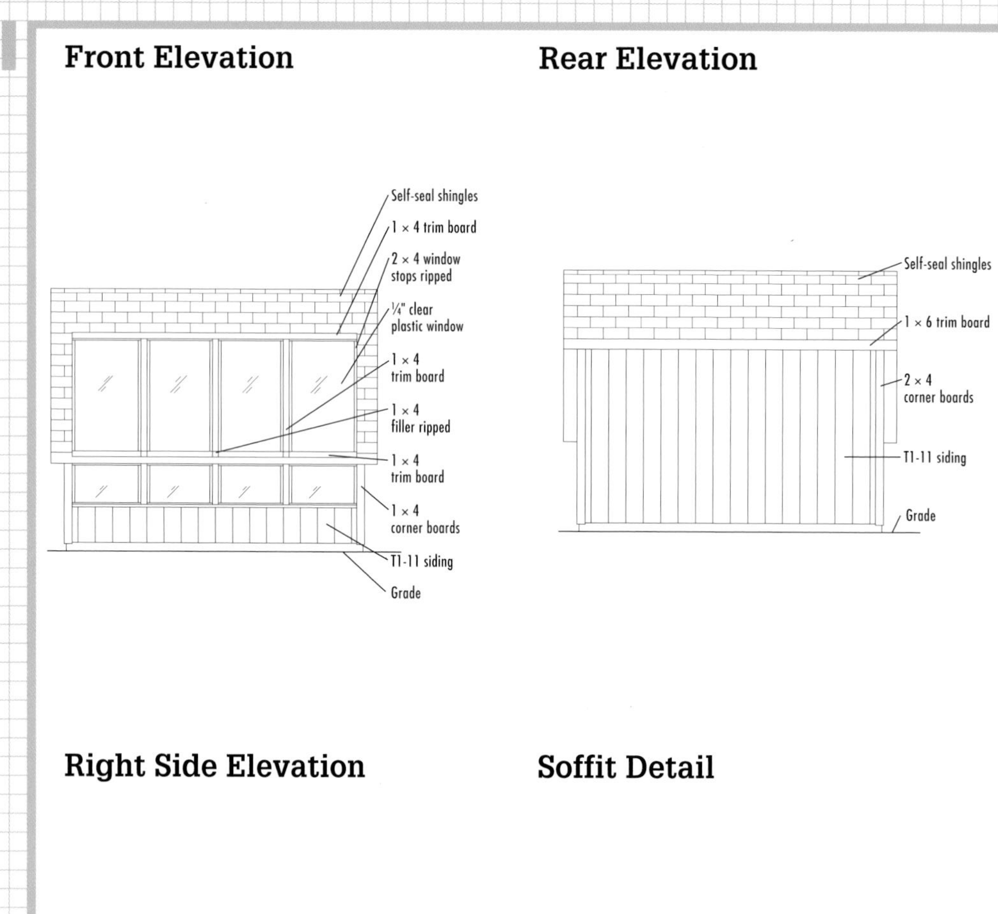

Right Side Elevation

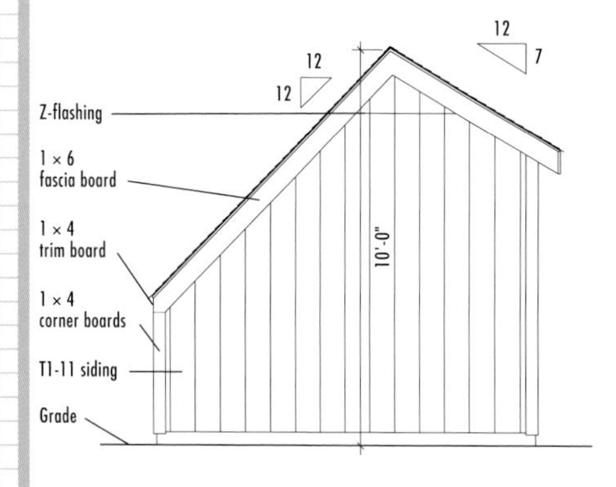

Soffit Detail

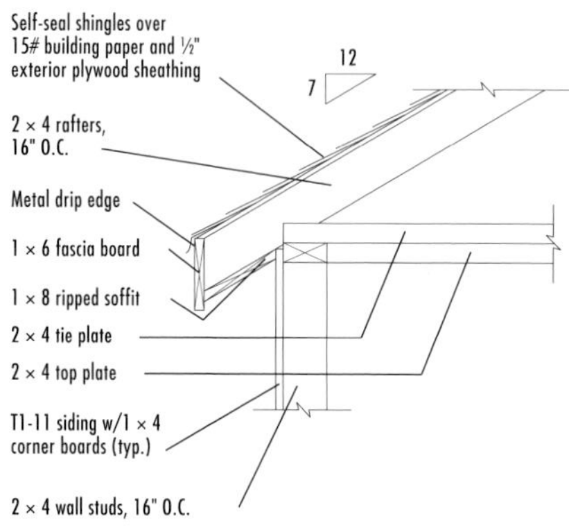

Door Construction (Front & Side View)

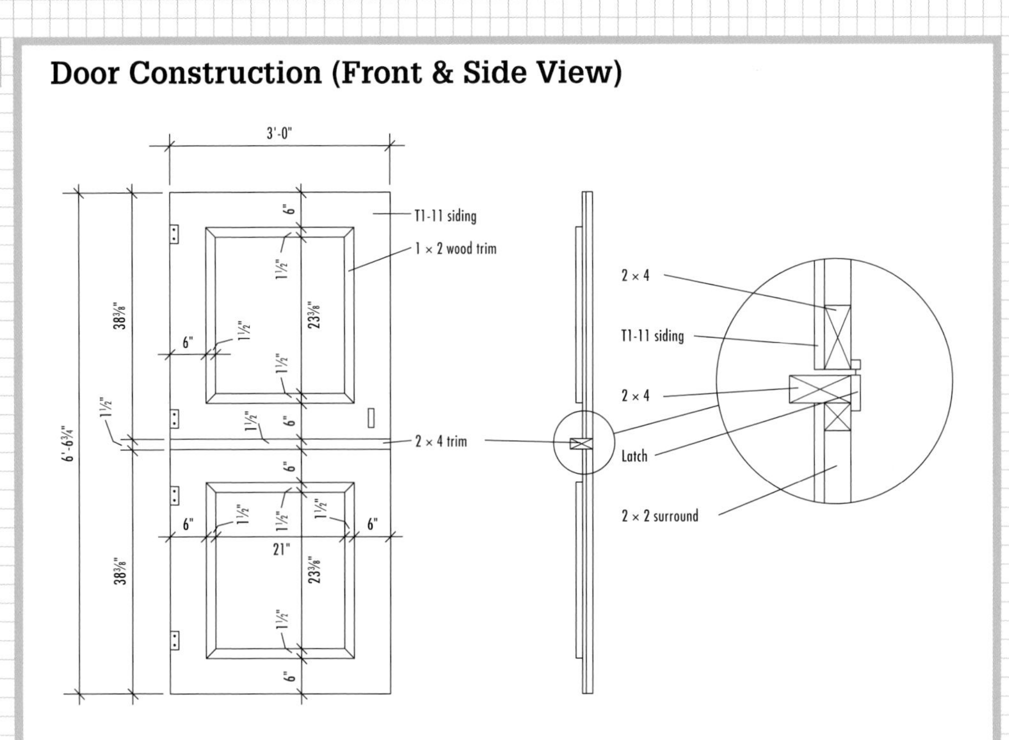

Door Construction (Door Jamb, Rear, Door Header)

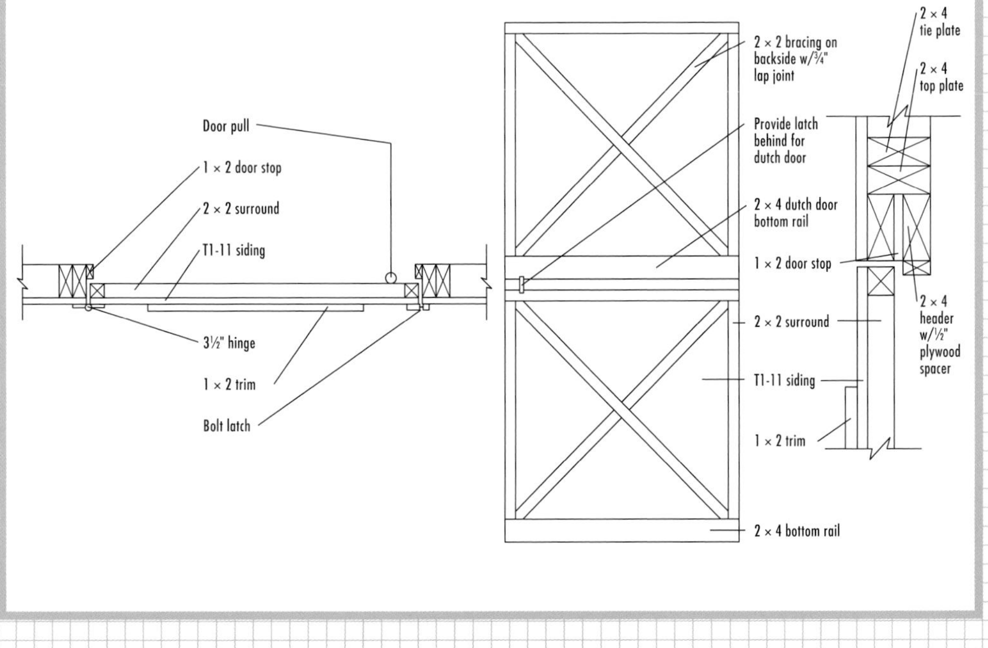

Header & Window Detail

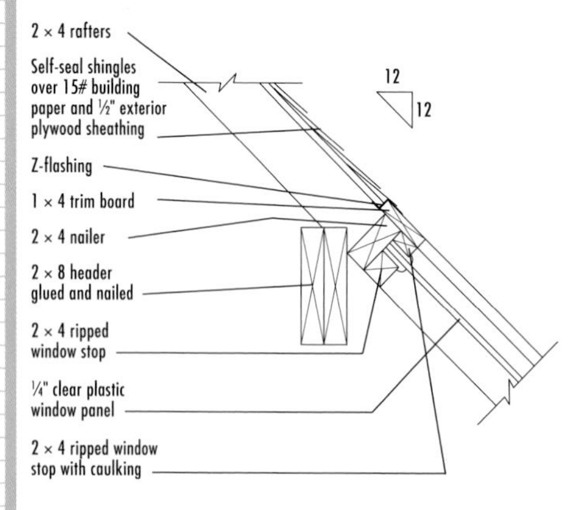

Window Section

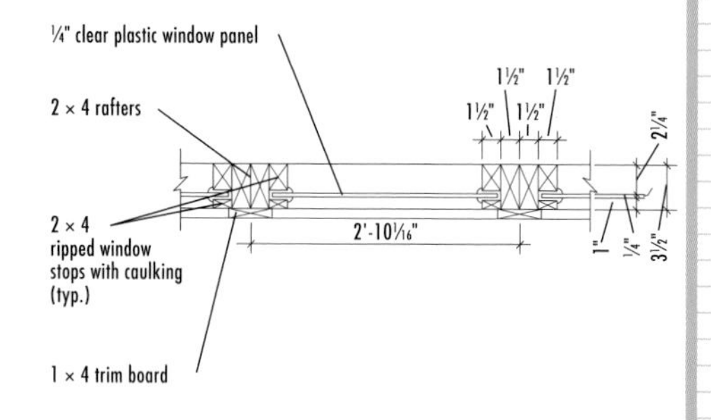

Window Detail

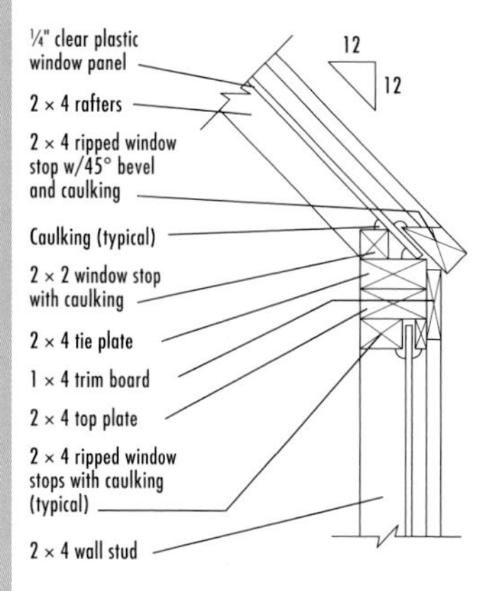

Table & Lower Window Detail

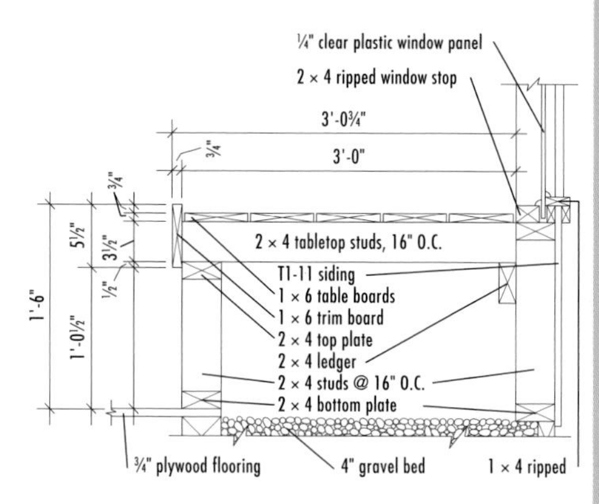

Rafter Templates

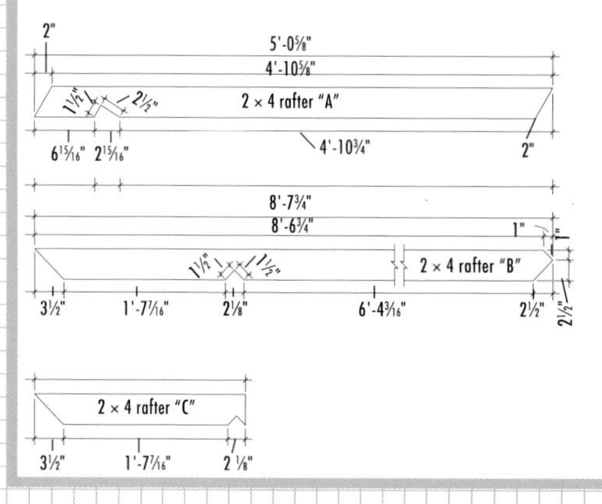

Rake Board Detail

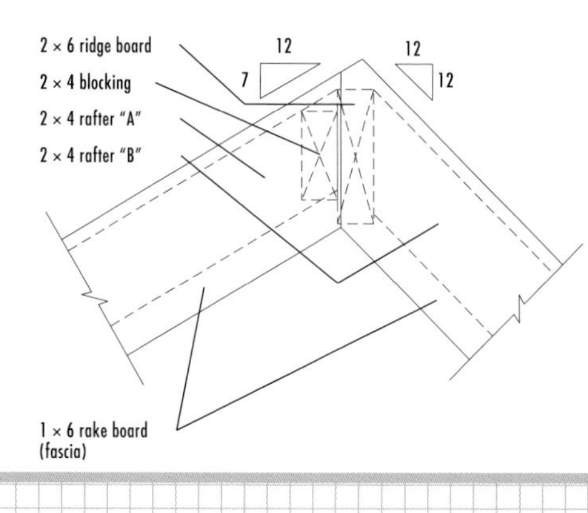

How to Build a Sunlight Garden Shed

Build the foundation following the basic steps used for a wooden skid foundation (pages 432 to 433). First, prepare a bed of compacted gravel. Make sure the bed is flat and level. Cut seven 4 × 4" × 10 ft. pressure-treated posts to 115¾" to serve as floor joists (see page 369). Position the joists as shown in the FLOOR FRAMING PLAN on page 252. Level each joist, and make sure all are level with one another, and the ends are flush. Add rim joists and blocking: Cut two 12-ft. 2 × 4s (142¾") for rim joists. Fasten the rim joists to the ends of the 4 × 4 joists (see the FLOOR FRAMING PLAN) with 16d galvanized common nails.

Cut ten 4 × 4 blocks to fit between the joists. Install six blocks 34½" from the front rim joist, and install four blocks 31½" from the rear. Toenail the blocks to the joists. All blocks, joists, and sills must be flush at the top.

To frame the rear wall, cut one top plate and one pressure-treated bottom plate (142¾"). Cut twelve studs (81"). Assemble the wall following the layout in the REAR FRAMING (page 253). Raise the wall and fasten it to the rear rim joist and the intermediate joists using 16d galvanized common nails. Brace the wall in position with 2 × 4 braces staked to the ground.

For the front wall, cut two top plates and one treated bottom plate (142¾"). Cut ten studs (35¾") and eight cripple studs (13¼"). Cut eight 2 × 4 windowsills (31¹⁄₁₆"). Assemble the wall following the layout in the FRONT FRAMING (page 253). Add the double top plate, but do not install the window stops at this time. Raise, attach, and brace the front wall.

(continued)

Cut lumber for the right side wall: one top plate (54⅞"), one treated bottom plate (111¾"), four studs (81"), and two header post studs (86⅞"); for the left side wall: top plate (54⅞"), bottom plate (111¾"), three studs (81"), two jack studs (79"), two posts (86⅞"), and a built-up 2 × 4 header (41¼"). Assemble and install the walls as shown in the RIGHT SIDE FRAMING and LEFT SIDE FRAMING (page 252). Add the doubled top plates (tie plates) along the rear and side walls. Install treated 2 × 4 nailing cleats to the joists and blocking as shown in the FLOOR FRAMING PLAN (page 252) and BUILDING SECTION (page 251).

Trim two sheets of ¾" plywood as needed and install them over the joists and blocking as shown in the FLOOR FRAMING PLAN, leaving open cavities along the front of the shed and a portion of the rear. Fasten the sheets with 8d galvanized common nails driven every 6" along the edges and 8" in the field. Fill the exposed foundation cavities with 4" of gravel and compact it thoroughly.

Construct the rafter header from two 2 × 8s cut to 142¾". Join the pieces with construction adhesive and pairs of 10d common nails driven every 24" on both sides. Set the header on top of the side wall posts and toenail it to the posts with four 16d common nails at each end.

Cut one of each "A" and "B" pattern rafters using the RAFTER TEMPLATES (page 256). Test-fit the rafters. The B rafter should rest squarely on the rafter header, and its bottom end should sit flush with the outside of the front wall. Adjust the rafter cuts as needed, then use the pattern rafters to mark and cut the remaining A and B rafters.

Cut a 2 × 6 ridge board (153¼"). Mark the rafter layout onto the ridge and front and rear wall plates following FRONT FRAMING and REAR FRAMING. Install the A and B rafters and ridge. Make sure the B rafters are spaced accurately so the windows will fit properly into their frames; see the WINDOW SECTION (page 256).

Cut a pattern "C" rafter, test-fit, and adjust as needed. Cut the remaining seven C rafters and install them. Measure and cut four 2 × 4 nailers (31¹⁄₁₆") to fit between the sets of B rafters (as shown). Position the nailers as shown in the HEADER & WINDOW DETAIL (page 256), toenail them to the B rafters, and endnail them to the C rafters.

Complete the rake portions of each side wall. Mark the stud layouts onto the bottom plate and onto the top plate of the square wall section; see RIGHT and LEFT SIDE FRAMING. Use a plumb bob to transfer the layout to the rafters. Measure for each stud, cutting the top ends of the studs under the B rafters at 45° and those under the A rafters at 30°. Toenail the studs to the plates and rafters. Add horizontal 2 × 4 nailers as shown in the framing drawings.

Create the inner and outer window stops from 10-ft.-long 2 × 4s. For stops at the sides and tops of the roof windows and all sides of the front wall windows, rip the inner stops to 2¼" wide and the outer stops to 1" wide; see the WINDOW SECTION and WINDOW DETAIL (page 256). For the bottom of each roof window, rip the inner stop to 1½"; bevel the edge of the outer stop at 45°.

(continued)

Install each window as follows: Attach inner stops as shown in the drawings using galvanized finish nails. Paint or varnish the rafters and stops for moisture protection. Apply a heavy bead of caulk at each location shown on the drawings (HEADER & WINDOW DETAIL, WINDOW SECTION/DETAIL, TABLE & LOWER WINDOW DETAIL). Set the glazing in place, add another bead of caulk, and attach the outer stops. Cover the rafters and stop edges with 1 × 4 trim.

Cover the walls with T1-11 siding, starting with the rear wall. Trim the sheets as needed so they extend from the bottom edges of the rafters down to at least 1" below the tops of the foundation timbers. On the side walls, add Z-flashing above the first row and continue the siding up to the rafters.

Install 1 × 6 fascia over the ends of the A rafters. Using scrap rafter material, cut the 2 × 4 lookouts (5¼"). On each outer B rafter, install one lookout at the bottom end and four more spaced 24" on center going up. On the A rafters, add a lookout at both ends and two spaced evenly in between. Install the 1 × 6 rake boards (fascia) as shown in the RAKE BOARD DETAIL (page 256).

Rip 1 × 6 boards to 5¼" width (some 1 × 6s may come milled to 5¼" already) for the gable soffits. Fasten the soffits to the lookouts with siding nails. Rip a 1 × 8 board for the soffit along the rear eave, beveling the edges at 30° to match the A rafter ends. Install the soffit.

Deck the roof with ½" plywood sheathing, starting at the bottom ends of the rafters. Install metal drip edge, building paper, and asphalt shingles following the steps on pages 60 to 67. If desired, add one or more roof vents during the shingle installation. Install Z-flashing and overlap shingles onto the 1 × 4 trim board above the roof windows, as shown in the HEADER & WINDOW DETAIL.

Construct the planting tables from 2 × 4 lumber and 1 × 6 boards, as shown in the TABLE & LOWER WINDOW DETAIL and BUILDING SECTION. The bottom plates of the table legs should be flush with the outside edges of the foundation blocking.

Build each of the two door panels using T1-11 siding, 2 × 2 bracing, a 2 × 4 bottom rail, and 1 × 2 trim on the front side; see the DOOR CONSTRUCTION drawings (page 255). The panels are identical except for a 2 × 4 sill added to the top of the lower panel. Install 1 × 2 stops at the sides and top of the door opening. Hang the doors with four hinges, leaving even gaps all around. Install a bolt latch for locking the two panels together.

Complete the trim details with 1 × 4 vertical corner boards, 1 × 4 horizontal trim above the front wall windows, and ripped 1 × 4 trim and 1 × 2 trim at the bottom of the front wall windows (see the TABLE & LOWER WINDOW DETAIL). Paint the siding and trim, or coat with exterior wood finish.

Shelters, Arbors & Gazebos

Outdoor recreational shelters are a unique blend of architecture and landscaping. While they define natural spaces and create comfortable rooms for outdoor living, they can be just as sculptural as they are practical.

Shelters, arbors, and gazebos—all classic backyard buildings—represent a range of outdoor architecture meant for everything from alfresco dining to afternoon naps. And, all ten of the projects in this chapter offer an opportunity for you to expand upon and enjoy your home landscape in a new way. Some tend toward the exotic, like the Lattice Gazebo, the design of which is inspired by Japanese teahouses, while others, such as the Rustic Summerhouse, utilize construction and finishing techniques that have been slowly perfected in the United States over the last two centuries.

Each project comes with complete plans and building instructions, but when it comes to finishing details, don't be afraid to embellish. Outdoor rooms should complement and accent your home and reflect the style and charm of your personal estate.

In this chapter:

- Rustic Summerhouse
- 3-season Gazebo
- Summer Pavilion
- Lattice Gazebo
- 8-sided Gazebo
- Pool Pavilion
- Patio Shelter
- Arbor Retreat
- Classical Pergola
- Corner Lounge

Rustic Summerhouse

Whether close to your home, adjoining a garden or pond, or elsewhere on your property, this charming hideout will quickly become a focal point for your friends and family when enjoying the outdoors.

The quaint and cozy Rustic Summerhouse is a perfect spot for enjoying a meal, relaxing with the sunset, escaping into a good book, or taking a quick break from gardening or yard work. The entirely screened-in room is bug-free, large enough to comfortably seat six, strong enough to support a hammock for two, and stable enough to house a hot tub or whirlpool spa. This charming summerhouse, made with traditional rough-hewn materials, is the perfect finishing touch to any backyard or estate.

The building is constructed using post-and-beam construction techniques that have been passed down through the centuries, ensuring that the finished product will stand the test of time. The design's eleven screens ensure excellent airflow, keeping the interior comfortable even on hot days. Attractively finished with board-and-batten siding and a corrugated metal roof, the Rustic Summerhouse's crowning achievement is the decorative cedar clapboard sunburst design above the door, handcrafted by each builder. Following the techniques discussed here to create this building design, your summerhouse will have its own completely unique finishing touch.

The materials are key to this building's quality. Hearty 2 × 6 flooring provides a solid base for even the most aggressive wear and tear, and rough-sawn 1 × 12 pine siding with two-inch battens gives the building its charm and character. Intermediate or expert builders may find this building to be quite quick to erect, whereas beginners may select this project as a fulfilling challenge with a fantastic payoff.

This cozy outdoor room is the perfect getaway or relaxation spot. Large enough to seat six for dinner, the Rustic Summerhouse is also the perfect retreat for one or two to sit and enjoy the view.

Tools, Materials & Cutting List

Chain saw
Circular saw
Hammer
Nail puller (crow bar)
Tape measure
Chalk line
Extension cord
Tin snips
Staple gun
Speed square
4-ft. level
Drill with ¼" drive bit
Utility knife
Plate compactor
Eye and ear protection
Plumb bob
Wood shims
Excavation tools
Reciprocating saw

Description	Qty/Size	Material
Foundation/Floor		
Foundation base	2.5 cu. yds.	Compactable stone gravel
Concrete blocks	6	4 × 8 × 16" blocks
Skids	2@14'	4 × 6 hemlock
Floor joists	8@10'	2 × 6 hemlock
Rim joists	2@14'	2 × 6 hemlock
Flooring	24@14'	2 × 6 tongue-and-groove spruce or pine
Wall Framing		
Rear top plate beam	14'	4 × 4 hemlock
Front top plate beams	2@62¾"	4 × 4 hemlock
Gable beams	2@10'	4 × 4 hemlock
Rear wall horizontal nailers	4@17¼"	2 × 4 hemlock
Front wall horizontal nailers	8@7⅛"	2 × 4 hemlock
Window studs	8@74"	2 × 4 hemlock
Front and rear wall window headers and sills	15@36½"	2 × 4 hemlock
Side wall window headers and sills	18@32½"	2 × 4 hemlock
Side wall spacer blocks	12@12"	1 × 4 hemlock
Wall, corner & door posts	14@74"	4 × 4 hemlock
Door studs	2@82"	2 × 4
Door header 1	93"	2 × 4
Door header 2	112"	2 × 4
Roof Framing		
Ridge board	14'	1 × 8 pine
Dormer ridge board	59½"	1 × 8 pine
Common rafters	12@66⅝"	2 × 6
Dormer common rafters	2@65⅝"	2 × 6
Valley rafters	2@87"	2 × 6
Jack 1 rafters	4@12½"	2 × 4
Jack 2 rafters	4@39½"	2 × 4
Collar ties	4@4'	1 × 6

Description	Qty/Size	Material
Exterior Finishes		
1 × 4 door, window, and corner trim, ridge & temporary bracing	24@12'	1 × 4 pine
Rear and front wall corner siding	6@90"	1 × 12
Rear and front wall siding – under windows	16@21"	1 × 12
Rear and front wall siding – above windows	12@12"	1 × 12
Rear and side wall siding – between windows	8@57"	1 × 6
Front wall siding – above windows	2@18" and 24"	1 × 12
Front wall siding	2@106"	1 × 12
Front wall siding	2@109"	1 × 8
Front wall siding – above door	2@23" and 27"	1 × 8
Front wall siding – above door	31"	1 × 8
Side wall corner siding	4@94"	1 × 6
Side wall siding – under windows	18@21"	1 × 12
Side wall siding – above windows	4@22", 28", 34", and 40"	1 × 12
Side wall siding – above windows	2@43"	1 × 12
Decorative sunburst boards	20@4'	6"-wide red cedar clapboard
Sunburst spacer	2@2'	1 × 6 pine
Sunburst spacer	2@75"	1 × 5 pine
Kick plate	40½"	1 × 8 pine
Fascia	5@16'	2 × 8 pine
Shadow board	5@16'	2 × 4 pine
Window trim	6@55½"	1 × 6 pine
Door casing	2@81", 1@38½"	1 × 5 pine
Doorstop	2@80", 1@36½"	1 × 2 pine
Rear wall battens	2@82½"	¾ × 2" pine*
Under-window battens	40@19"	¾ × 2" pine*
Gable battens	4@11", 17", 23", 29", and 35"	¾ × 2" pine*

Description	Qty/Size	Material
Roofing		
Rear-side sheathing	5@14'	1 × 12 pine
Rear-side sheathing – top ridge piece	14'	1 × 6 pine
Front-side sheathing	2@33", 42¾", 55", 65½", and 76"	1 × 12 pine
Front-side sheathing – top ridge pieces	2@75" and 81½"	1 × 6 pine
Dormer sheathing	2@10½", 20½", 32", 42½", and 53"	1 × 12 pine
Dormer sheathing – top ridge pieces	2@59"	1 × 6 pine
36"-wide roofing	11@12'	Corrugated metal
Ridge cap	22 ft.	
24"-wide valley flashing	20 ft.	Aluminum
Windows		
2'8" screens	6	
3' screens	5	

Description	Qty/Size	Material
Door		
3' heavy-duty screen door	1	
Hardware		
16d common nails	15 lbs	
16d galvanized common nails	12 lbs	
8d galvanized common nails	17 lbs	
6d stainless steel nails	2 lbs	
1⅝" zinc screws	1 lb	
1½" metal roofing screws with rubber gasket, painted to match your roof color	450	
Metal roofing nails or staples	1 lb	
Black spring screen door hinges	2	
2½" decorative black screen door hinges	2	
Black screen door knob and latch	1	

**Battens ripped on-site from 2" × 4" lumber*

Floor Plan

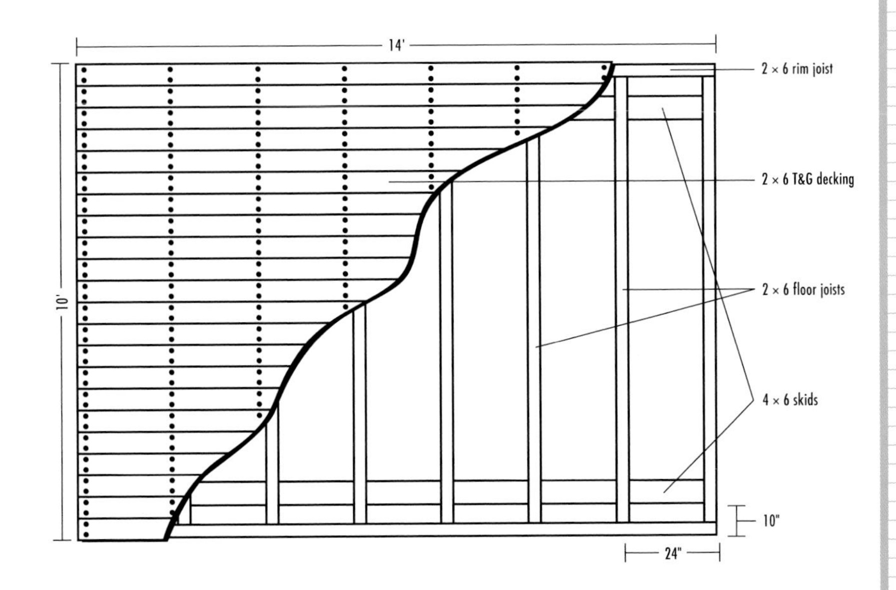

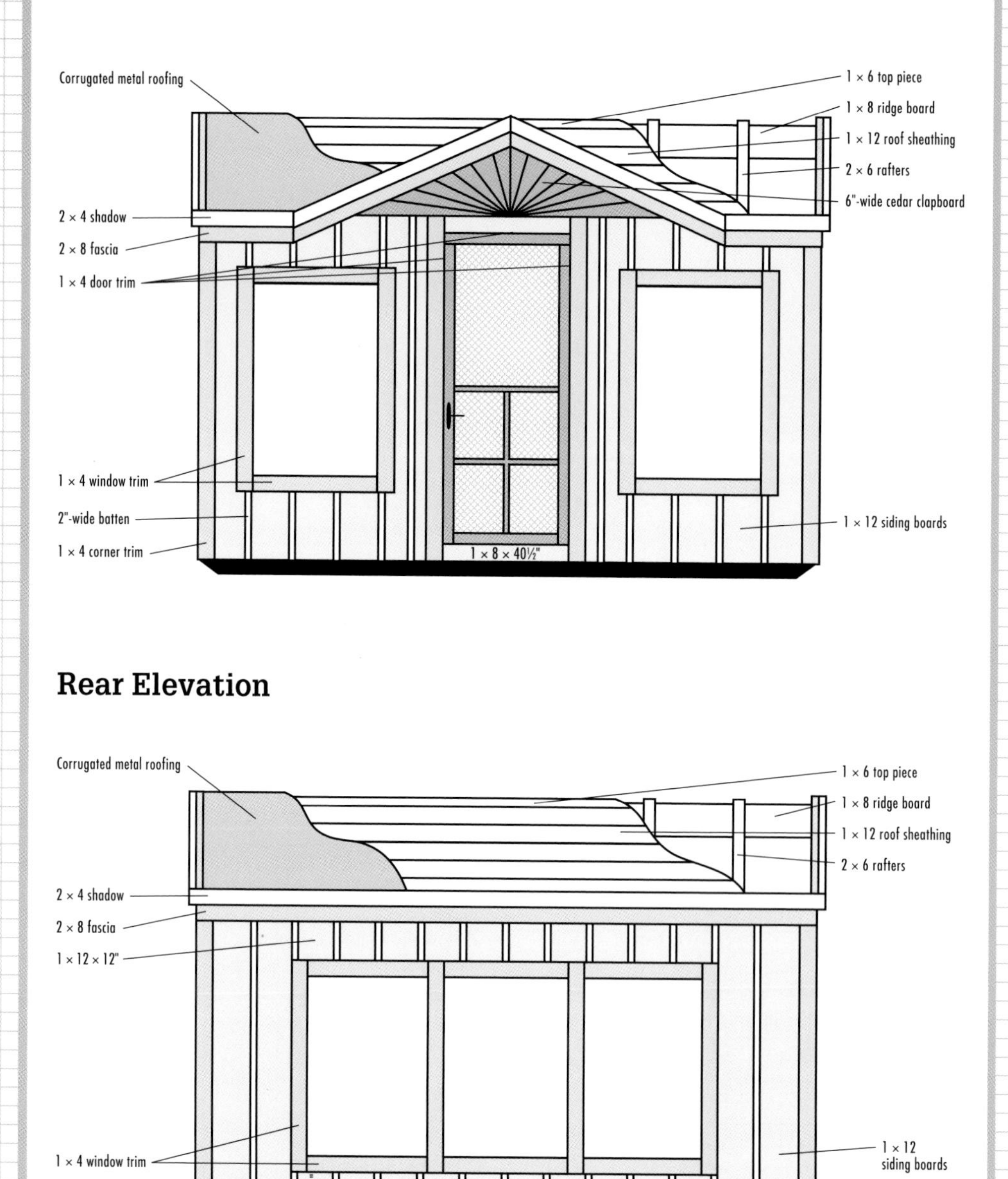
Front Elevation
Corrugated metal roofing
1 × 6 top piece
1 × 8 ridge board
1 × 12 roof sheathing
2 × 6 rafters
6"-wide cedar clapboard
2 × 4 shadow
2 × 8 fascia
1 × 4 door trim
1 × 4 window trim
2"-wide batten
1 × 4 corner trim
1 × 8 × 40½"
1 × 12 siding boards
Rear Elevation
Corrugated metal roofing
1 × 6 top piece
1 × 8 ridge board
1 × 12 roof sheathing
2 × 6 rafters
2 × 4 shadow
2 × 8 fascia
1 × 12 × 12"
1 × 4 window trim
2"-wide batten
1 × 4 corner trim
1 × 12 × 21"
1 × 12
siding boards
14'

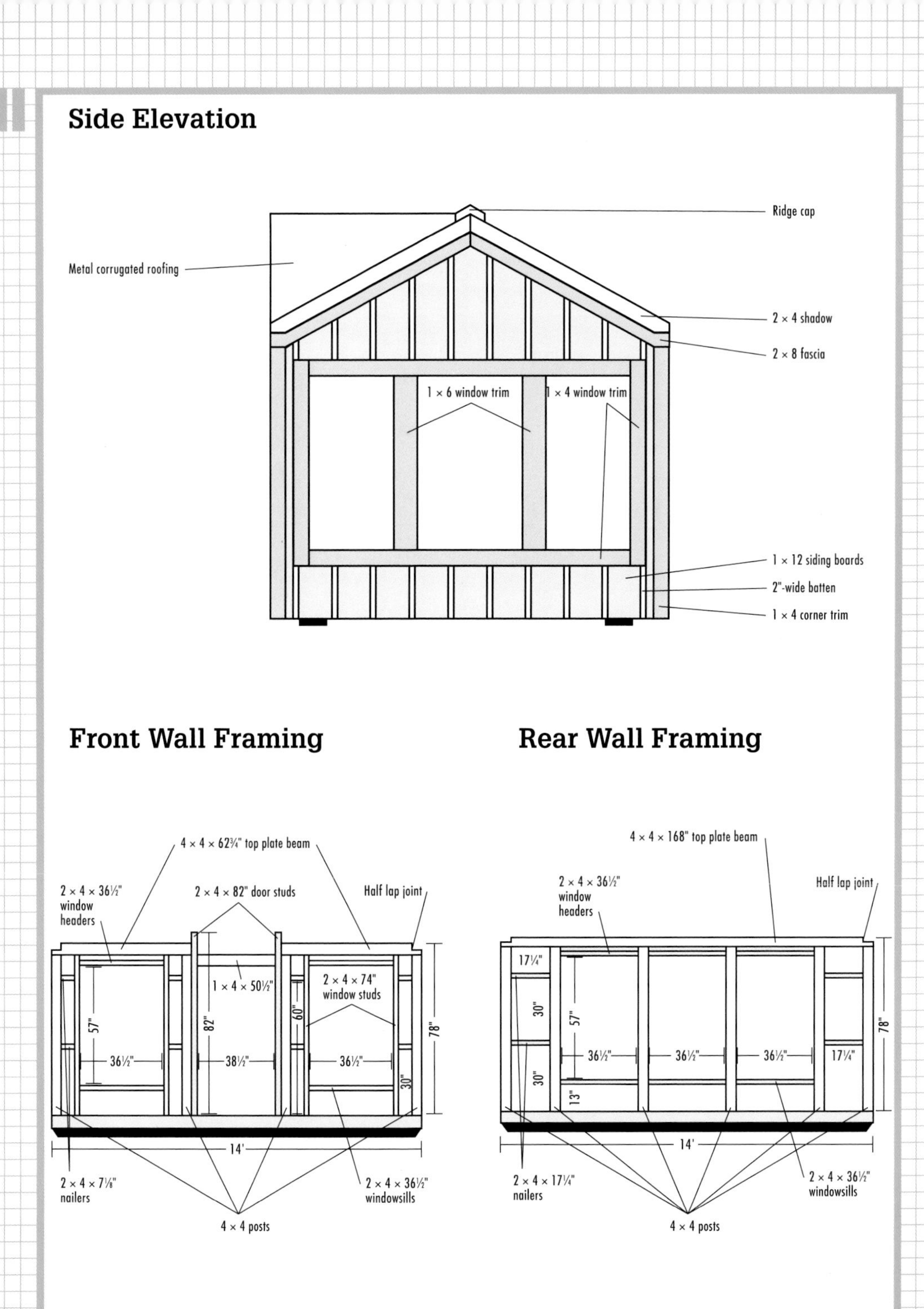

Side Elevation
Ridge cap
Metal corrugated roofing
2 × 4 shadow
2 × 8 fascia
1 × 6 window trim
1 × 4 window trim
1 × 12 siding boards
2"-wide batten
1 × 4 corner trim
Front Wall Framing
4 × 4 × 62¾" top plate beam
2 × 4 × 36½" window headers
2 × 4 × 82" door studs
Half lap joint
1 × 4 × 50½"
2 × 4 × 74" window studs
57"
82"
60"
78"
36½"
38½"
36½"
30"
14'
2 × 4 × 7⅛" nailers
2 × 4 × 36½" windowsills
4 × 4 posts
Rear Wall Framing
4 × 4 × 168" top plate beam
2 × 4 × 36½" window headers
Half lap joint
17¼"
30"
57"
30"
13"
36½"
36½"
36½"
17¼"
78"
14'
2 × 4 × 17¼" nailers
2 × 4 × 36½" windowsills
4 × 4 posts

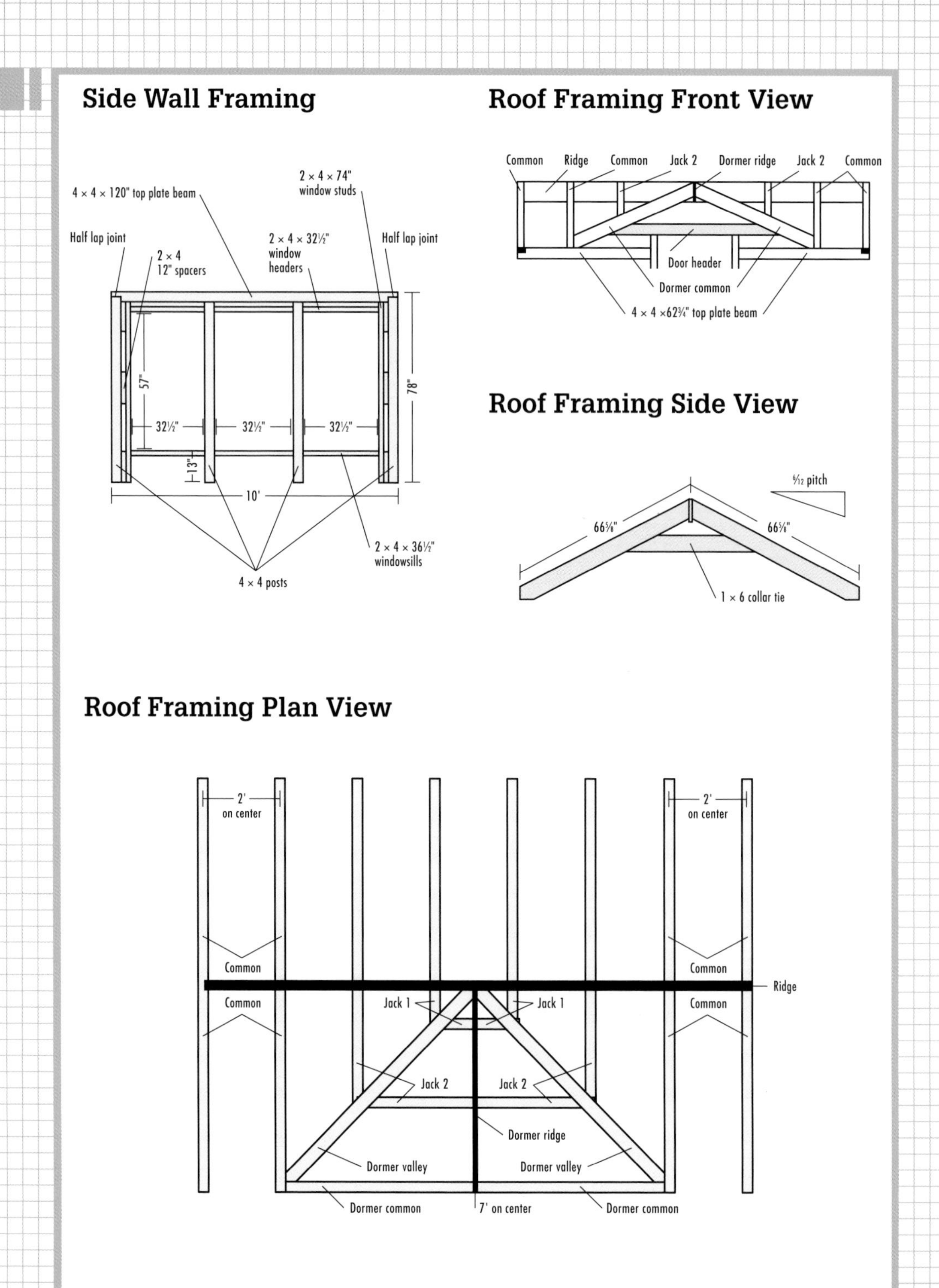

Side Wall Framing
4 × 4 × 120" top plate beam
2 × 4 × 74"
window studs
Half lap joint
2 × 4
12" spacers
2 × 4 × 32½"
window
headers
Half lap joint
57"
78"
32½"
32½"
32½"
13"
10'
2 × 4 × 36½"
windowsills
4 × 4 posts
Roof Framing Front View
Common
Ridge
Common
Jack 2
Dormer ridge
Jack 2
Common
Door header
Dormer common
4 × 4 ×62¾" top plate beam
Roof Framing Side View
6/12 pitch
66⅝"
66⅝"
1 × 6 collar tie
Roof Framing Plan View
2'
on center
2'
on center
Common
Common
Ridge
Common
Common
Jack 1
Jack 1
Jack 2
Jack 2
Dormer ridge
Dormer valley
Dormer valley
Dormer common
7' on center
Dormer common

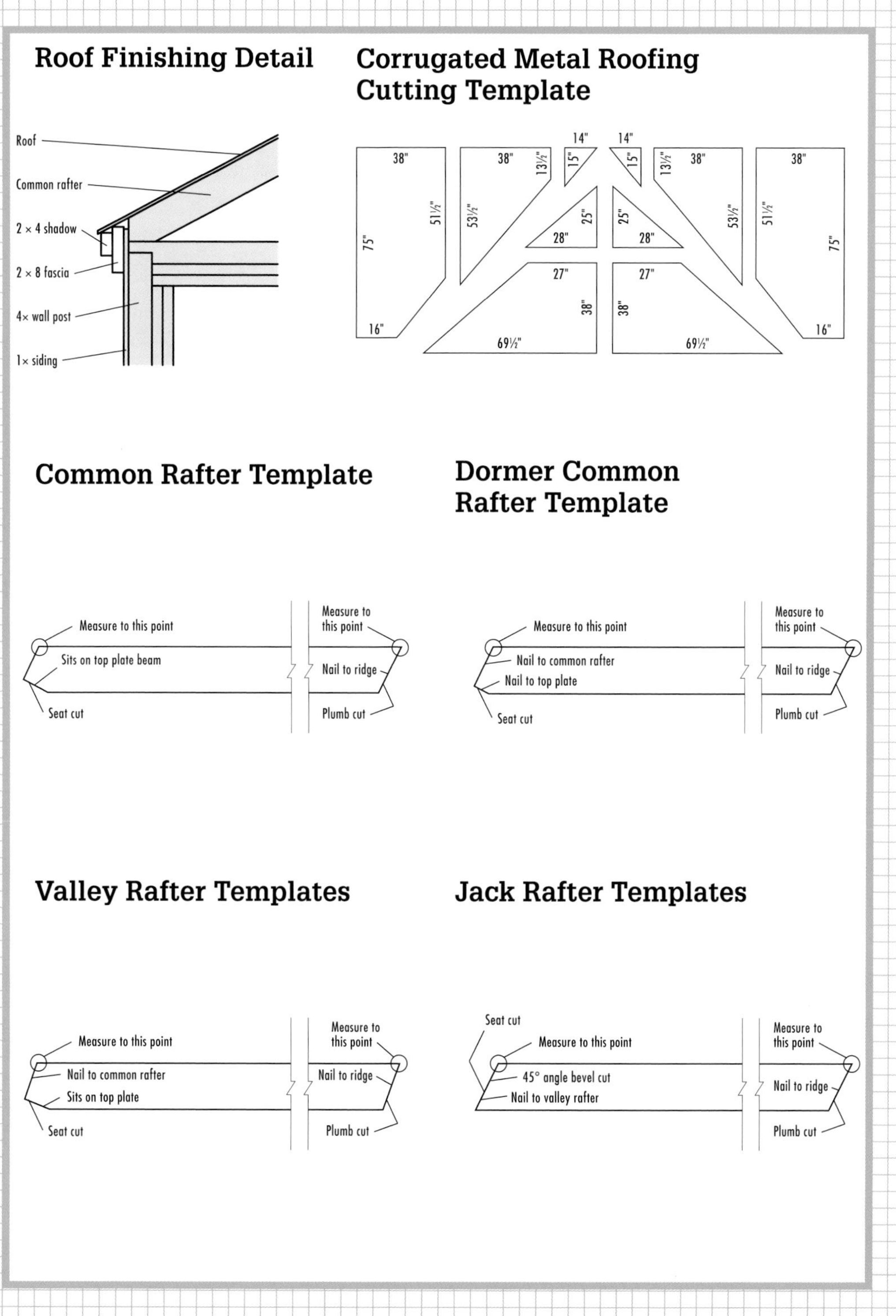

Roof Finishing Detail
Roof
Common rafter
2 × 4 shadow
2 × 8 fascia
4× wall post
1× siding
Corrugated Metal Roofing Cutting Template
38"
75"
51½"
16"
38"
13½"
53½"
14"
15"
25"
28"
27"
38"
69½"
Common Rafter Template
Measure to this point
Sits on top plate beam
Seat cut
Measure to this point
Nail to ridge
Plumb cut
Dormer Common Rafter Template
Measure to this point
Nail to common rafter
Nail to top plate
Seat cut
Measure to this point
Nail to ridge
Plumb cut
Valley Rafter Templates
Measure to this point
Nail to common rafter
Sits on top plate
Seat cut
Measure to this point
Nail to ridge
Plumb cut
Jack Rafter Templates
Seat cut
Measure to this point
45° angle bevel cut
Nail to valley rafter
Measure to this point
Nail to ridge
Plumb cut

How to Build a Rustic Summerhouse

Prepare the 12 × 16-ft. foundation site with a level, 4"-deep layer of compacted gravel. Measure 12" in from the edges of the foundation site on all sides and outline the building's footprint. Position six concrete blocks approximately 6" within the outlines at the corners to form a rough 9 × 13-ft. footprint for the building to sit upon.

Prepare and position the skids. Cut the 4 × 6 skids (see page 369) to exactly 168" and cut rough 45° angles on the bottom edges of both ends of each skid with a reciprocating saw. Position the skids on the blocks with the angled cut facing down, roughly 10" inside of the outline, checking that they are level and parallel with one another. Cut eight 2 × 6 floor joists at 120" and cut two 2 × 6 rim joists at 168".

Lay out the floor joists perpendicular to the skids, starting at one end and positioning them every 24" on-center, overhanging the skids equally on each side. Nail the rim joists to the floor joists using three 16d common nails at each end. Check for level. Use wood shims under the skids to increase deck height, if necessary. Square the deck frame by measuring across the diagonals. When both corner-to-corner measurements are equal, the floor is square. Secure the squared deck by toenailing each joist to the skids.

Install 2 × 6 tongue-and-groove decking perpendicular to the floor joists. Hold the first piece of decking flush to the rim joists and cut it to fit. Nail decking to each joist with two 16d galvanized common nails. Rip the last piece to fit before installing.

Align the 4 × 4 corner posts (cut to length at 74") flush to the edges of each corner. Toenail each corner post to the deck on both outer sides with three 16d common nails. Use additional nails on the interior sides of the posts, if necessary.

Cut half-lap joints into the 4 × 4 beams with a circular saw and mark wall post and framing locations on beams as follows: for the 14-ft.-long rear top plate beam (see Diagrams, pages 267 to 271).

Place the rear wall top plate beam on the rear wall corner posts, with the half-lap joint and layout markings facing up. Endnail through the half-laps and into the top of the corner posts using three 16d common nails at each post.

(continued)

Set both center posts for the front wall. Measuring from the end of the deck, mark the post locations on the deck floor. Set 4 × 4 posts at these locations and toenail them to the deck on the outside using three 16d common nails each.

Position each front top plate beam, with the half-lap joint facing up over the corner post (the end over the center posts will not have a lap). Endnail the front top plate beams into the top of the corner posts and toenail into the center posts with three 16d common nails at each end (inset).

Attach 120"-long 4 × 4 side wall on the building sides, fitting the half-lap joints face down into the half-lap joints on the front and rear top plate beams. Endnail the side wall top plate into the front and rear top plate beams using three 16d common nails at each end.

Set the wall rear and side wall posts at the locations marked on the top plate beams; check for plumb and toenail to the deck and the top plate beam. If necessary, toenail additional nails on the interior sides of the posts to secure them in place. Check the deck for level in multiple places.

Level and attach 1 × 4 temporary front wall bracing inside the door opening on the inside of the building. Position this board high enough to walk under during the rest of construction and do not remove it until construction is complete.

Tack temporary 1 × 4 bracing diagonally on the inside of the side walls between a corner post and the top plate beam. Hold the corner posts perfectly plumb, then position bracing so as not to interfere with attaching the siding or the rafters later on. Do not remove the bracing until construction is complete.

Install horizontal nailers on the rear wall; cut four 2 × 4 nailers to 17¼". Measure up from the deck floor between a rear wall corner post and the first interior post and mark nailer locations at 30" and 60". Attach nailers to both posts with their bottom faces flush to these marks by toenailing one 16d common nail into each edge and two nails into one face.

Frame the rear wall windows. Install the first window headers tight against the top plate beam in the three gaps between interior posts along the rear wall. Nail the headers directly into the top plate beam. Install the second header tight to the first in the same manner.

(continued)

Install the rear wall windowsills. Measure up from the deck and make a mark at 13" on all interior posts. Toenail the windowsills to the posts with the top face flush to this mark using one 16d common nail on each edge and two on a face.

Begin framing the side walls. Cut 2 × 4 spacers at 12"—these will be nailed directly to the corner posts facing the side wall. Attach one spacer at the top of the post, one at the bottom, and one centered between them; attach with two 8d common nails on each end. Install 74" window studs tight to the spacers; endnail studs to the spacers and toenail to the deck and to the top plate beam.

Frame the side-wall windows by installing a double window header and a windowsill in each gap. Install double window headers as in step 15 and windowsills as in step 16. Install temporary diagonal bracing to all walls as in step 13 (inset).

Install overhead bracing. Plumb all walls and adjust if necessary. Install a long 1 × 4 brace (approximately 10-ft.) between each side wall and the front wall. Attach the brace to the interior front wall posts (flush with the inside edge) and the rear wall top plate beam, directly above the opposite wall post, side-wall framing.

Install the front wall window framing. From each side, measure from the corner and mark positions of window studs on the deck (inset). Cut four window studs to 74" and install at the marks; toenail to the top plate beam and decking. Measuring from the deck, make a mark at 30" and at 60" on each corner post and window stud. Cut 2 × 4 nailers to 7⅛" and install between the corner posts and window studs with their bottom edge flush to these marks. Install the double window header and window sill between the window studs as in steps 15 and 16.

Prepare the door frame by installing two 82"-long 2 × 4 studs tight to the inside of each door post using two 16d common nails at the bottom, top, and center of each stud.

Install the rear-wall siding. Use 1 × 12 pine boards and set siding to hang below rim joists ¼" and at least 4" above the top plate. Attach siding with two 8d galvanized nails at each floor joist, nailer, and top plate. When the wall is finished, snap a chalk line 4" above the top plate and use a circular saw to trim siding along the chalk line (inset). *Note: For this load-bearing wall, do not use siding smaller than 1 × 6.*

Mark and cut a pattern rafter (inset). Test-fit the pattern rafters and make any necessary adjustments before cutting the rest. Crown the 1 × 8 ridge and mark rafter locations every 24" on-center. Attach two 2 × 6 front rafters, one to each end of the side wall. Endnail the ridge to the two front rafters. Attach the rafters on the opposing side wall to the top plate first and then to the ridge. Plumb the rafters.

(continued)

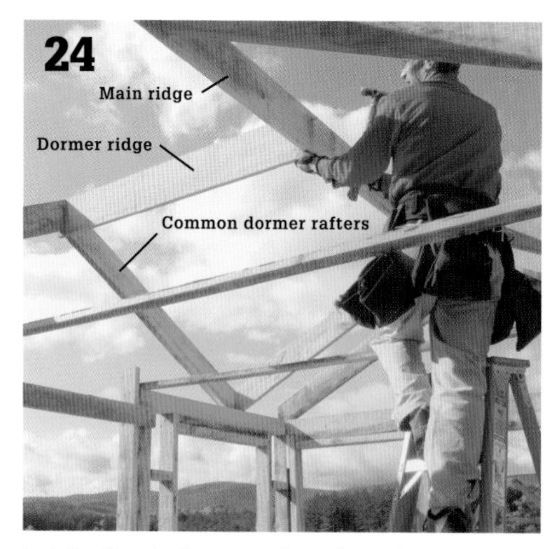

Install rafters to the rear main wall top plate beam and the main ridge at 6-ft. and 8-ft. Install 2 × 6 common dormer rafters to the front wall; first toenail each seat cut end to the front wall top plate beam at 2-ft. and 12-ft, then endnail each through the end of the 1 × 8 dormer ridge. Last, attach the dormer ridge to the main ridge at 7-ft. on-center. Endnail the dormer ridge through the main ridge using three 8d common nails.

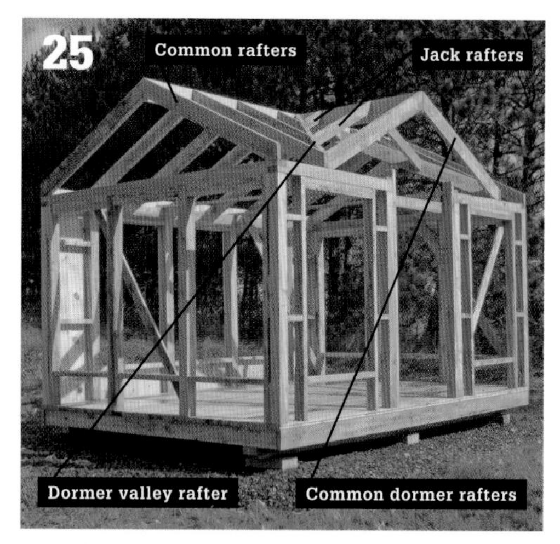

Install the remaining rafters. Attach front and rear rafters to the main ridge and to the front and rear top plate beams at 2-ft. and 12-ft. on-center. Then install dormer valley rafters to the ridge at the junction of the two ridge poles and to the front wall top plate beam. Install the remainder of the rear common rafters and jack rafters as shown in ROOF FRAMING PLAN VIEW, making sure to install each rafter and its opposite together to maintain a straight ridge pole.

Brace the roof. Once all rafters are installed and you've double-checked for level, nail temporary bracing from the top of the main ridge at a 45° angle to a block attached to a floor joist to hold the building square. Do not remove the temporary bracing until construction is complete.

Finishing framing the door. Cut a 2 × 4 header at 93", and miter-cut the ends at 63° to fit within the dormer rafters. Position the door header on top of the door studs, tight within the dormer rafters, and toenail in place. Cut a second door header at 112" (with the same 63° miter-cut ends); position the second door header behind the first so it extends to the outer edge of the dormer rafters. Endnail to the dormer rafters on both ends.

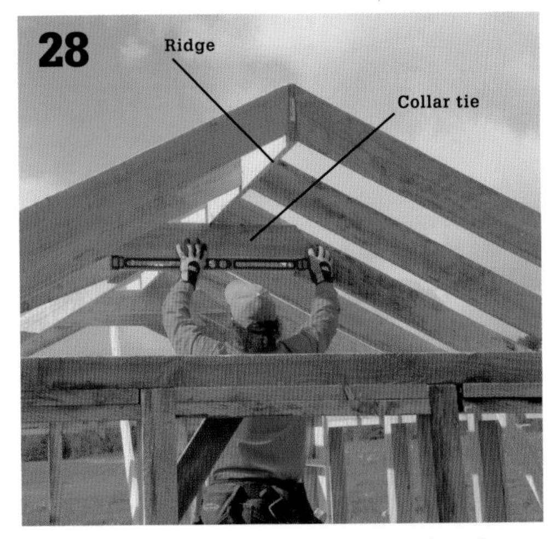

Cut four 1 × 6 collar ties at 48" and miter-cut the ends at 26½°. Install collar ties tight to the ridge on the 2-ft., 4-ft., 10-ft., and 12-ft. rafters with five 8d common nails per rafter. Check each tie for level before fastening it.

Install 1 × 12 siding on the side and front walls, setting it to hang below the rim joists ¼". Cut each siding piece to fit flush to the rafters and attach at the rim joist, horizontal nailer, top plate and rafters with four 8d galvanized common nails at each position. Allow front wall siding to extend above the top plate at least 4". Trim siding on the ends of the front wall as in step 22.

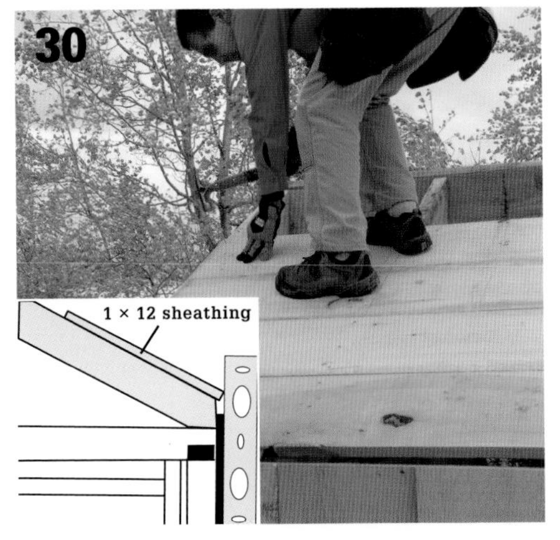

Install 1 × 12 roof decking, beginning at the eave and working your way toward the ridge. Position the first piece flush with the rafter edge on the gable side; to position the decking on the eave side, hold a level or straightedge flush with the siding and place the first piece flush with the level. Attach decking with four 8d nails at each rafter. Install the rest of the boards flush with the first piece, cutting to fit.

Install the decorative sunburst above the door. Nail a level 1 × 6 piece of clapboard 3" above the door, flush to the subfascia on both sides. Blindnail to the siding using 6d stainless steel nails. Install the next piece with one end at the center of the first piece and the other end fanned up about 3" toward the roof, then install the opposing piece. Continue to install clapboard in this fashion as you move up toward the center; pieces will overlap at the bottom, reaching a thickness of about 6" when the sunburst is complete. Cap the sunburst by cutting and installing a piece to cover the central seam.

(continued)

Apply the 2 × 8 pine fascia and 2 × 4 shadow. Install fascia to the two side walls using the common rafter templates to match the angle of the rafters. Attach with three 16d galvanized common nails spaced every 24" on-center. Install fascia on the front and rear walls by nailing to a spacer at each rafter location with three 16d galvanized common nails. Install shadow board using the same technique, nailing it in place with two 16d galvanized common nails at each rafter (or every 24" on-center on the side walls).

Install the door, window, and corner trim. First, install the vertical 1 × 4 corner trim flush on the gable ends. Install the front and rear wall corner trim flush to the gable corner trim. Next, install the 1 × 5 door jamb trim to cover the siding and door framing on the interior of the door. Nail this in place. Finally, attach 1 × 4 exterior window and door trim to the siding. Set the window trim to overhang 1" inside the frame and door trim flush with the jamb trim.

Rip 2" pine stock into strips ¾- to ⅝"-thick to create the siding battens. Install the battens over the seams where two siding boards meet. Cut battens to fit up to the fascia and above and below the window trim. Attach them using 8d galvanized common nails following the nailing pattern of the siding. Avoid nailing into knots.

Install the corrugated metal roofing. First, install metal flashing in the valleys of the dormer, starting at the eaves using roofing nails spaced every 12". Leave the flashing at least 2" long at both the ridge and the eave overhang. Cut metal roofing according to the CORRUGATED METAL ROOFING CUTTING TEMPLATE on page 271; to cut roofing in the flats, score with a utility knife on the white side and flex until it snaps. To cut over ridges, use a circular saw with a metal abrasive blade. Install the first piece, allowing it to hang over the shadow board by 2" on both the gable end and the bearing wall. Screw the metal roofing into the sheathing with 1½" metal roofing screws with a rubber gasket, positioning the screws in the center of the flats and on all seams. Cut metal flashing with a utility knife as you install metal roofing over it. *Tip: Nail into the ridge of the roofing material only where two pieces of metal overlap.*

Install the ridge cap, fastening it into the ridges of the roofing material only. (Fastening the ridge cap in the flats will cause the roofing material to dent.) Attach the ridge with metal roofing screws every 8 to 12" on-center. Trim the ridge cap for length.

Install the screen door and window screens. Attach the door according to the manufacturer's instructions. Use adjustable spring hinges, if desired. Install a door knob and latch. Install window screens so they set tight against the exterior trim. You may need to have window screens custom-made. Allow the building to cure for one season and then paint or stain as desired.

3-season Gazebo

A large, windowed gazebo is the ultimate outdoor room. In summer, fully screened openings usher cooling breezes through the shaded interior. During the cooler days of spring and fall, slide-up storm windows provide comfort without limiting the full, 360-degree view. Because the gazebo is enclosed, you can decorate the interior and keep it furnished year-round, or use the space for off-season storage.

The gazebo featured in this project has a classic hexagonal floor plan. Five of the walls are framed identically and are designed with standard-sized combination storm windows and standard fixed utility windows. The sixth wall contains a standard pre-hung storm door and a utility window. You can follow the plan's specifications for window and door sizes or choose custom sizes and alter the framing accordingly. Either way, it's a good idea to buy the units and have them on hand for measuring before you frame the walls.

As with the 8-sided Gazebo (page 320), making the many angled cuts on this project is much easier with a power miter saw, preferably a compound saw.

This gazebo design has combination windows with screens and glass, but you could opt to fill the wall openings only with screens.

Tools, Materials & Cutting List

Wood stakes
Mason's string
Circular saw
Batter boards
Tape measure
Plumb bob
Excavation tools
Straight board
Framing square
Hammer
Line level
Level
Construction adhesive
Caulk gun
Ratchet wrench
Chalk line
Drill with bits
Table saw
15# building paper
Cedar shingles
T-bevel
Eye and ear protection

Description (No. finished pieces)	Quantity/Size	Material
Foundation		
Concrete	Field measure	3,000 PSI concrete
Concrete tube forms	7 — field measure for length	12"-diameter cardboard forms
Compactable gravel	2 cubic feet	
Framing		
Main posts (6)	6 @ 10'	4 × 6
Floor support posts (6)	1 @ 8'	4 × 4
Center pier pad (2–3 pieces)	1 @ 3'	
Exterior-grade plywood as needed for shim material		2 × 8
Perimeter and interior floor beams (18)	18 @ 8'	2 × 8
Floor joists (9)	9 @ 8'	2 × 8
Perimeter roof beams (6)	6 @ 8'	4 × 6
Roof hub (1)	1 @ 1'	6 × 6
Hip rafters (6)	6 @ 10'	2 × 8
Purlins (6)	2 @ 10'	2 × 8
Intermediate rafters (18)	9 @ 12'	2 × 8
Trim nailers (24)	3 @ 8', plus cutoffs from intermediate rafters	2 × 4
Window & door frames	Field measure	4 × 4
Corner studs (12)	12 @ 10'	2 × 4
Truss top chord (4)	4 @ 12'	2 × 12
Truss bottom chord (8)	8 @ 8'	2 × 6
Truss strut (4)	1 @ 10'	2 × 4
Hub (1)	1 @ 51"	4 × 4
Support joists (4 outer, 4 inner)	4 @ 14', 4 @ 10'	2 × 4
Slat braces (4)	1 @ 8', 1 @ 3'	2 × 4
Slats	Field measure	2 × 2
Floor Decking	39 @ 8'	5/4 × 6 decking boards
Roofing (roof covering)		
Roof sheathing	9 @ 4' × 8'	¾" exterior-grade plywood
Shingles and 15# building paper	Coverage for 220 square feet, plus ridge caps	

Description (No. finished pieces)	Quantity/Size	Material
Stairs		
Stringers (3) and stair pad (1)	2 @ 8'	Pressure-treated 2 × 12
Treads (6)	3 @ 10'	5/4 × 8 decking boards
Risers (optional, 3)	1 @ 10', 1 @ 6'	1 × 6
Wall Finishes & Trim		
Top-of-wall trim	6 @ 8', 6 @ 8'	1 × 12, 1 × 10
Window/door header trim	Field measure	1× lumber
Exterior sheathing/trim	Field measure	
Interior sheathing/trim	Field measure	
Optional skirt framing	12 @ 8'	2 × 4
Optional skirt sheathing/trim	Field measure	
Framing Connectors		
Post bases, main posts	6, with standoff plates and anchoring hardware	Simpson ABU46, or similar approved connector
Post bases, floor support posts	6, with standoff plates and anchoring hardware	Simpson AB44, or similar approved connector
Beam hangers	3, with recommended fasteners	Simpson LS50Z, or similar approved connector
Hurricane ties	6, with recommended fasteners	Simpson H8, or similar approved connector
Rafter connectors	6, with recommended fasteners	Simpson FB26, or similar approved connector
Stair stringer framing connectors	6, with recommended fasteners	Simpson L30, or similar approved connector
Hardware & Fasteners		
⅝" × 12" J-bolts	7, with washers and nuts	
½" × 12" J-bolts	6, with washers and nuts	
¼" × 10" galvanized carriage bolts	12, with washers and nuts	
3½" x ¼" galvanized lag bolt	6, with washers and nuts	
Construction adhesive		
10d galvanized common nails		
16d galvanized common nails		
2¼" deck screws		
2½" deck screws		
8d galvanized box nails		
Heavy-duty staples		
Roofing nails		
3½" galvanized wood screws		
¼" × 6" galvanized lag screws, with washers		

** Wall Frames are made up of the corner studs and door and window frames.*

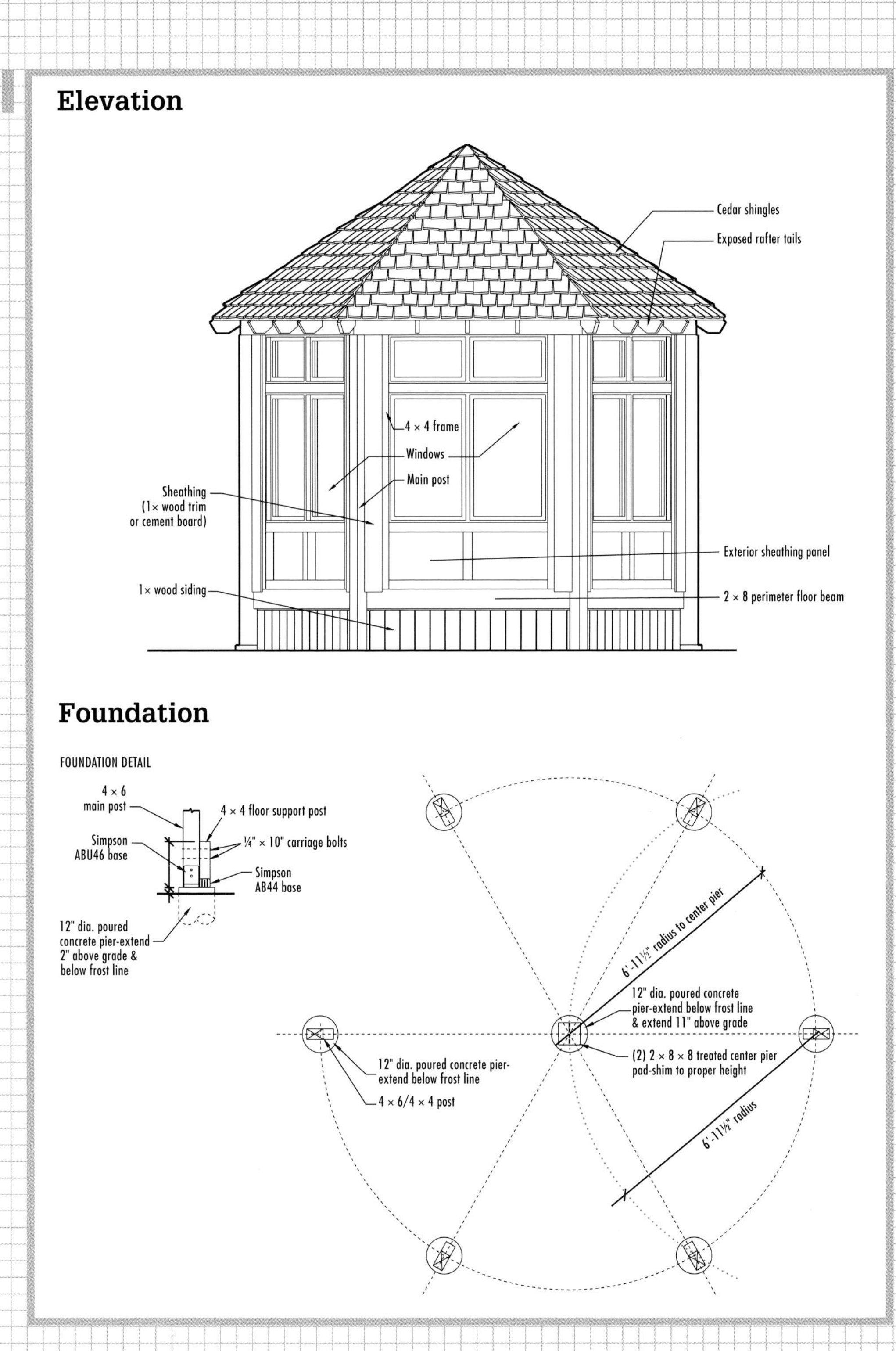

Elevation
Cedar shingles
Exposed rafter tails
4 × 4 frame
Windows
Main post
Sheathing (1× wood trim or cement board)
Exterior sheathing panel
1× wood siding
2 × 8 perimeter floor beam
Foundation
FOUNDATION DETAIL
4 × 6 main post
4 × 4 floor support post
Simpson ABU46 base
¼" × 10" carriage bolts
Simpson AB44 base
12" dia. poured concrete pier-extend 2" above grade & below frost line
6'-11½" radius to center pier
12" dia. poured concrete pier-extend below frost line & extend 11" above grade
(2) 2 × 8 × 8 treated center pier pad-shim to proper height
12" dia. poured concrete pier-extend below frost line
4 × 6/4 × 4 post
6'-11½" radius

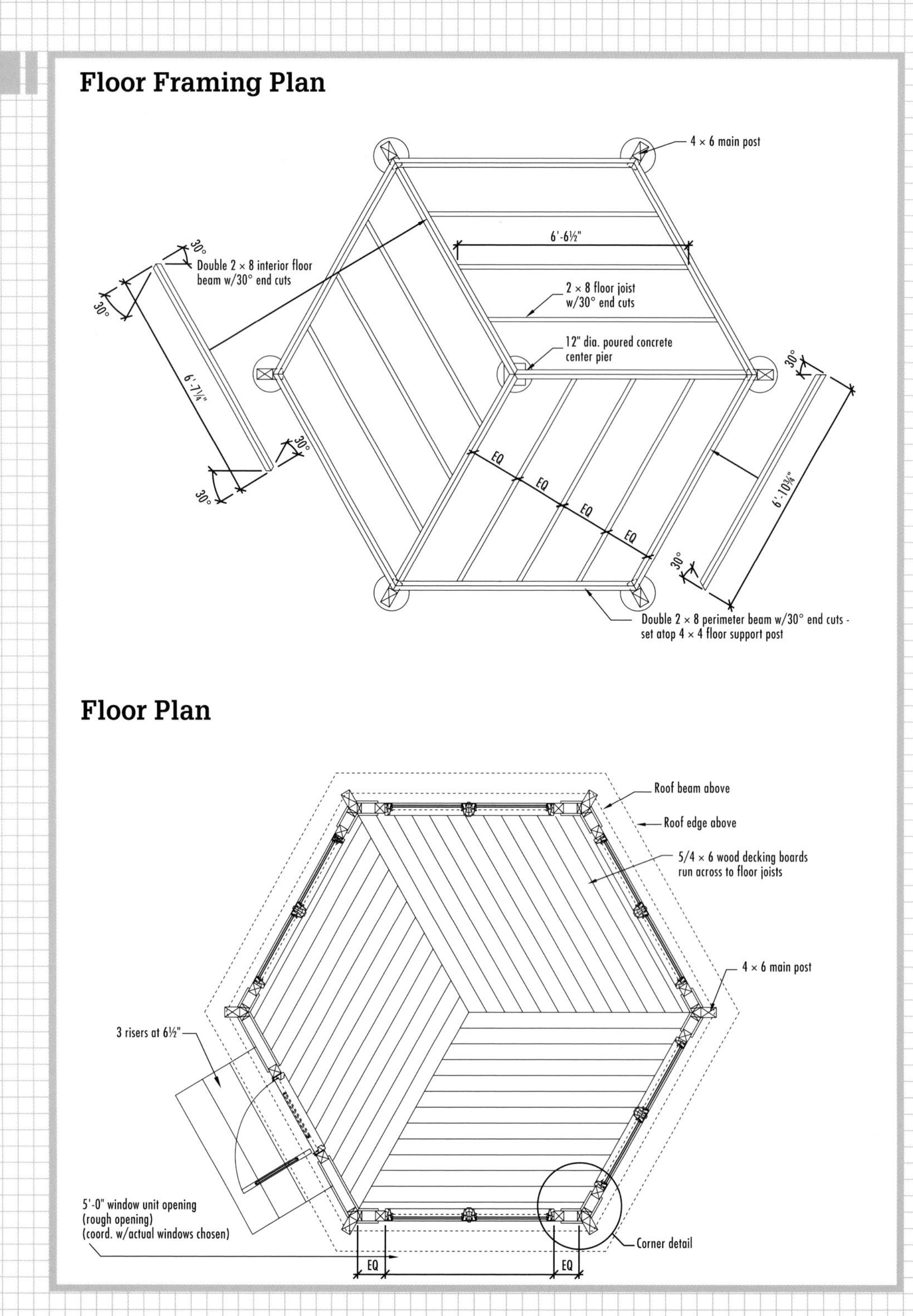

Floor Framing Plan
4 × 6 main post
6'-6½"
Double 2 × 8 interior floor beam w/30° end cuts
30°
30°
6'-7¼"
30°
30°
2 × 8 floor joist w/30° end cuts
12" dia. poured concrete center pier
30°
6'-10¾"
30°
EQ
EQ
EQ
EQ
Double 2 × 8 perimeter beam w/30° end cuts - set atop 4 × 4 floor support post
Floor Plan
Roof beam above
Roof edge above
5/4 × 6 wood decking boards run across to floor joists
4 × 6 main post
3 risers at 6½"
5'-0" window unit opening (rough opening) (coord. w/actual windows chosen)
Corner detail
EQ
EQ

Building Section

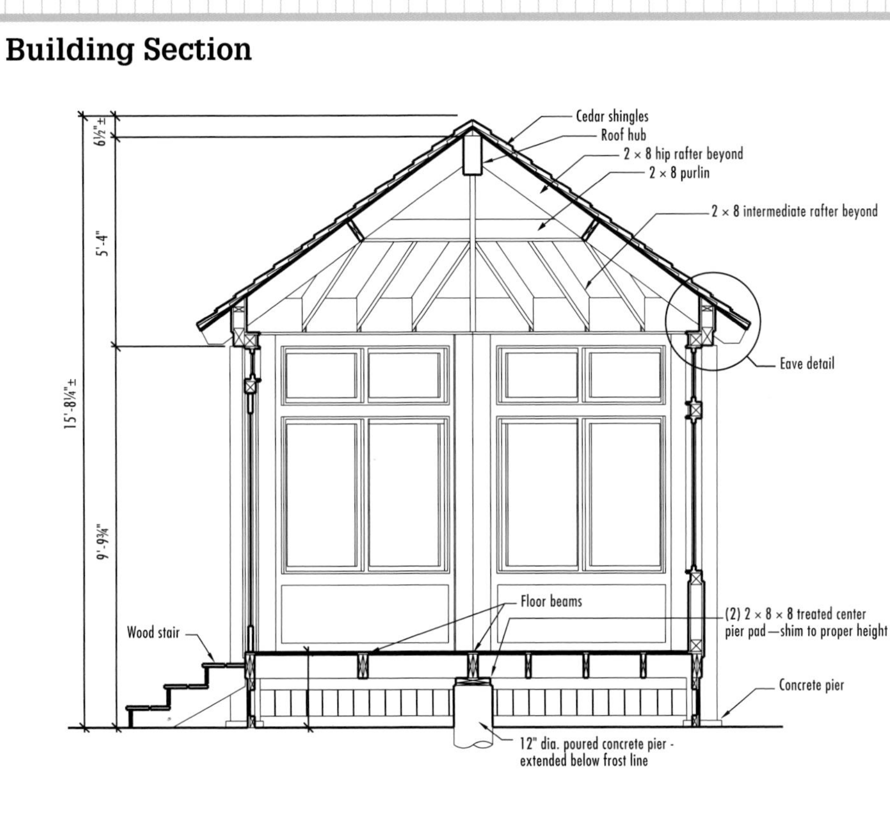

Window Frame Detail

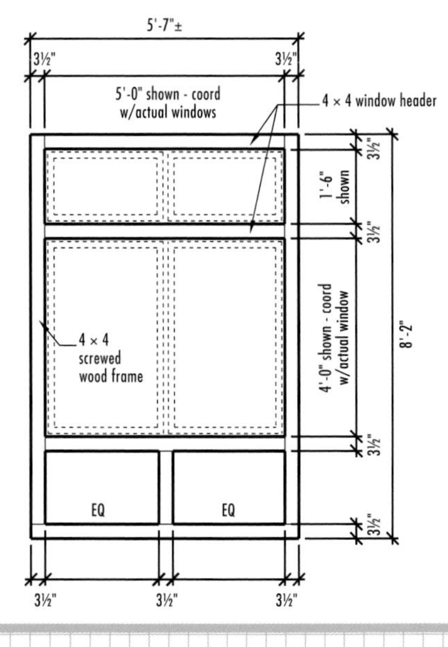

Door Frame Detail

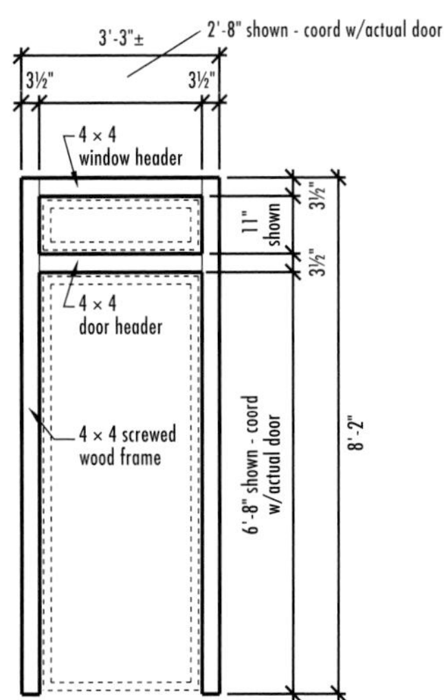

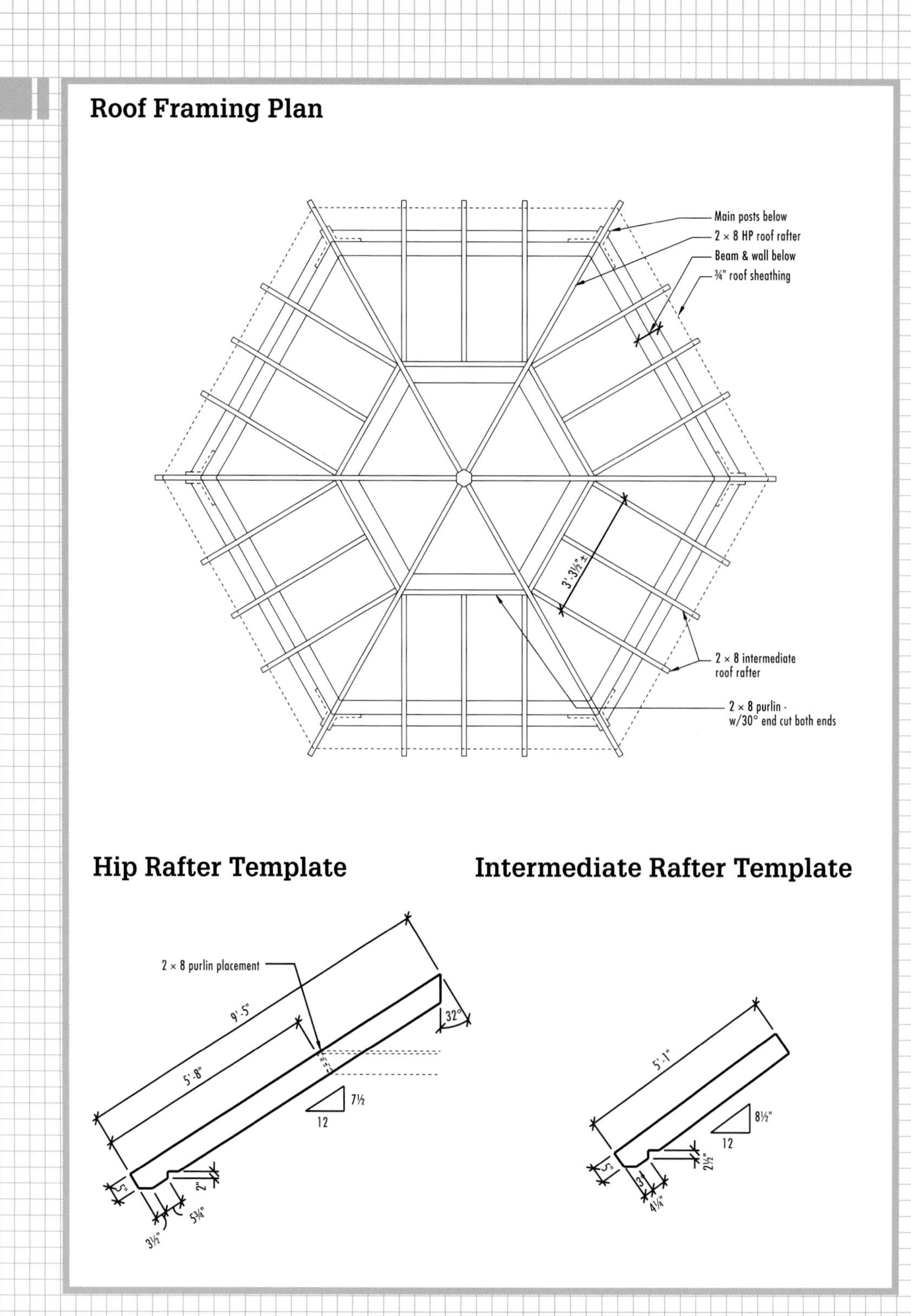

Roof Framing Plan
Main posts below
2 × 8 HP roof rafter
Beam & wall below
¾" roof sheathing
3'-3½" ±
2 × 8 intermediate roof rafter
2 × 8 purlin - w/30° end cut both ends
Hip Rafter Template
2 × 8 purlin placement
9'-5"
5'-8"
32°
7½
12
5"
2"
5¾"
3½"
Intermediate Rafter Template
5'-1"
8½"
12
5"
2½"
3"
4¼"

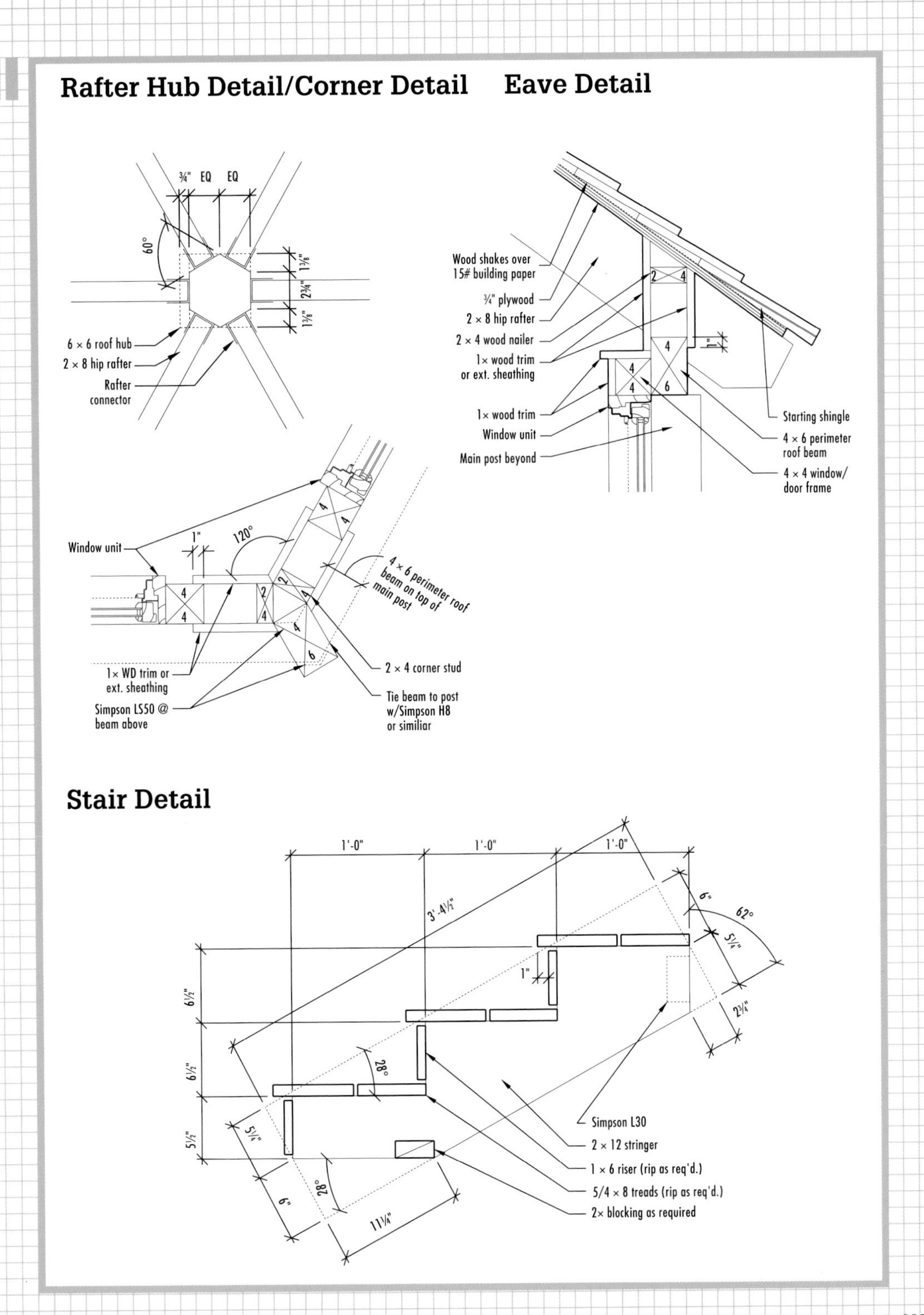

Rafter Hub Detail/Corner Detail
Eave Detail
¾" EQ EQ
60°
1⅜"
2¾"
1⅜"
6 × 6 roof hub
2 × 8 hip rafter
Rafter connector
Wood shakes over 15# building paper
¾" plywood
2 × 8 hip rafter
2 × 4 wood nailer
1× wood trim or ext. sheathing
1× wood trim
Window unit
Main post beyond
1"
Starting shingle
4 × 6 perimeter roof beam
4 × 4 window/ door frame
Window unit
1"
120°
4 × 6 perimeter roof beam on top of main post
1× WD trim or ext. sheathing
Simpson LS50 @ beam above
2 × 4 corner stud
Tie beam to post w/Simpson H8 or similiar
Stair Detail
1'-0"
1'-0"
1'-0"
3'-4½"
6"
62°
5¼"
2¾"
1"
6½"
6½"
5½"
28°
5¼"
6"
28°
11¼"
Simpson L30
2 × 12 stringer
1 × 6 riser (rip as req'd.)
5/4 × 8 treads (rip as req'd.)
2× blocking as required

How to Build a 3-season Gazebo

STEP 1: POUR THE CONCRETE PIERS

See pages 434 to 437 for basic instructions on laying out and pouring concrete pier footings. Detailed steps for laying out a hexagon are given below. Use 12"-diameter concrete tube forms for the six outer piers and the center pier.

Tip: There's a convenient mathematical rule that makes it easy to lay out a hexagon: the centerpoint and all six outer points are equidistant. Therefore, if you measure the same distance from the center and one outer point, the intersection of those measurements is the location of a second outer point, and so on.

A. Drive a stake into the ground at the gazebo's centerpoint, then drive a nail into the center of the stake. Set up batter boards on opposing sides of the gazebo footprint, and run a mason's line that passes directly over the centerpoint.

B. Mark the string at the centerpoint, then measure out in both directions and mark the string at 83½". Drive a stake and a nail at each outer mark using a plumb bob to transfer the string markings to the ground. These points represent the centers of two outer piers.

C. With two helpers, pull one tape measure from the centerpoint and one from an outer stake. Cross the tapes so they meet at 83½"—at that intersection, drive a stake and nail to represent a third outer pier. Repeat the process to lay out the three remaining piers. Each time, measure from the centerpoint and one of the original two outer piers to avoid compounding inaccuracy. Untie the mason's string.

D. Dig the holes and set the concrete forms, following the steps on page 436 and the requirements of your local Building Code. The outer forms should extend 2" above the ground; the center form should extend 11" above the ground.

E. Set up batterboards behind all of the piers. Run three mason's strings over the centers of opposing piers, making sure they all cross over the centerpoint (on the center pier). *Note: The strings stay in place for the concrete pour, so make sure they're high enough to allow easy access to the forms. Use the measuring technique from sub step C to mark the centers of the piers onto the strings. Confirm that each concrete form is centered on the layout.*

F. Make two more marks on the strings to indicate the J-bolt locations: one at 80¾" from the centerpoint and one at 85¼" from the centerpoint.

G. Pour and screed the concrete. Into each outer pier, set a ½ × 12" J-bolt at the 80¾" string marking. Set a ⅝ × 12" J-bolt at the 85¼" mark. Position the bolts so they are plumb and extend ¾ to 1" above the concrete (follow the post base manufacturer's specifications). Set a ⅝ × 12" J-bolt into the center of the center pier so it extends 5" above the concrete.

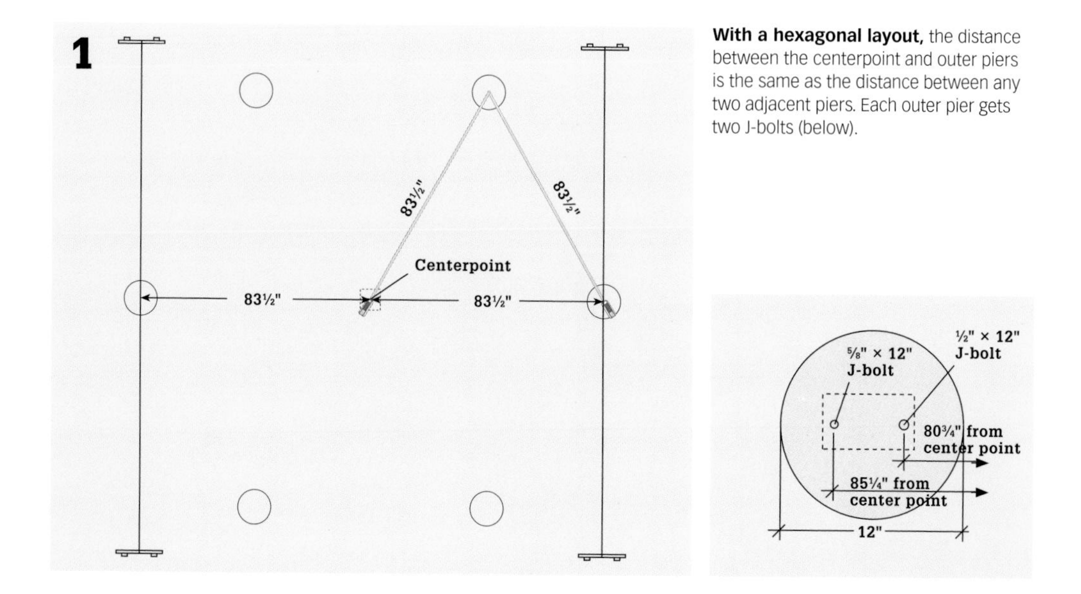

With a hexagonal layout, the distance between the centerpoint and outer piers is the same as the distance between any two adjacent piers. Each outer pier gets two J-bolts (below).

STEP 2: SET THE POSTS

A. Use a straight board with one end cut perfectly square to mark reference lines for squaring the post bases: Set the board flat against the J-bolts on the center pier and each outer pier and mark along the board across the top of the outer pier.

B. Center a 4 × 6 post base over each outer J-bolt and use a framing square to make sure the base is square to the reference line. Secure the bases with washers and nuts; use the hexagon measuring technique to make sure all points are equidistant. Add the provided standoff plate to each base.

C. Leaving the 4 × 6 main posts long, set them into the bases, and position them so their outside edges are 88" from the centerpoint. Tack each post in place with a nail, then install cross bracing so the post is perfectly plumb. Re-check for equidistant placement, then fasten the posts as recommended by the base manufacturer.

D. On the inside face of one of the main posts, make a mark 15" above the top of the pier. Using a mason's string and a line level, transfer this height mark to the other main posts.

E. Install the 4 × 4 post bases (with standoffs) so the floor support posts will be flush against the inside edges of the main posts. Measure from the standoff to the height mark and cut each floor support post to fit.

F. Anchor the support posts to their bases using the recommended fasteners. Then, anchor each support post to a main post with two ¼ × 10" carriage bolts, as shown in the FOUNDATION on page 285.

STEP 3: ADD THE CENTER PIER PAD

A. Run a level line across the tops of two opposing floor support posts. Measure from the top of the center pier to find the thickness of the center pier pad.

B. Create a pad from 2 × 8 lumber, adding exterior-grade plywood as needed for shim material to achieve the proper thickness. Assemble the pad with construction adhesive and nails.

C. Drill a counterbored hole in the pad's center for the J-bolt, and anchor the pad to the center pier. Cut off any excess bolt so it's flush with the top of the pad.

Mark a line across the tops of the piers; this helps ensure all post bases are facing the gazebo's center.

Countersink the J-bolt nut and washer into the wood pier pad, so the floor beams sit flush on top.

(continued)

STEP 4: FRAME THE FLOOR

A. Build the six double 2 × 8 perimeter floor beams following the FLOOR FRAMING PLAN on page 286. Check the boards for crowning, making sure the crowned edges are up; then cut the beam ends at 30° so they break on the centers of the floor support posts. Join the two boards for each beam with construction adhesive and 10d common nails.

B. Install the perimeter beams by toenailing into the main posts and floor support posts with 16d common nails. Also nail the beam ends together.

C. Build the three double 2 × 8 interior floor beams like the perimeter beams. Cut the ends at 30°, too.

D. Fasten the interior beams to the center pier pad and to one another with 16d nails. Anchor the outer ends of the beams to the perimeter beams with beam hangers using the recommended fasteners.

E. Mark the floor joist layout onto the beams following the FLOOR FRAMING PLAN; space the joists equally across each section of the floor frame.

F. Measure and cut each joist to fit, beveling the ends at 30°. Install the joists crown-up, driving three 16d common nails into the beams at each end. Make sure the joists are flush with the tops of the floor beams.

STEP 5: INSTALL THE FLOOR DECKING

As shown in the FLOOR PLAN (on page 286), the 5/4 × 6 decking is laid nearly perpendicular to the floor joists. Each of the three flooring sections starts with a full-width piece aligned with the center of an interior floor beam. Install the decking one section at a time.

A. Starting with any one of the three flooring sections: Cut the outside end of the first decking board at 30° to fit flush with the outside face of the perimeter beam. The pointed end should touch the main post. Let the inside end of the decking board run long over the center of the floor frame.

B. Align the decking board with the center of the interior floor beam (the seam between the two beam boards), and fasten the decking to the beams and joists with two 2¼" deck screws driven through pilot holes.

Join the interior beams to the perimeter beams with framing connectors, making sure the tops of the beams are flush.

Cut along chalk lines to trim the ends of the decking boards along the floor beams.

C. Install the remaining boards in the section, running the ends long over the interior and perimeter beams. Fit the boards tightly together to prevent gaps that could let in insects. Rip the last board to fit so it breaks over the center of the interior beam.

D. Trim the decking boards in the first section so the outside ends are flush with the outside edge of the perimeter beam and the inside ends are aligned with the center of the interior beam. Snap a chalk line to ensure a straight cut, and use a circular saw set to cut just through the decking.

E. Complete the remaining sections of flooring using the same procedure. For the final section, precut the inside ends of the boards at 30° to fit against the first board in the first section.

STEP 6: INSTALL THE ROOF BEAMS

A. Measure up from the floor deck and mark one of the main posts at 94¼". Transfer that height mark to the remaining posts using a mason's string and a line level. Cut off the posts at the height marks (see page 369).

B. Cut the 4 × 6 roof beams to span across the tops of the main posts, as shown in the ROOF FRAMING PLAN on page 288, the EAVE DETAIL and the CORNER DETAIL on page 289. Miter the ends at 30° so the joints break over the centers of the posts.

C. Toenail the beams to the main posts with 16d common nails, then anchor the beams together with hurricane ties. Reinforce the beam-post connection with adjustable-angle framing connectors.

STEP 7: INSTALL THE HIP RAFTERS

A. Cut the 6 × 6 roof hub to length at 12".

B. On a table saw, trim off ¾" from one side so the hub measures 4¾" × 5½". Set the saw blade to 30° and make four full-length cuts to create six facets at 2¾", as shown in the RAFTER HUB DETAIL on page 289.

C. Select two straight 2 × 8s to use for the pattern hip rafters. Check the boards for crowning, then cut them following the HIP RAFTER TEMPLATE on page 288. The slope for the hip rafters is 7½ -in-12.

Anchor the beam assembly to the main posts with hurricane ties, and then tie the roof beams together with metal angles.

Install the hip rafters so their top ends are flush with the top of the hub. Toenail the rafters to the sides of the hub for extra support (inset).

(continued)

D. Test-fit the rafters and hub on the gazebo frame. Make any necessary adjustments for a good fit. Use one of the rafters to mark the remaining four hip rafters, and make the cuts.

E. Install the hip rafters so their top ends are flush with the top of the hub and their bottom ends fall over the joints of the roof beams. Fasten the rafters to the hub using 3½ × ¼" lag bolts. Also, toenail the sides of the hip rafters to the hub with 2½" deck screws.

STEP 8: INSTALL THE PURLINS & INTERMEDIATE RAFTERS

A. Measure up from the ends of the hip rafters and mark the side faces at 68". These marks represent the bottom faces of the 2 × 8 purlins.

B. Cut the six 2 × 8 purlins to fit between the rafters, beveling the ends at 30°.

C. Nail the purlins to the rafters with 16d common nails, so their faces are perpendicular to the rafter faces and all edges are flush along the top.

D. Mark the layout for the intermediate rafters onto the purlins and roof beams; follow the ROOF FRAMING PLAN on page 288.

E. Cut a pattern intermediate rafter from a 2 × 8, following the INTERMEDIATE RAFTER TEMPLATE on page 288. Test-fit the rafter against the purlins and roof beams, and make any adjustments necessary for a good fit.

F. Using the pattern rafter, mark and cut the 17 remaining intermediate rafters.

G. Install the rafters on their layout marks using 16d common nails.

STEP 9: SHEATH & SHINGLE THE ROOF

A. Starting at the eave and working up, cut ¾" plywood to span between the centers of the hip rafters.

Note: Starting with a 4 × 8-ft. sheet of plywood means that about 2–3" of the rafter tails will be seen, as shown in the ROOF FRAMING PLAN. If desired, you can slightly adjust the amount of exposure.

B. Fasten the sheathing with 8d box nails driven every 6" along the perimeter and every 12" along intermediate rafters and purlins.

Endnail or toenail the intermediate rafters to the purlins. Toenail the rafters to the roof beams.

Bevel the edges of both shingles for each cap, and alternate the overlap between courses.

C. Cut and install the remaining sheathing. At the roof peak, cut the pieces to a point so they enclose the hub.

D. Install 15# building paper and cedar shingles, following the steps on pages 440 to 441. Install the shingles so their top ends run long over the ridges of the roof, then trim them off with a saw.

E. Cap the ridges with custom-beveled shingle caps or 1× cedar boards. Use a T-bevel to find the angle of the ridge (see Step 4 on page 441), then bevel the edges of the cap shingles on a table saw. Alternate the overlap with each cap.

STEP 10: FRAME THE WALLS

The wall frames consist of 2 × 4 corner studs at the ends of each wall section and 4 × 4 frames for the window and door openings. You can build the window and door frames on the ground, then tip them up and secure them to the gazebo framework.

Follow the WINDOW FRAME DETAIL and the DOOR FRAME DETAIL, page 287, for the basic layout of the frames. Modify any dimensions as needed to fit your window and door units.

A. At each corner inside the gazebo, measure up from the bottom edges of the roof beams and make a mark at 3½". Cut the 12 2 × 4 corner studs to fit between the floor and the marks on the roof beams.

B. Position the studs, as shown in the FLOOR PLAN on page 286, making sure they are flush to the outside edges of the floor decking. Fasten the studs to the floor and roof beams with 16d common nails.

C. Cut the 4 × 4 members for each window and door frame using the rough opening dimensions specified by the window/door manufacturer. When the frames are installed, the bottom faces of the top horizontal pieces should be flush with the bottom faces of the roof beams, as shown in the EAVE DETAIL, page 289.

D. Assemble the frames with 3½" wood screws driven at an angle through pilot holes. Drive four screws at each joint, locating the screws on the faces that will be least visible after the windows and door are installed. For example, on the horizontal header and sill pieces, drive two screws through the top and bottom faces and into the vertical frame members.

When installed, the tops of the window and door frames are level with the tops of the corner studs.

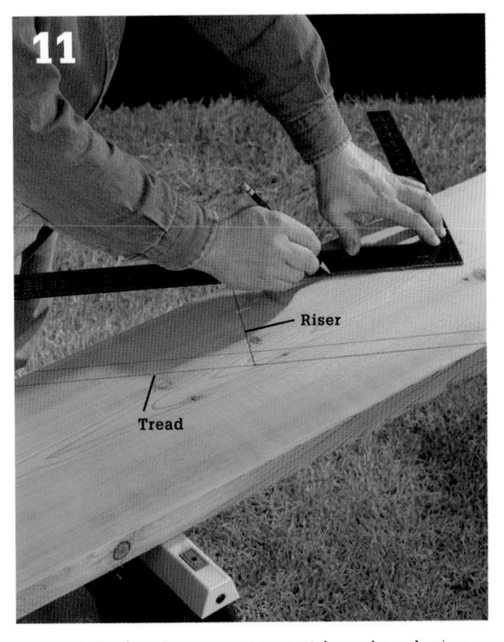

Position the framing square on a stringer board using the tread and riser dimensions, then trace along the square to mark the cutting lines.

(continued)

E. Center each window and door frame within its wall section. Measure the diagonals to make sure the frame is square, then fasten the frame to the roof beam and floor with ¼ × 6" lag screws. Countersink the screws below the surface of the frame pieces.

STEP 11: BUILD THE STAIRS

A. Use a framing square to lay out the first 2 × 12 stair stringer; follow the STAIR DETAIL on page 289. Starting at one end of the board, position the framing square along the board's top edge. Align the 12" mark on the square's blade (long part) and the 6½" mark on the tongue (short part) with the edge of the board. Trace along the outer edges of the blade and tongue, then use the square to extend the blade marking across the width of the board. The tongue mark represents the first riser.

B. Measure down 1" from the blade mark and make another line parallel to it—this is the cutting line for the bottom of the stringer (the 1" offset compensates for the thickness of the treads on the first step).

C. Continue the step layout, starting at the point where the first riser mark meets the top edge of the board. Mark the top cutting line by extending the third tread mark across the board's width. Mark the top end cut 12" from the top riser.

D. Cut the stringer and test-fit it against the gazebo. Make any adjustments necessary for a good fit, then use the stringer as a pattern to mark the remaining two stringers, and make the cuts.

E. Anchor the top ends of the stringers to the floor beam using framing connectors. Secure the bottom ends as required for your specific application, such as with 2× blocking nailed between the stringers and anchored to a concrete pad or spiked into the ground. You can cut the blocking from leftover 2 × 12 stringer material.

F. Cut the stair treads to fit the stringer assembly, overhanging the risers as desired. Start with a full-width 5/4 × 8 tread at the front of each step, then rip the next piece to fit. Fasten the treads with 2¼" deck screws. If desired, add 1× riser boards to enclose the back of each step.

STEP 12: ADD THE WALL FINISHES

How you sheath and trim the walls of your gazebo is up to you and will likely be determined by the style of the windows and door. The trim details shown in the plans are merely suggestions for enclosing the walls and covering some of the framing for a more finished appearance.

The openings in the wall frames can be covered with exterior-grade plywood, T1-11 siding, or another type of siding. The spaces above the roof beams should be covered with 1× trim or exterior sheathing and secured at the top to 2 × 4 nailers installed between the rafters; see the EAVE DETAIL on page 289. Also, see the CORNER DETAIL on page 289 for interior/exterior corner trim ideas.

You also have the option of adding a skirt below the floor frame between the main posts. To do this, install 2 × 4 nailers between the posts, then add exterior sheathing, as shown in BUILDING SECTION drawing, on page 287.

Overlap the framing with trim or sheathing by at least 1" to make room for nailing and to help weatherproof the structure. Consider filling in the cavities below windows with siding installed with molding, as shown here.

Summer Pavilion

The ideal summer pavilion combines the open-air shelter of a screened porch with the remote seclusion of an outbuilding. Add a beautiful view, and you have the very essence of leisure. Built on a concrete slab foundation, this project is designed for years of outdoor exposure. The fully screened front wall stands about nine feet high and offers a sweeping view of your favorite scenery. The lower-entry side has solid corner walls for structural support and added privacy. Dual screened doors ensure plenty of airflow for the hottest summer days.

As shown in the plans, the Summer Pavilion's roof is covered with translucent polycarbonate panels that let in plentiful light while sheltering the interior from showers. However, other roof coverings might better suit your climate and use of the space. For example, you might choose a more opaque material, or hang fabric beneath the panels to help keep the interior cool. Whatever type of roofing you use, make sure it complies with the local Building Code standards for year-round exposure (e.g., snow loads).

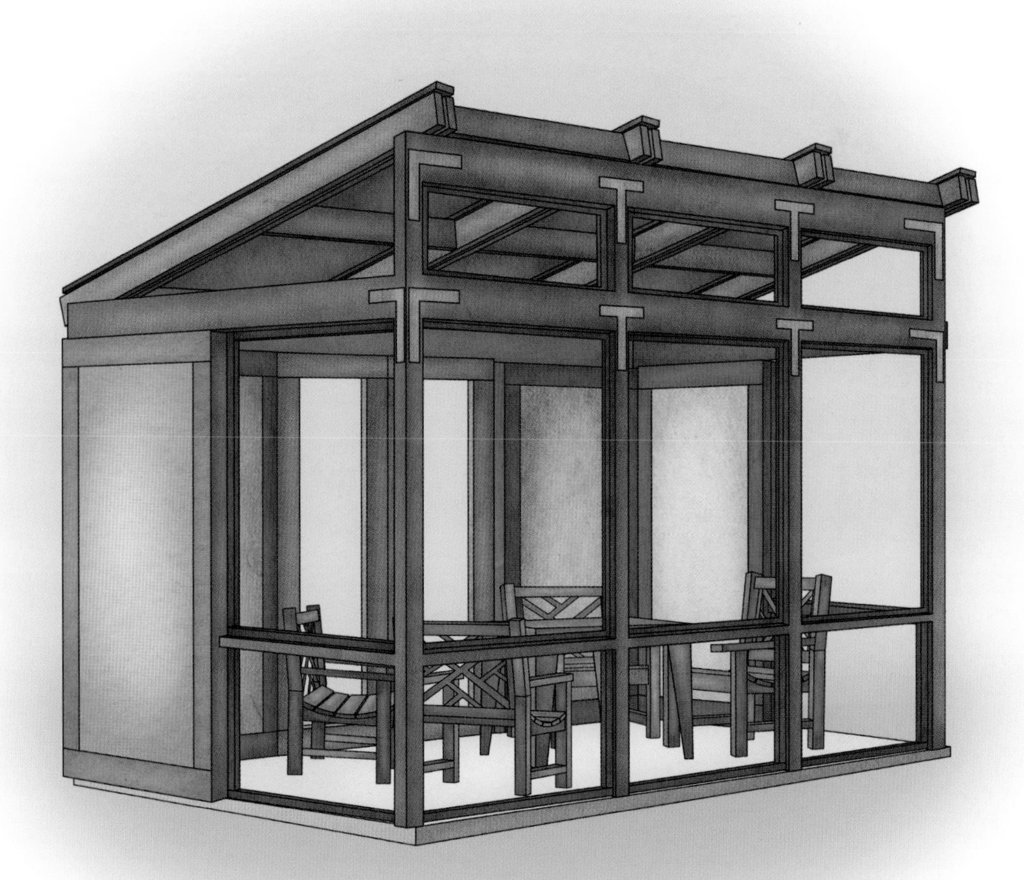

A pavilion like this is a great place to spend warm evenings with friends—eating dinner, playing games, or enjoying good conversation.

Tools, Materials & Cutting List

Hammer
Tape measure
Circular saw
Level
⅝"-diameter threaded rod
Drill with bits
Epoxy (recommended by post manufacturer)
Post base standoff
Mason's string
Chain saw or reciprocating saw
Caulk gun
Construction adhesive
Stapler and staples
Glazing tape or adhesive sealant
Aluminum tape
Eye and ear protection

Description (No. finished pieces)	Quantity/Size	Material
Foundation		
Concrete	1.63 cubic yards	3,000 PSI concrete
Gravel	1.25 cubic yards	Compactable gravel
Mesh	100 square feet	6" × 6": W1.4 × W1.4 welded wire mesh
Form materials	See page 28	
Wall Framing		
Corner wall framing (22 studs, 4 bottom plates, 4 top plates)	26 @ 8'	2 × 4 (pressure-treated lumber for plates)
Front wall posts (4)	4 @ 9'	4 × 4
Front wall top beam (1)	1 @ 13'	4 × 8
Front wall intermediate beams (3)	1 @ 8', 1 @ 4'	4 × 6
Side wall beams (2)	2 @ 8'	4 × 6
Rear wall beam (1)	1 @ 13'	4 × 6
Corner Wall Finish		
Corner wall siding (and interior finish)	8 sheets @ 4 × 8'	⅝" T1-11 plywood siding
Horizontal trim (8, exterior only)	4 @ 8'	1 × 6
Vertical trim (12, including interior)	12 @ 8'	1 × 4
End of wall trim (4)	4 @ 8'	1 × 6
Top cap trim	Field measure	1× lumber
Roof Framing		
Roof beam members (8)	8 @ 9'	2 × 6
Roof beam spacers (4)	4 @ 9'	1 × 6
Battens (4)	4 @ 9'	1 × 4 composite decking material
Blocking (6)	3 @ 8'	2 × 6
Cross beams (9)	3 @ 12'	2 × 6

Description (No. finished pieces)	Quantity/Size	Material
Screens		
Sills (5)	2 @ 8', 1 @ 4'	2 × 4
Screen molding (58, cut to fit)	58 @ 8'	¾" quarter-round
Screen (15, cut to fit)	4'-wide roll × 55' long	
Adhesive rubber weatherstripping	225 linear feet	
Roof panels (3)	3 @ 4 × 8'	Multiwall polycarbonate panel
Screen doors	2 doors or prehung unit; field measure	
Hardware & Fasteners		
⅝" × 8" J-bolts	8, with washers & nuts	
16d galvanized common nails		
8d galvanized finish nails		
8d galvanized box nails		
6d galvanized box nails		
6d galvanized finish nails		
16d galvanized finish nails or 3½" deck screws		
Heavy-duty staples		
2" deck screws		
Post bases	4, with recommended anchoring hardware	Simpson CPS4 or similar approved base with standoff
Post-to-beam T connectors	8, with recommended anchoring hardware	Simpson OT or similar approved connector
Beam-to-beam angles	4, with recommended anchoring hardware	Simpson HL35PC or similar approved connector
Post-to-beam L connectors	6, with recommended anchoring hardware	Simpson OL or similar approved connector
Roof beam to wall beam connectors	8, with recommended fasteners	Simpson H2.5 or similar approved connector
Aluminum tape		
Glazing tape or adhesive sealant		

Front Elevation

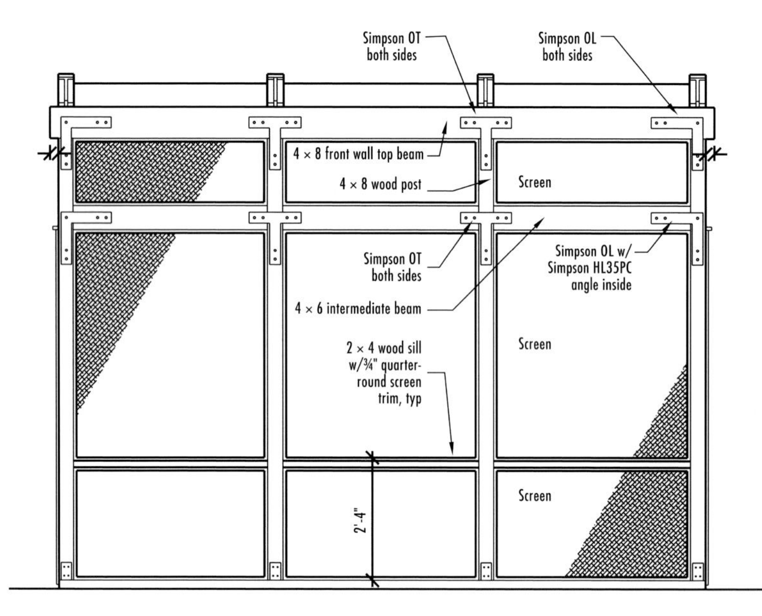

Rear Elevation

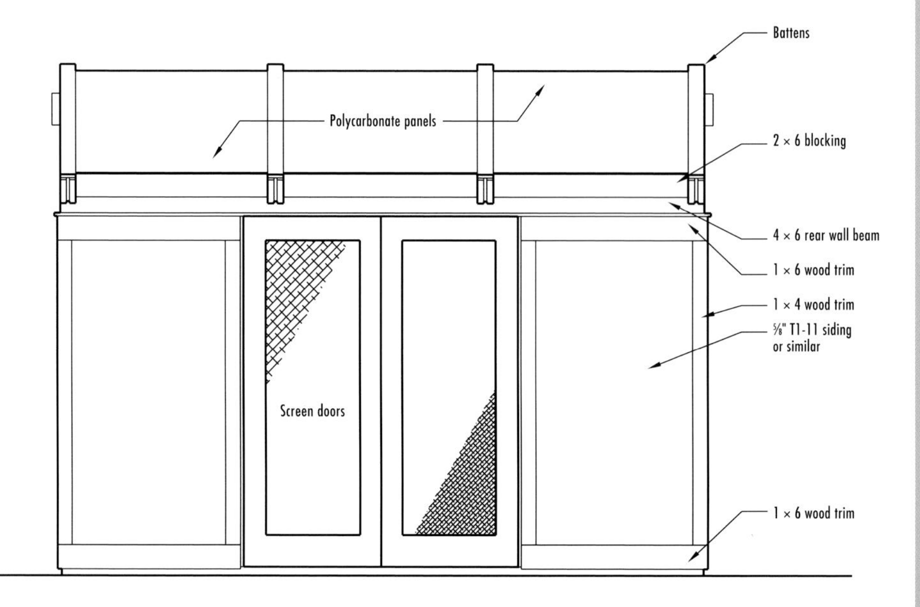

Roof Beam Template/Side Elevation

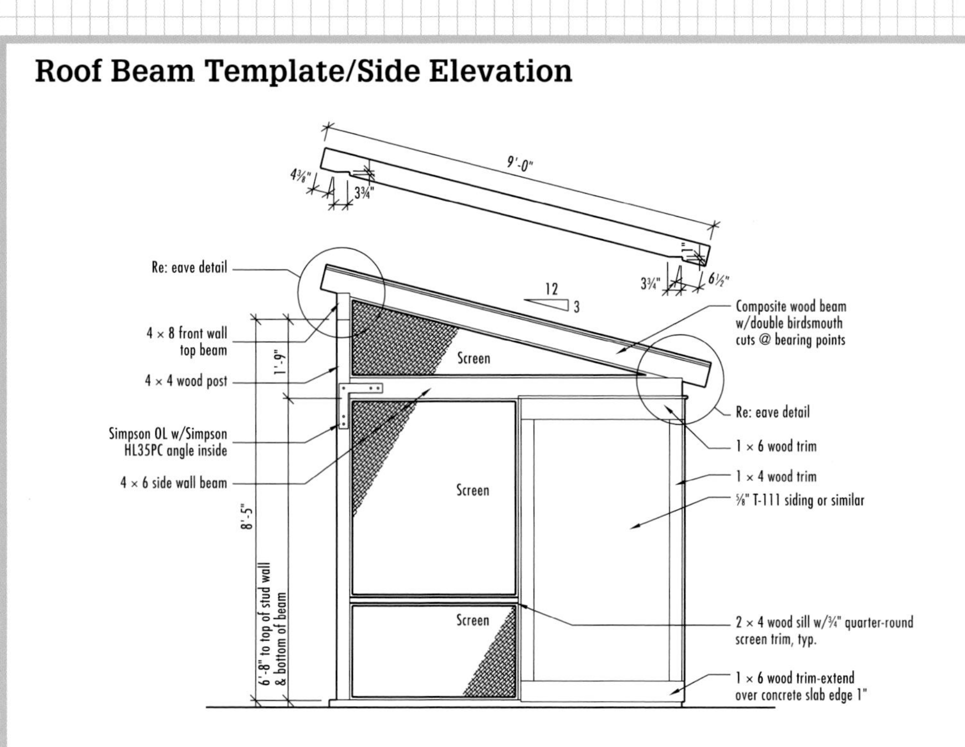

Foundation Plan

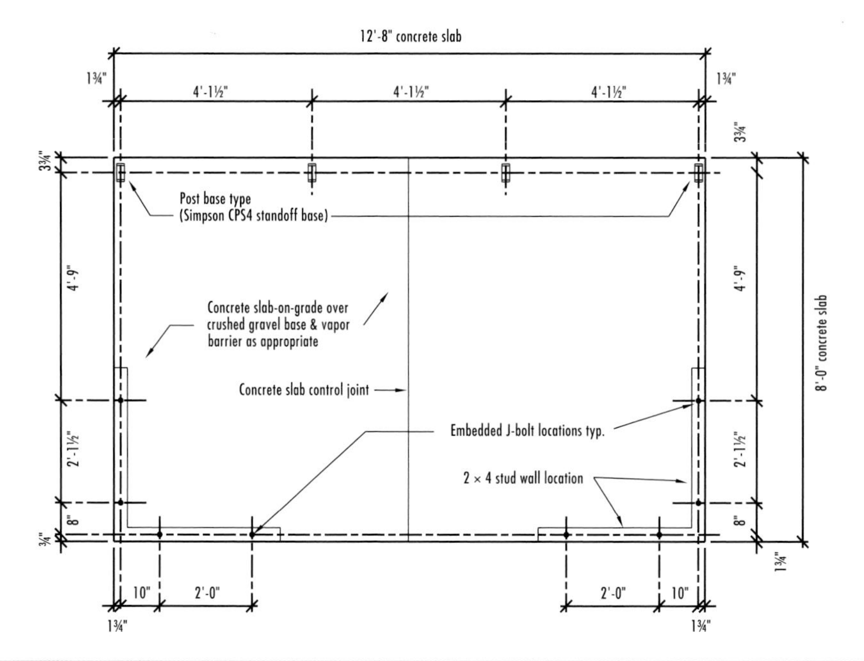

Wall Framing

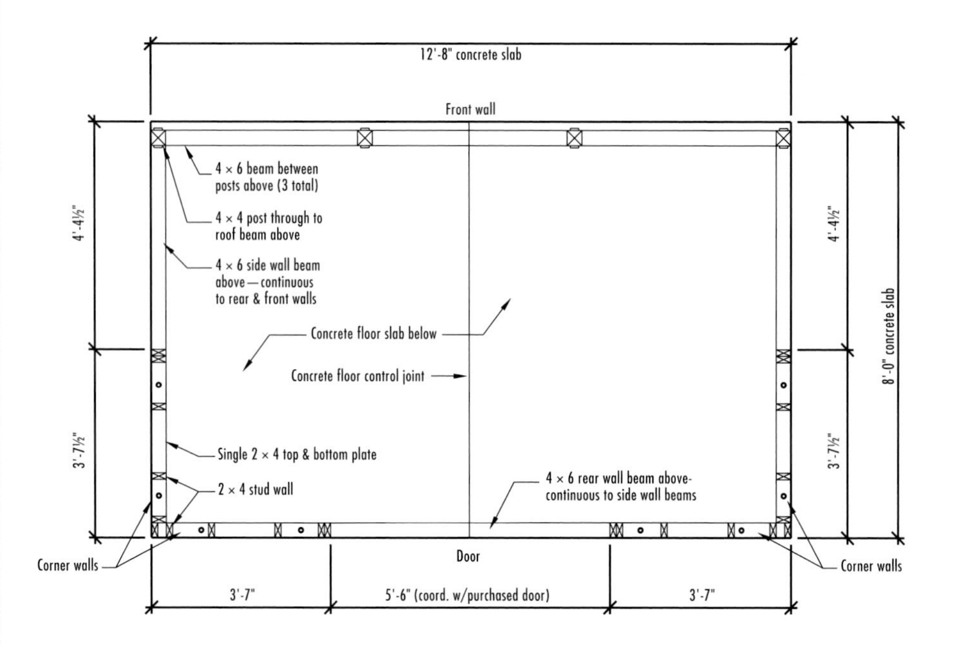

Floor and Roof Framing

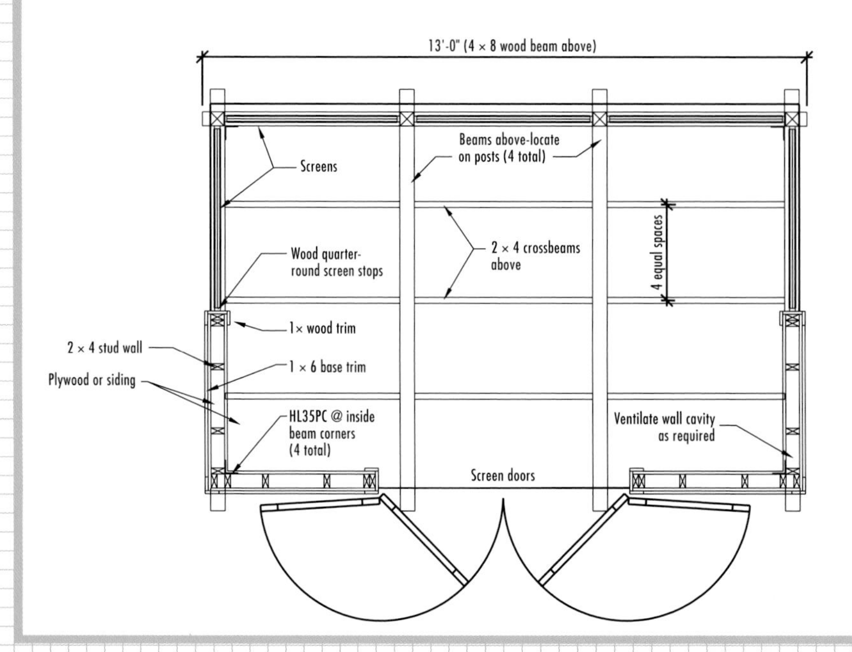

Roof Plan

Eave Details

Roof Beam Detail

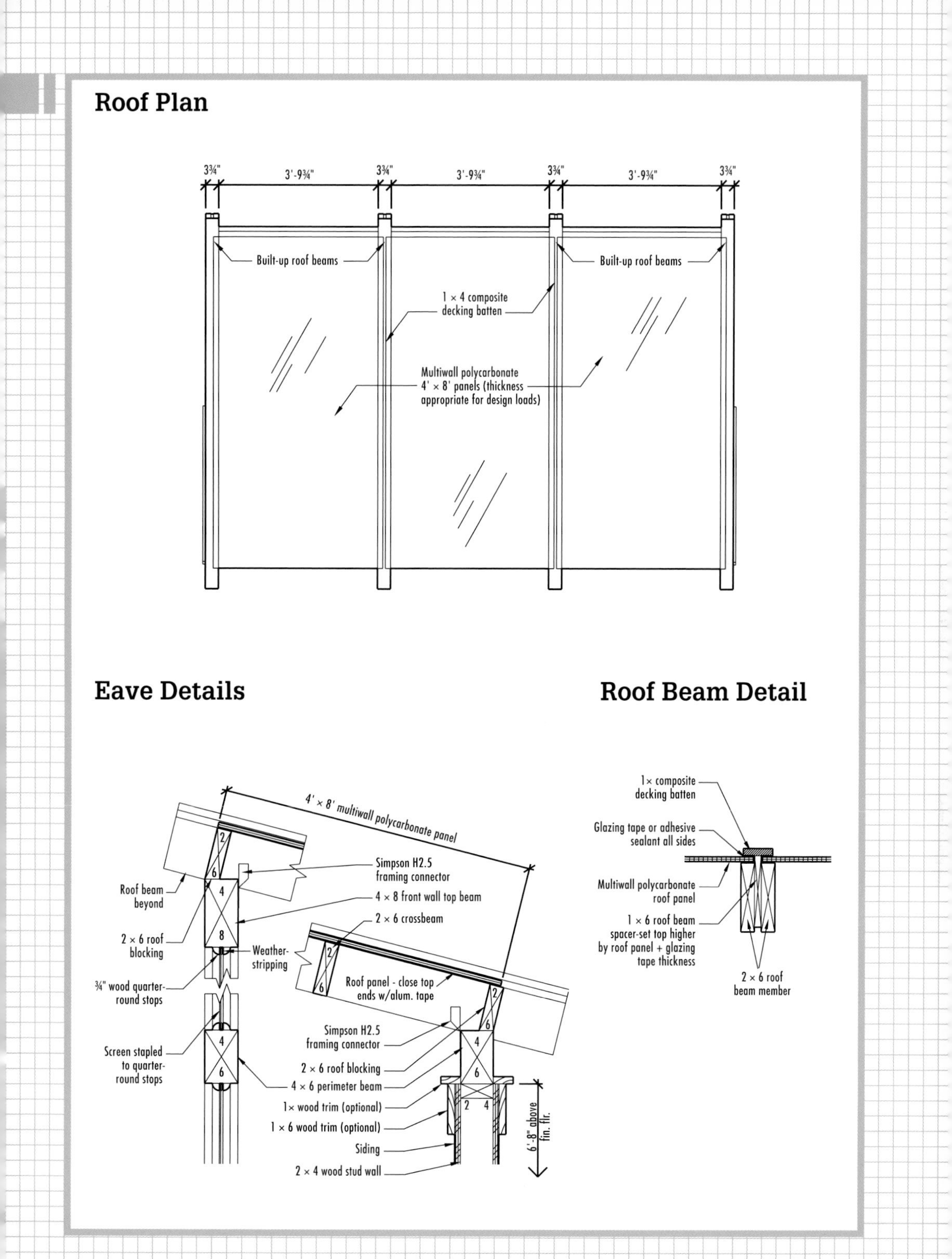

How to Build a Summer Pavilion

STEP 1: BUILD THE FOUNDATION

See page 28 for step-by-step instructions on pouring and finishing a concrete slab foundation. The finished slab should measure 96 × 152", as shown in the FOUNDATION PLAN on page 301.

Set eight ⅝ × 8" J-bolts in the concrete, following the layout shown in the FOUNDATION PLAN; these are for anchoring the bottom plates of the corner walls. The bolts should extend 2½" from the slab. Also create a control joint down the center of the slab, as shown in the FOUNDATION PLAN.

Note: If you live in an area that gets heavy rainfall, you might want to slope the slab slightly toward one side for drainage.

STEP 2: FRAME THE CORNER WALLS

Unless you plan to build custom screen doors, it's best to buy the doors (or a prehung door unit) and have them on hand before framing the walls—it's the only way to ensure your door opening will be the right size.

You can frame the four corner walls on the ground, then raise them onto the foundation using standard wall framing techniques.

A. Mark the rough opening for the screen doors so it is centered along the rear edge of the foundation; see WALL FRAMING on page 302. Size the opening according to the door manufacturer's instructions, but be sure to account for 1× lumber trim at each side; see FLOOR & ROOF FRAMING on page 302.

B. Measure from your layout marks to determine the lengths of the bottom and top plates of the rear wall frames. Cut two 2 × 4 bottom plates and two top plates using this dimension.

C. For the side walls, cut two bottom plates and two top plates at 40".

D. Cut 22 studs at 77".

E. Mark the stud layouts onto the plates following the WALL FRAMING plan. Mark the J-bolt locations onto the bottom plates, then drill holes for the bolts.

F. Assemble the walls with 16d galvanized common nails. Raise each rear wall, setting the bottom plate over the J-bolts. Position the wall so it's flush with the foundation, and secure the plates to the J-bolts with washers and nuts.

G. Raise the side walls and anchor them to the J-bolts. Make sure all of the walls are plumb. Fasten the side walls to the rear walls, driving 16d nails through the end studs.

STEP 3: SET THE POSTS

A. Cut four pieces of ⅝"-diameter threaded rod, following the specifications of the post base manufacturer.

B. At the bottom end of each post, drill a ¾-diameter × 10" hole in the center of the post. Clean dust and debris out of the holes, then fill each hole halfway with the base manufacturer's recommended epoxy. Insert a piece of threaded rod into each hole, and let the epoxy cure.

C. Attach a post base standoff to each post using the recommended fasteners.

D. Mark the centers of the four front wall posts, following the FOUNDATION PLAN. At each point, drill a ¾"-diameter hole into the slab, following the manufacturer's specifications for anchor depth.

E. Clean the anchor holes and fill them halfway with epoxy. Set the posts by inserting the threaded rods into the foundation holes. Temporarily brace the posts so they are perfectly plumb, then let the epoxy cure.

F. Mark one of the posts at 101" above the slab. Use a level or mason's line to transfer the height mark to the remaining posts, then cut the posts (see page 369).

STEP 4: SHEATH THE CORNER WALLS

You can cover the outsides of the corner walls with any outdoor siding material. However, if you use traditional board siding, you must first sheath the walls with ½" or thicker plywood to give the wall frames rigidity. The project as shown here calls for ⅝" T1-11 plywood siding, which takes care of the structural support and siding at once.

A. For each wall, cut the siding to width so it's flush to the framing; cut it to length so it's flush with the top of the top plate and extends 1" below the top of the foundation slab. Overlap the siding where it meets at the outside corners.

B. Fasten the siding to the wall studs and plates with 8d galvanized finish nails; nail every 6" along the perimeter and every 12" in the field of the sheet.

C. Sheath the insides of the walls with the same plywood siding or another exterior-grade wall finish.

Build the concrete slab foundation following the requirements of your local Building Code.

Secure the wall frames to the foundation J-bolts, then fasten the walls together along the full lengths of the end studs.

Set the posts into the slab with anchoring rods—brace them so they are perfectly plumb. *Note: Post shown semi-transparent here for clarity.*

Install the plywood siding so it's flush to the framing and overhangs the foundation by 1".

(continued)

STEP 5: INSTALL THE WALL BEAMS

Note: You will need at least one helper for this step.

A. Cut the 4 × 6 side-wall beams to length at 90½".

B. Making sure the crowned edge is up, set each beam on top of the side corner wall so its front end is butted against the corner post and its back end is flush with the rear wall; see SIDE ELEVATION on page 301. Fasten the beam to the wall's top plate with 16d nails.

C. Anchor the beam to the corner post using a post-beam connector on the outside only—you will install a connector on the inside after the front wall beams are in place. Position and fasten the connector following the manufacturer's specifications.

D. Cut the rear wall 4 × 6 beam to length at 145". Position the beam—crown up—between the ends of the side-wall beams so all beams are flush at the top. Anchor the beams together using beam angles and the manufacturer's recommended hardware, then nail the rear beam to the rear walls.

E. Cut the 4 × 8 front wall top beam to length at 156". Set the beam on top of the posts so it overhangs the end posts by 2" and is flush to the faces of all of the posts; see FRONT ELEVATION on page 300. Anchor the beam with pairs of post-beam connectors at each post.

F. Cut the three 4 × 6 front wall intermediate beams to fit snugly between the posts. Position the beams so they are aligned with the side-wall beams and anchor them to the posts with pairs of post-beam connectors. On the inside of the walls, use angle connectors to anchor the two outer intermediate wall beams to the side-wall beams.

Above left: Side-wall beam to corner post connection.

Above right: Side-wall beams to rear wall beam connection.

Left: Front wall beam to post connections.

STEP 6: BUILD & INSTALL THE ROOF BEAMS

The roof beams are built-up beams made with two 2 × 6s sandwiched over a 1 × 6 spacer, which protrudes above the tops of the 2 × 6s to accommodate the roof panels; see ROOF BEAM DETAIL on page 303.

A. Cut eight 2 × 6s and four 1 × 6s to length at 108".

B. Select a straight 2 × 6 to use for the pattern beam member. Make the birdsmouth cuts following the ROOF BEAM TEMPLATE on page 301 (see pages 44 to 48 for help with marking and cutting birdsmouths).

C. Set the pattern on the front and rear wall beams to test-fit the cuts. Make any necessary adjustments so the birdsmouths fit flush against the beams.

D. Use the pattern to mark the birdsmouths on the remaining beam members and the 1 × 6s, then make the cuts. Using construction adhesive and nails, construct the beam so the 1 × 6 extends far enough to accommodate the roof panels, plus a little extra space for glazing tape or adhesive sealant (see Step 10, on page 309). Nail the pieces together from both sides using pairs of 16d galvanized common nails driven every 12". Drive the nails at a slight angle so they won't protrude through the opposite side.

E. Mark the layout of the roof beams onto the front and rear wall beams, following the FLOOR AND ROOF FRAMING plan.

F. Set the roof beams on their layout marks and fasten them to the wall beams with two 16d nails on each side. Then reinforce each joint with a framing connector.

Construct built-up beams with adhesive and nails (inset). Secure the roof beams to the wall beams with framing connectors fastened with nails.

STEP 7: TRIM THE CORNER WALLS

The four corner walls have 1 × 6 trim along the top and bottom and 1 × 4 trim at the sides. The wall ends are capped with custom-cut 1 × 6 trim. Finally, a 1× cap is added to finish off the walls along the beams. See SIDE ELEVATION, REAR ELEVATION, and EAVE DETAILS. Installing trim on the interior wall as shown is optional.

A. Cap the ends of the walls by ripping 1 × 6 trim to width so it covers the edges of the siding. Install the trim flush to the top and bottom edges of the siding using 8d galvanized box nails.

B. On the side walls, cut 1 × 6 trim boards to span from the outside corners of the rear walls and overlap the end-wall trim. Install the trim flush with the top and bottom edges of the siding.

With the exception of the top cap trim, install the trim so all of the joints are flush.

(continued)

C. On the rear walls, cut and install 1 × 6 trim boards starting at the door opening and overlapping the side-wall trim.

D. Cut and install 1 × 4 trim to fit vertically between the 1 × 6s. Overlap the 1 × 4s at the outside corners, as shown in the FLOOR AND ROOF FRAMING plan.

E. Rip 1× boards to cap the tops of the walls; see EAVE DETAILS. Size the cap trim to overhang the 1 × 6s as much as you like. Install the cap trim with 6d box nails.

STEP 8: BUILD THE SCREENS

A. Cut 2 × 4 sills to fit snugly between pairs of posts along the front and side walls (on the side walls, the sills span between the corner post and framed corner wall).

B. Install the sills so their top faces are 28" above the slab. Fasten the sills to the posts and side walls with 16d finish nails or 3" deck screws.

Note: All of the screens are built on-site with each custom-fit to its opening. The simple procedure is the same for every screen.

C. On the inside perimeter of each screened opening, draw a reference line 1⅝" from the outside of the wall. Miter-cut ¾" quarter-round molding to fit around the opening, fitting the pieces into the opening as you work. Cut two frames of molding for each opening; see EAVE DETAILS.

D. Install the outside frame of molding so its flat edge is on the reference lines using 6d finish nails driven through pilot holes.

E. Using a flexible screen material of your choice, staple the screen to the flat side of the molding, keeping the screen taught and smooth. Trim excess screen after completing the stapling.

F. Cover the edge of the screen with adhesive rubber weatherstripping, positioned so it will be hidden behind the molding. Install the second frame of molding so it's tight against the weatherstripping, further securing the edges of the screen.

Secure the screen with staples and weatherstripping sandwiched between quarter-round molding.

STEP 9: INSTALL THE ROOF BLOCKING & CROSSBEAMS

A. Cut 2 × 6 blocking to fit snugly between the roof beams at the outer edges of the wall beams (six pieces total); see EAVE DETAILS.

B. Rip the bottom edge of each piece of blocking at 14° so the top edge is flush with the tops of the roof beam 2 × 6s, as shown in the EAVE DETAILS. Install the blocking between the roof beams with 16d common nails.

C. Cut nine 2 × 6 crossbeams to fit between the roof beams, as shown in the FLOOR AND ROOF FRAMING plan. Mark the crossbeam layout onto the roof beams.

D. Position the crossbeams at an angle so they are perpendicular to the roof beams and their top edges are flush with the roof beam 2 × 6s. Fasten the crossbeams with 16d nails.

STEP 10: INSTALL THE ROOF PANELS & DOORS

The roof framing is sized to accept uncut 4 × 8-ft. sheets of multiwall polycarbonate panels. Make sure the panels you use are rated for the spans as shown, or install additional blocking and/or crossbeams as needed. The battens that hold down the roof panels are made of composite decking material to prevent rot.

A. Cut four 1 × 4 battens to length at 108".

B. Apply aluminum tape to enclose the ends of the roof panel cells, following the manufacturer's directions.

C. Apply glazing tape or adhesive sealant to the edges of each roof panel or the roof framing, according to the manufacturer's directions. Also add tape or sealant along the top edges of the crossbeams and roof blocking.

D. Set each panel on the roof beams so its front edge is flush with the front of the blocking, as shown in EAVE DETAILS.

E. With the panels in place, center the battens over the beams, drill pilot holes, and fasten them to the beams with 2" deck screws. Make sure the panels are held firmly by the battens.

F. Install the screen doors following the manufacturer's instructions.

Fasten the blocking and crossbeams to the roof beams with angled nails or screws.

Center the battens over the roof beams to provide equal overlap on adjacent roof panels.

Lattice Gazebo

A garden retreat surrounded by lattice is a study in dynamic views, both inside and out. The sun's movement fills the interior with ever-changing light patterns made even more compelling by vines twisting along the walls and roof. And with only partial separation from the elements, the sense of enclosure is blended with the sounds and smells of the outdoors.

Borrowing elements from Japanese teahouses and other Asian sources, this design has a gently sloping roofline that brings lightness to its broad, sweeping form. Rounded openings give the walls an uncommon decorative quality and provide visual contrast to the grid pattern of the lattice. In daylight, the shadowed interior offers a pleasant retreat for people as well as sun-shy plants. At night, when lighted from inside, the gazebo glows like a paper lantern.

The simple, symmetrical design of this project makes it easy to alter its size. As shown in the plans, the gazebo is big enough for a patio table and chairs. To create a smaller version, first resize the footprint, then decide how tall you want the structure, being mindful of proper proportions. From there, the pieces can be measured and cut to fit.

This gazebo can easily be resized to fit a large or small space on your property and is an elegant way to build your garden up and around you, adding color and dimension to your landscape.

Tools, Materials & Cutting List

Gravel
Excavation tools
Mason's string
Level
Tape measure
Line level
Chain saw or reciprocating saw
Circular saw
Cardboard
Jigsaw or bandsaw
Sandpaper
Clamps
Ratchet wrench
Drill with bits
Compound miter saw
Eye and ear protection

Description (No. finished pieces)	Quantity/Size	Material
Posts & Foundation		
Posts (4)	4 @ field measure	4 × 4
Concrete	Field measure	3,000 PSI concrete
Compactable gravel	Field measure	
Walls		
Top and bottom rails (14 full-length, 4 door-wall bottom rails)	15 @ 10'	2 × 4
Window headers (6) and sills (6), door headers (2), and jambs (16)	23 @ 10'	2 × 4
Window and door brackets (14)	3 @ 10', 1 @ 6'	1 × 12
Lattice	Field measure	¾" (total panel thickness) manufactured wood lattice
Roof		
Truss top chord (4)	4 @ 12'	2 × 12
Truss bottom chord (8)	8 @ 8'	2 × 6
Truss strut (4)	1 @ 10'	2 × 4
Hub (1)	1 @ 51"	4 × 4
Support joists (4 outer, 4 inner)	4 @ 14', 4 @ 10'	2 × 4
Slat braces (4)	1 @ 8', 1 @ 3'	2 × 4
Slats	Field measure	2 × 2
Hardware & Fasteners		
3" galvanized wood screws		
1⅝" galvanized wood screws		
1½" galvanized wood screws		
⅜ × 6" galvanized carriage bolts	16, with washers and nuts	
¼ × 7" galvanized lag screws	4, with washers	
¼ × 6" galvanized lag screws	12, with washers	
2½" deck screws		

Note: Truss assembly consists of 2 bottom chords, 1 top chord, and 1 strut.

Front Elevation

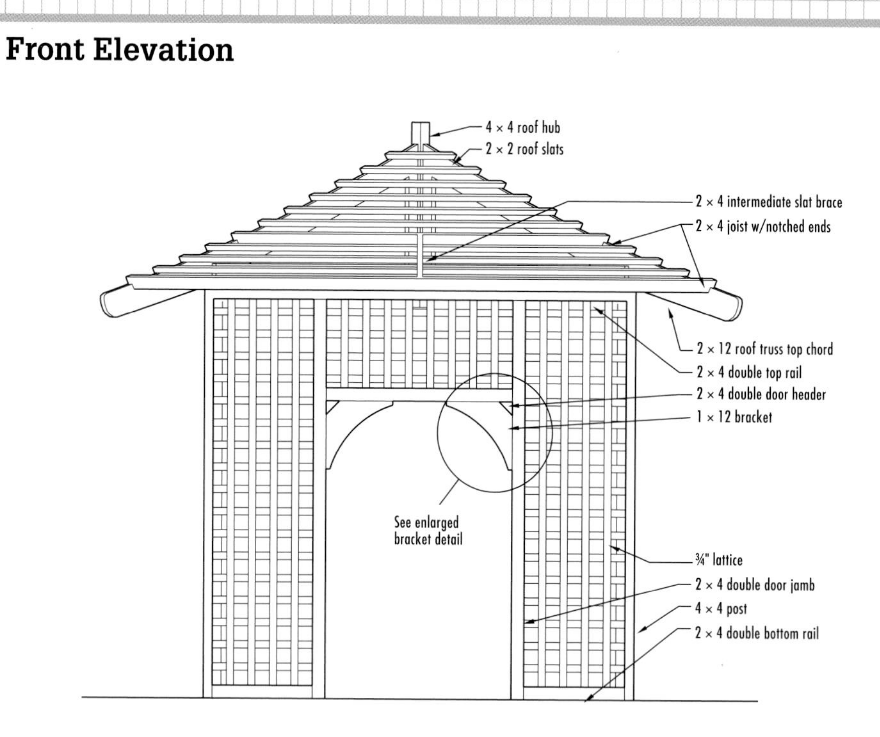

Window Bracket/Door Arch

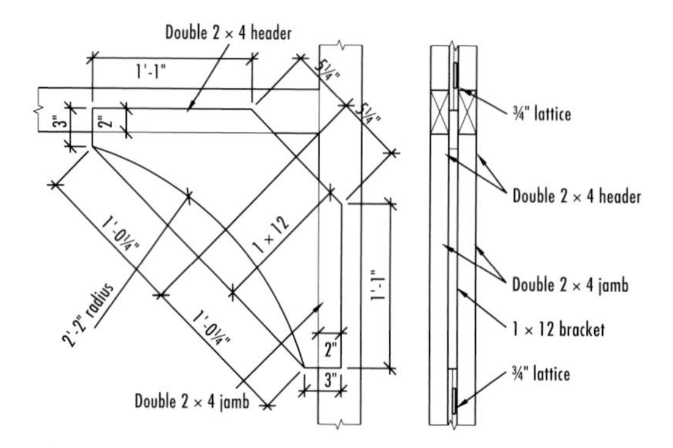

Side Elevation

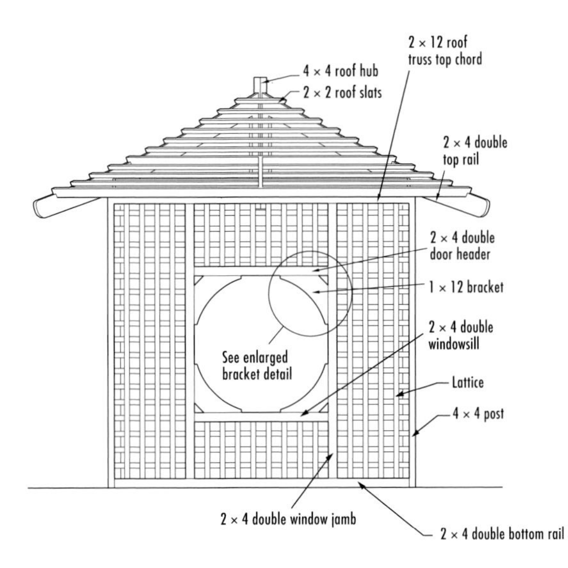

Floor Plan

Top/Bottom Rails Detail

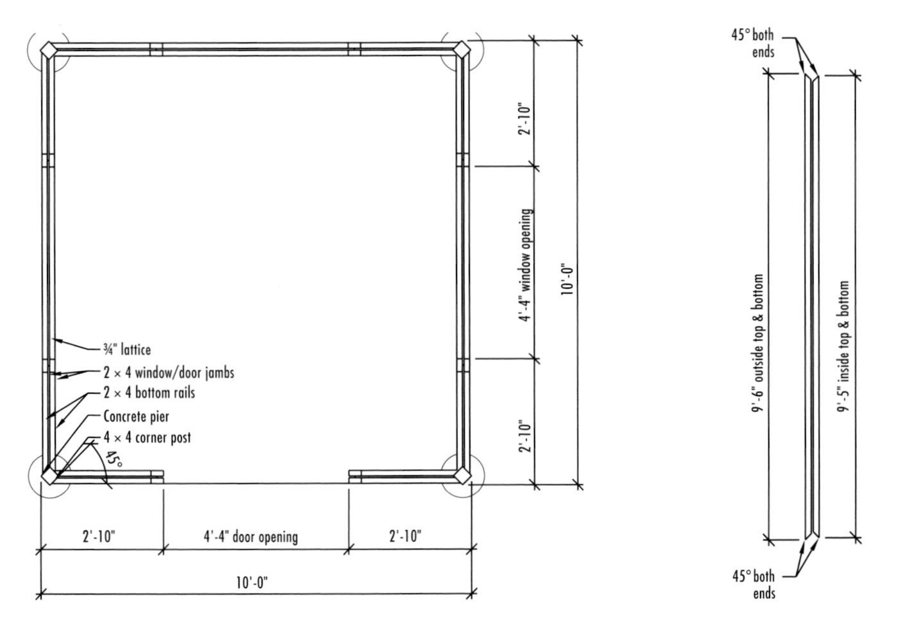

Framing Elevation

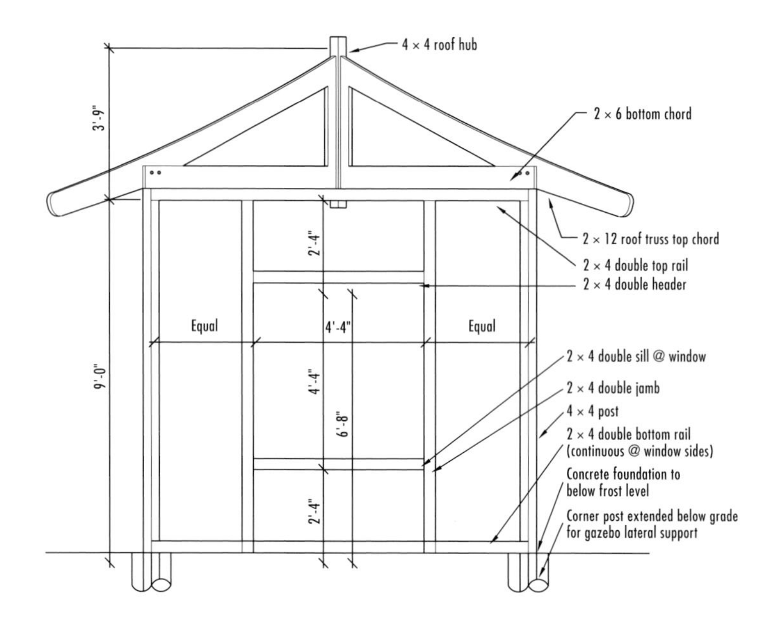

Corner Detail

Roof Plan

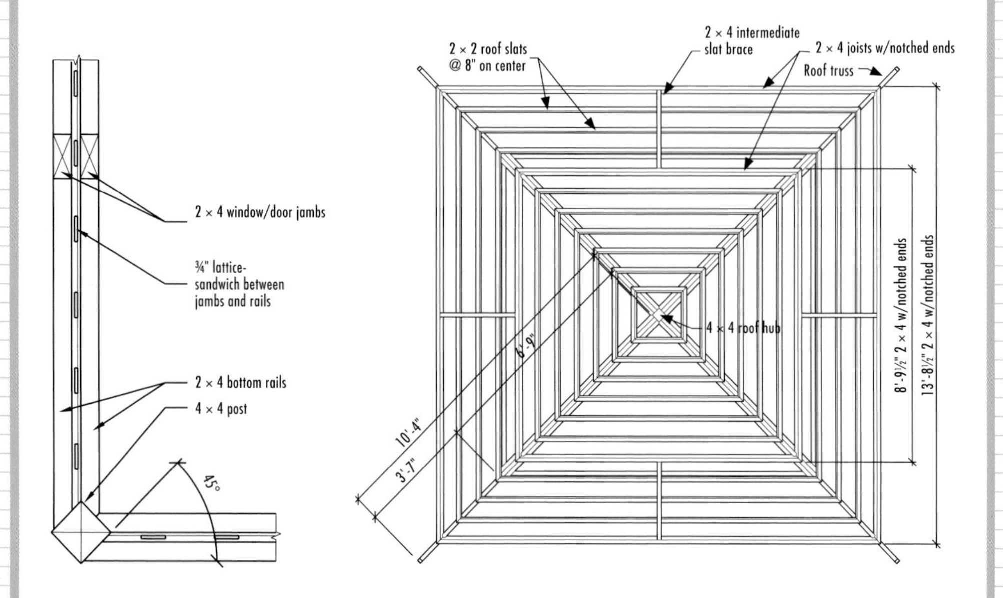

Slat Section

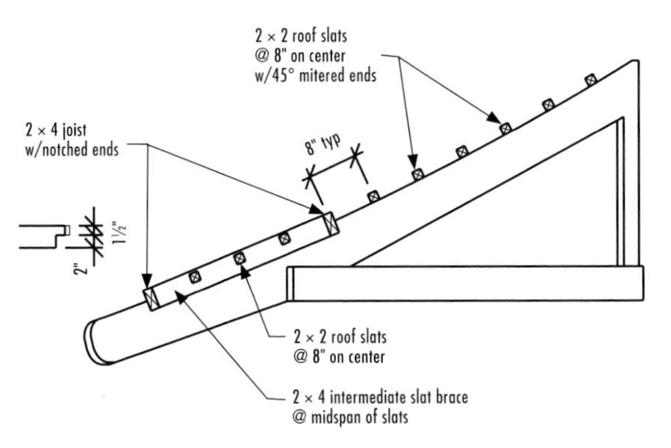

Roof Truss Template

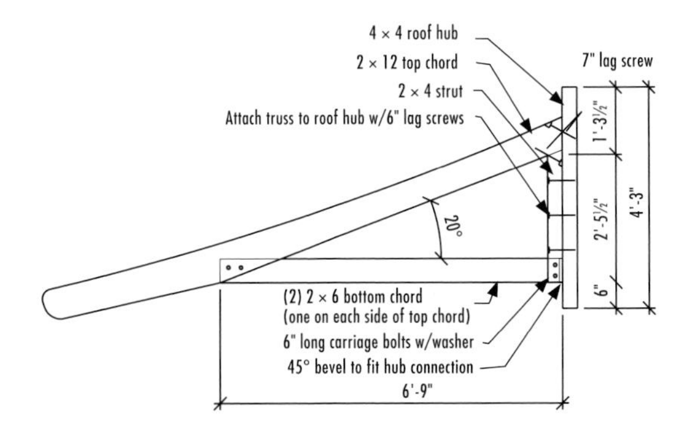

Truss Top Chord Template

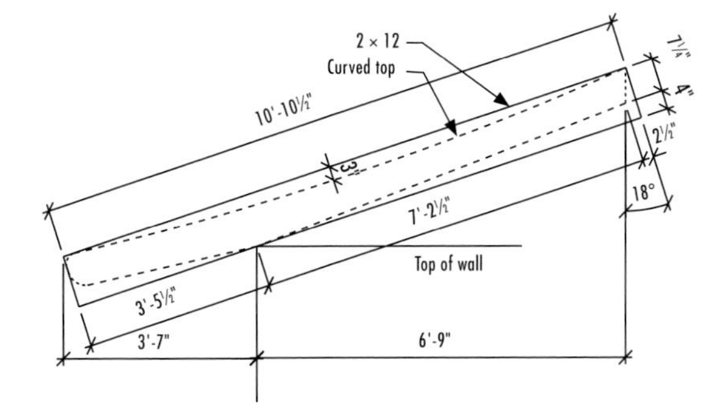

How to Build a Lattice Gazebo

STEP 1: INSTALL THE POSTS

The four 4 × 4 posts are buried in concrete; see pages 438 to 439 for a detailed procedure. The depth and diameter of the post and the surrounding concrete pier must meet the requirements of your local Building Code and extend below the frost line (as a minimum, the posts should be buried 30" deep). Treat the bottom ends of the posts for rot-resistance before setting them.

A. Lay out the four post locations onto the ground, following the FLOOR PLAN on page 313. Dig the postholes and add a 6" layer of gravel to each for drainage.

B. Set up mason's lines to lay out the precise post locations, following the FLOOR PLAN. Set the posts in the holes and secure them with cross bracing so they are perfectly plumb and turned at 45° to the square layout. Measure the diagonals between posts to check for squareness: the layout is perfectly square when the diagonal measurements are equal.

C. Pour the concrete and let it dry completely.

D. Measure up from the ground and mark one of the posts at 108". Use a mason's string and a line level to transfer that height mark to the other three posts, then cut the posts to height (see page 369). Cut carefully so the post tops are flat and level.

Use mason's lines to set up the square post layout and to offset the posts at 45°.

STEP 2: BUILD THE OUTER WALL FRAMES

In this step, you complete the outer layer of the 2 × 4 wall framing. The main wall structures are made of 2 × 4 frame pieces sandwiched over lattice panels, as shown in the CORNER DETAIL (page 314). Because lattice varies in thickness, test-fit some scrap 2 × 4 frame pieces and lattice to determine how the assembly will fit against the posts.

A. Cut the four 2 × 4 bottom rails to fit between the posts along the sides and rear of the gazebo; cut the ends at 45° to fit flush against the post faces. You will cut and install the front wall, bottom rails in sub-step F. See TOP/BOTTOM RAILS DETAIL on page 313.

B. Position the bottom rails against the posts at the desired height above the ground, and fasten them to the posts with 3" wood screws.

C. Cut the top rails to fit between the posts on all four sides of the gazebo. Install the rails so their top edges are flush with the tops of the posts.

D. Mark the window and door openings onto the rails, following the FLOOR PLAN. All of the openings are centered on the walls and span 52".

Install the bottom rail at the desired height. Most likely it will sit close to, if not touching, the ground.

E. Cut the 2 × 4 window jambs to fit snugly between the top and bottom rails. Fasten them to the rails on the layout marks. For each window, cut a 2 × 4 header and a 2 × 4 sill to fit between the jambs. Install them at the desired height to create a square opening.

F. To frame the door opening, cut two jambs to reach from the top rail to the bottom of the bottom rail position. Cut two 2 × 4 bottom rails (left from sub step A) to span from the posts to the inside edges of the jambs, as shown in the FRONT ELEVATION (page 312). Also cut a 2 × 4 door header to length at 52". Assemble the door opening, as shown in the FRAMING ELEVATION, so the top of the opening is 80" from the ground on the bottom edges of the bottom rails.

STEP 3: ADD THE WINDOW & DOOR BRACKETS

A. Make a cardboard template for marking the bracket profiles, following WINDOW BRACKET/ DOOR ARCH on page 312.

B. Use the template to mark 14 brackets onto 1 × 12s, then cut the pieces with a jigsaw or bandsaw, and sand the cuts smooth.

C. Draw reference lines 2" in from the angled edges of each bracket; this designates the overlap onto the 2 × 4 wall framing, as shown in WINDOW BRACKET/DOOR ARCH.

D. Position the brackets at the corners of the window and door openings: Each window gets four brackets, while the door gets two at the top of the opening. Fasten the brackets to the frames with 1⅝" wood screws. Drive the screws from inside the gazebo, so the inner wall framing will hide the screw heads.

STEP 4: INSTALL THE LATTICE & INNER WALL FRAMES

You can install the lattice in any configuration you like using square- or diamond-patterned panels.

A. Working outward from the posts, cut the lattice panel for each framed section. Overlap the

Mark the bracket radius with a 26"-long string or board, pivoting from a centerpoint.

(continued)

2 × 4 wall framing by at least 3" along the rails and around the window and door openings; overlap about 1½" along the jambs above and below the openings. For best appearance, use a clean factory edge where the lattice meets the posts. You can also bevel those edges so you don't see the end grain of the lattice slats.

B. Fasten the lattice to the framing with 1½" wood screws—drive them through pilot holes to prevent splitting. Use mason's string or layout marks to help keep the lattice panels aligned with one another.

C. Cut and install the inner wall frames, following the same basic procedure used in Step 2 to install the outer frames. Bevel the ends of the rails where they meet the posts; if necessary, clip the rail ends so they fit together, as shown in the CORNER DETAIL. Fasten the inner framing parts together at the corner joints, as you did with the outer frames, and then fasten the inner and outer framing together with 3" wood screws.

STEP 5: BUILD THE ROOF TRUSSES

A. Select a straight, 12-ft. 2 × 12 to use as the pattern for the top chords. Check the board for crowning, and mark the top edge. Draw the outline of the chord onto the board, following the TRUSS TOP CHORD TEMPLATE (page 315). Make the cuts, and sand the curved edges smooth.

B. Using the cut board as a pattern, trace the outline onto a second 2 × 12 and make the cuts. For each of the two chords, cut two 2 × 6 bottom chords and one 2 × 4 strut, following the ROOF TRUSS TEMPLATE, on page 315.

C. Assemble the trusses as shown in the TEMPLATE: Sandwich two bottom chords over each top chord and strut and fasten at each end with two ⅜ × 6" carriage bolts. Fasten the strut to the top chord with a ¼ × 7" lag screw driven at an angle through a counterbored pilot hole.

D. Cut the 4 × 4 roof hub to length at 51". If desired, shape the ends to a point with four equal bevel cuts. Test-fit the trusses and hub on the gazebo. The outside ends of the bottom chords should be aligned with the outside edges of the posts. Make any adjustments necessary for a good fit.

E. Disassemble one of the trusses and use the parts as patterns to mark the remaining truss pieces. Cut the parts and assemble the remaining trusses.

If desired, bevel-cut the edges of lattice that meet the posts using a circular saw set at 45°.

Bend a long piece of trim or other flexible board against nails to make the top chord arch.

STEP 6: COMPLETE THE ROOF FRAME

A. Position two opposing trusses on the posts with their struts centered on the hub. The hub should extend 6" below the bottom chords. Drill pilot holes into the strut. Fasten three ¼ × 6" lag screws through each strut and into the hub (see ROOF TRUSS TEMPLATE). Offset the screws on opposing sides so they don't run into one another. Also fasten through the top chords into the hub with a lag screw driven at an angle through a counterbored pilot hole.

B. Install the two remaining trusses.

C. Mark the top edges of the rooftop chords at 8" and 40" in from their outside ends—these marks represent the outside faces of the 2 × 4 support joists; see the ROOF PLAN on page 314 and the SLAT SECTION on page 315. If you prefer slat spacing different than 8" on center, adjust the positions of the joists as desired.

D. Cut the 2 × 4 joists to span between the centers of the chords at the marks. Miter and notch the ends of the joists, as shown in the SLAT SECTION.

E. Install the joists on the chords with 2½" deck screws.

F. Cut the four 2 × 4 slat braces to fit between the joists at their centerpoints. Fasten the braces between the joists with screws.

STEP 7: INSTALL THE ROOF SLATS

A. Mark the layout of the roof slats onto the top edges of the truss chords and slat braces using 8" on-center spacing (or different spacing, as desired); see ROOF PLAN on page 314. Also mark the centers of the chords to facilitate measuring for the slats.

B. Cut 2 × 2 slats to span from the centers of the chords to the side faces of the slat braces.

Note: Compound-miter the chord ends of the slats. Fasten the slats with 2½" deck screws driven through pilot holes. The tops of the slats should be flush with the tops of the braces.

C. Continue installing slats up to the roof peak, mitering the ends so they meet at the centers of the truss chords.

If desired, countersink the lag screws into the truss struts to hide the screw heads.

Fasten the slats to the trusses using 2½" deck screws driven through pilot holes.

8-sided Gazebo

The traditional octagonal gazebo has timeless elegance. Viewed from any angle, its eight symmetrical sides give it an eye-catching, sculptural quality—a perfect centerpiece for the landscape. Inside the gazebo, an elevated floor adds a sense of loftiness, while open-air sides make it a great spot for escaping the sun and catching summer breezes.

The gazebo featured in this project measures nine feet across and has cedar decking laid in an octagonal pattern. Look up from the inside and you see an attractive panel ceiling of 1 × 6 cedar boards. The lattice panels help enclose the interior while giving the entire structure a light, airy feel. If you prefer more ornamentation, you can easily omit the upper lattice panels and add decorative brackets or scrollwork.

Not surprisingly, this project involves lots of angled cuts. If you've been looking for a reason to buy a compound miter saw, this is your ticket. It makes the project much, much easier.

The octagonal gazebo is a classic ornamental getaway. Finish your gazebo with similar coloring and style as your home to draw the two structures together and to help create a cohesive personal landscape.

Tools, Materials & Cutting List

Batter boards
Mason's string
Tape measure
Masking tape
Plumb bob
Stakes
Excavation tools
Gravel
Framing square
Hammer
Level
Line level
Circular saw
Chisel
Belt sander
Drill with bits
Caulk gun
Hand tamper
Concrete mixing tools
Stiff-bristle broom
Concrete edger
Reciprocating saw or handsaw
Table saw
Power miter saw (optional)
Chalk line
Eye and ear protection

Description (No. finished pieces)	Quantity/Size	Material
Foundation		
Concrete	Field measure	3,000 PSI concrete
Concrete tube forms	1 @ 16"-dia., 8 @ 12"-dia.	
Compactable gravel	2.5 cu. ft.	
Framing		
Posts (8)	8 @ 10'	6 × 6 cedar
Perimeter beams (16)	8 @ 8'	2 × 6 pressure-treated
Double joists (8)	8 @ 10'	2 × 6 pressure-treated
Angled joists (16)	8 @ 8'	2 × 6 pressure-treated
Roof beams (8)	4 @ 10'	6 × 8 cedar
Hip rafters (8)	8 @ 8'	2 × 8 cedar
Intermediate rafters (16)	8 @ 10'	2 × 6 cedar
Purlins (8)	2 @ 8'	2 × 8 cedar
Collar ties (4)	4 @ 10'	2 × 6 cedar
Rafter hub (1)	1 @ 2'	8 × 8 cedar
Wood sphere	1 @ 10"-dia., with dowel screw	
Pad (center pier) (2)	Cut from stringers	2 × 12 pressure-treated
Framing Anchors		
Joists to posts	8, with nails	Simpson U26-2
Angled joists to perim. beams	16, with nails	Simpson U26
Angled joists to double joists	16, with nails	Simpson LSU26
Anchor bolts	9 @ ⅝ × 12"	Galvanized J-bolt
Posts to piers	8, with fasteners	Simpson ABU66
Perimeter beams to posts	32	½ × 6" lag screws & washers
Metal anchors—rafters to beams	24	Simpson H1
Metal hangers—rafters to hub	8, with nails	Simpson FB26
Posts to roof beams	8, with fasteners	Simpson 1212T
Beams to roof beams	8, with nails	3 × 12" × 14-gauge galv. plate
Stair stringers to perimeter beam	6, with nails	Simpson L50

Description (No. finished pieces)	Quantity/Size	Material
Stairs		
Compactible gravel	4.5 cu. ft.	
Concrete form	2 @ 8'-0"	2 × 4
Stair pad	7 @ 60-lb. bags	Concrete mix
Stringers (3)	2 @ 8'-0"	2 × 12 pressure-treated
Stair treads (6)	2 @ 10'-0"	2 × 6 cedar
Stair risers (3)	1 @ 10'-0"	1 × 8 cedar
Finishing Lumber		
Decking	15 @ 8'-0", 6 @ 10'-0"	2 × 6 cedar
Deck starter	1 @ 1'-0"	2 × 8 cedar
Fascia	4 @ 10'-0"	2 × 4 cedar
Lattice	4 panels @ 4 × 8'	Cedar lattice
Stops	15 @ 8'-0" (horizontal), 10 @ 10'-0" (vertical)	5/4 × 5/4 cedar
Rails (37)	19 @ 8'-0"	2 × 4 cedar
Roofing		
Roof sheathing	26 @ 8'-0", 14 @ 10'-0"	1 × 6 T&G V-joint cedar
Asphalt shingles	256 sq. ft.	
15# building paper	300 sq. ft.	
Metal drip edge	36 linear ft.	
Galvanized flashing	3 linear ft.	
Roofing cement	1 tube	
Fasteners		
½ × 6" lag screws	16, with washers	
16d common nails		
16d galvanized common nails		
16d galvanized box nails		
16d galvanized casing nails		
10d galvanized common nails		
8d galvanized box nails		
8d galvanized finish nails		
3d galvanized finish nails		
Roofing nails, to fit roofing material		
1½" galvanized joist hanger nails		
Masonry screws or nails		
3" deck screws		
Construction adhesive		

Front Elevation

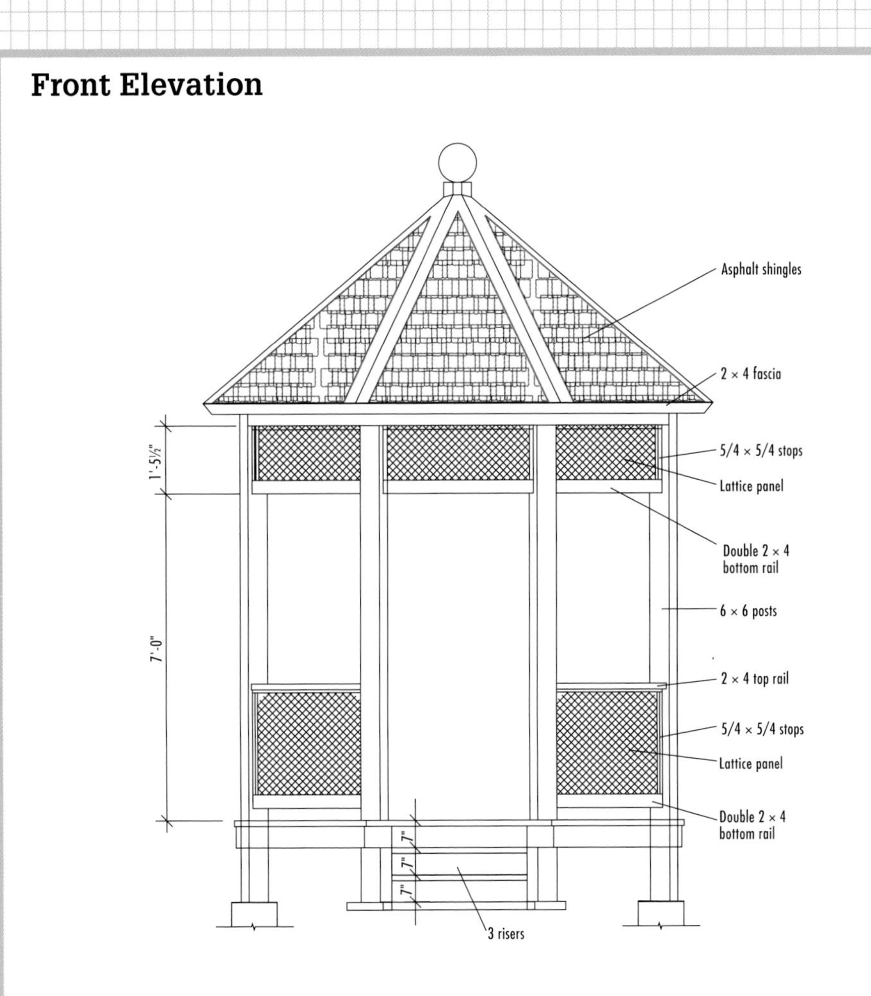

Center Pier Detail

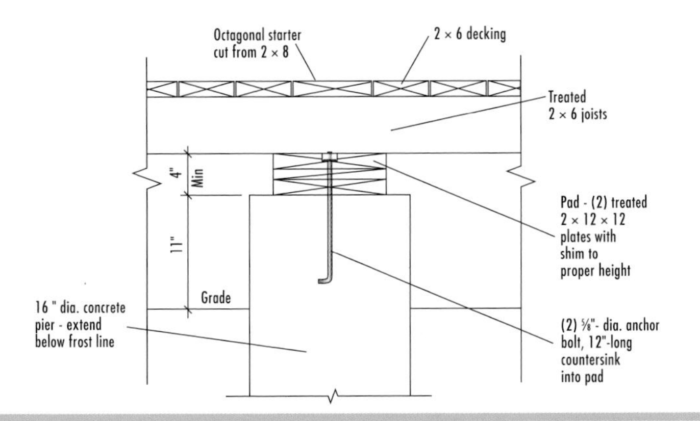

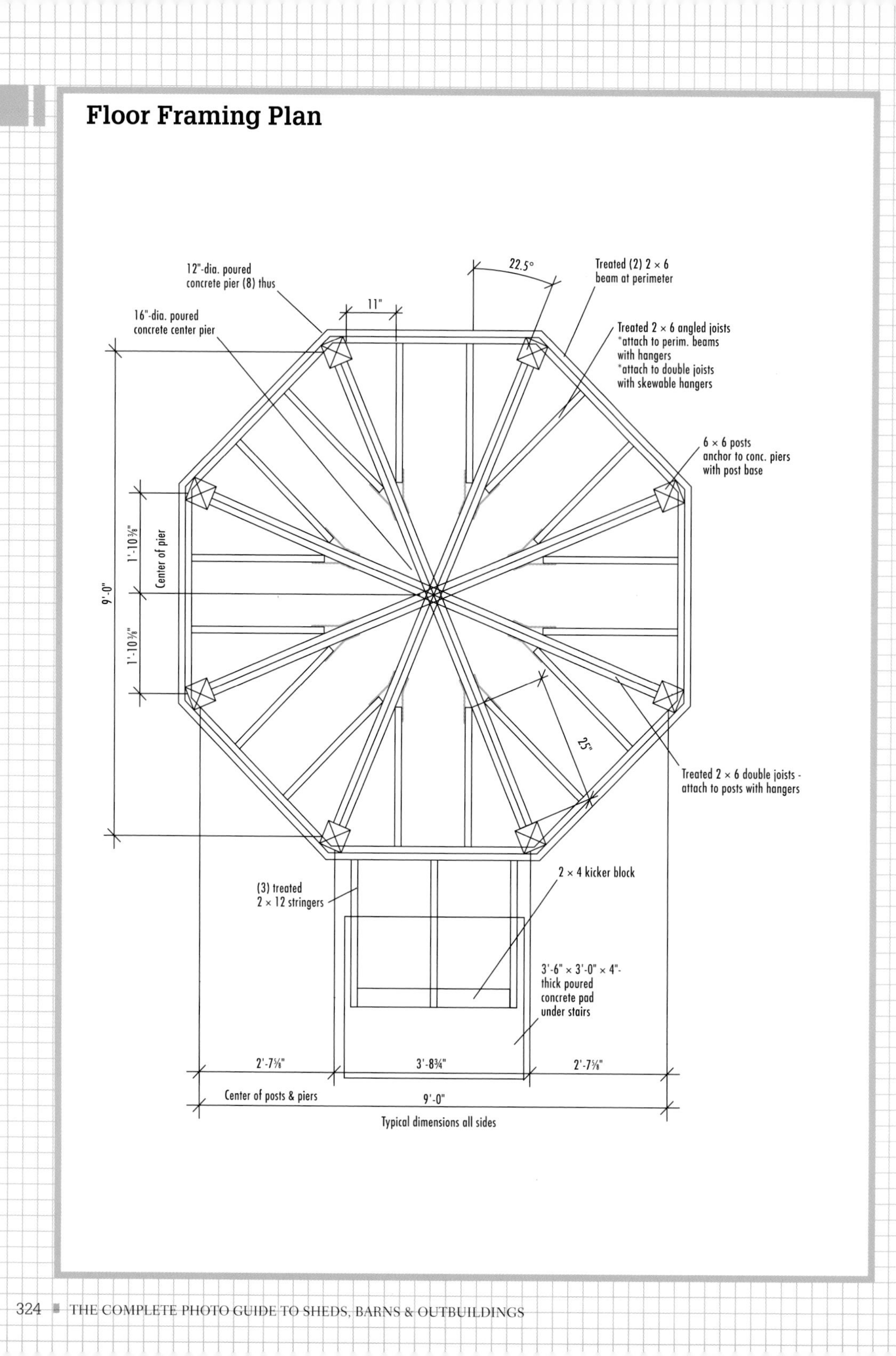
Floor Framing Plan
12"-dia. poured concrete pier (8) thus
16"-dia. poured concrete center pier
11"
22.5°
Treated (2) 2 × 6 beam at perimeter
Treated 2 × 6 angled joists *attach to perim. beams with hangers *attach to double joists with skewable hangers
6 × 6 posts anchor to conc. piers with post base
9'-0"
1'-10⅜"
1'-10⅜"
Center of pier
25"
Treated 2 × 6 double joists - attach to posts with hangers
2 × 4 kicker block
(3) treated 2 × 12 stringers
3'-6" × 3'-0" × 4"-thick poured concrete pad under stairs
2'-7⅝"
3'-8¾"
2'-7⅝"
Center of posts & piers
9'-0"
Typical dimensions all sides

Building Section

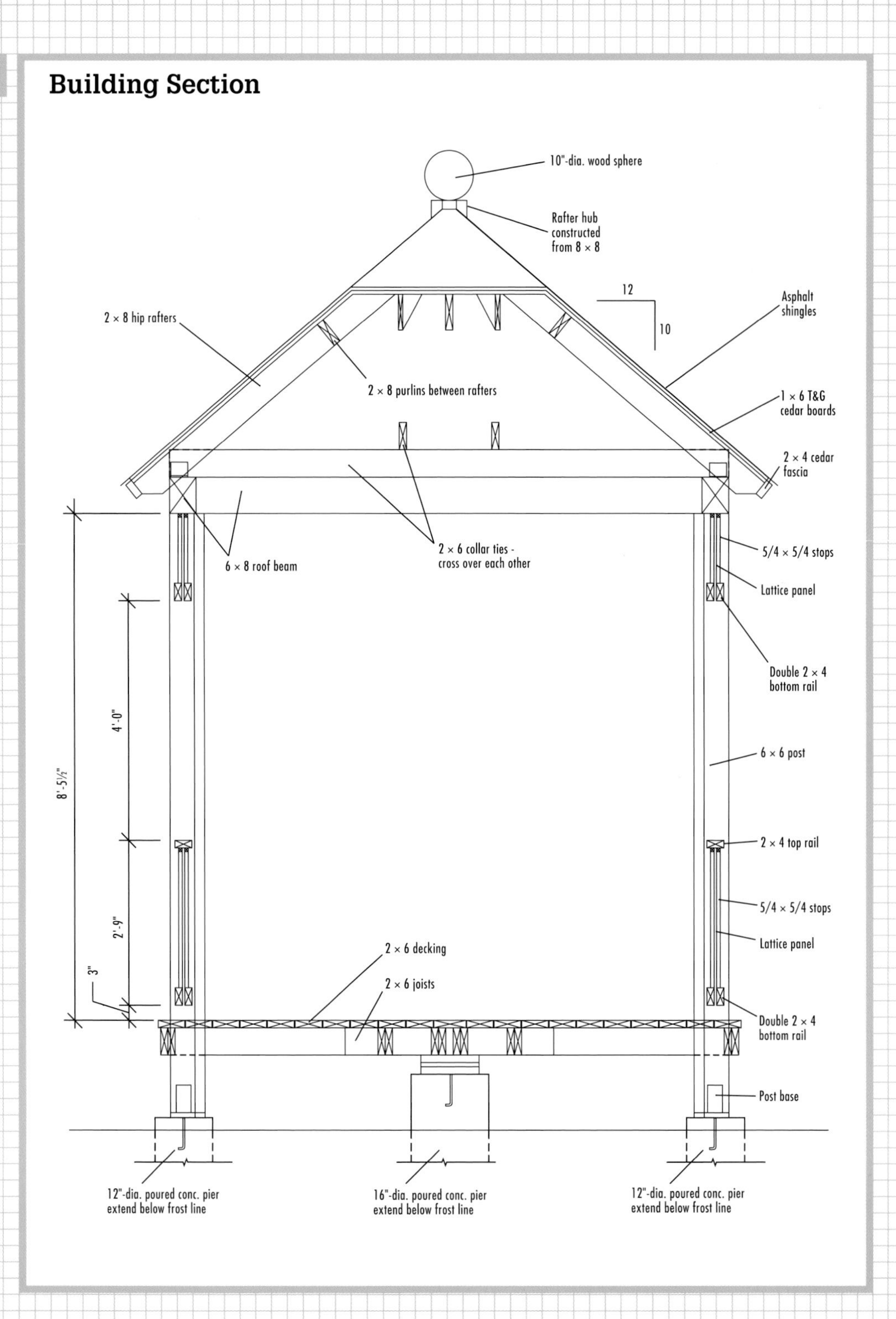

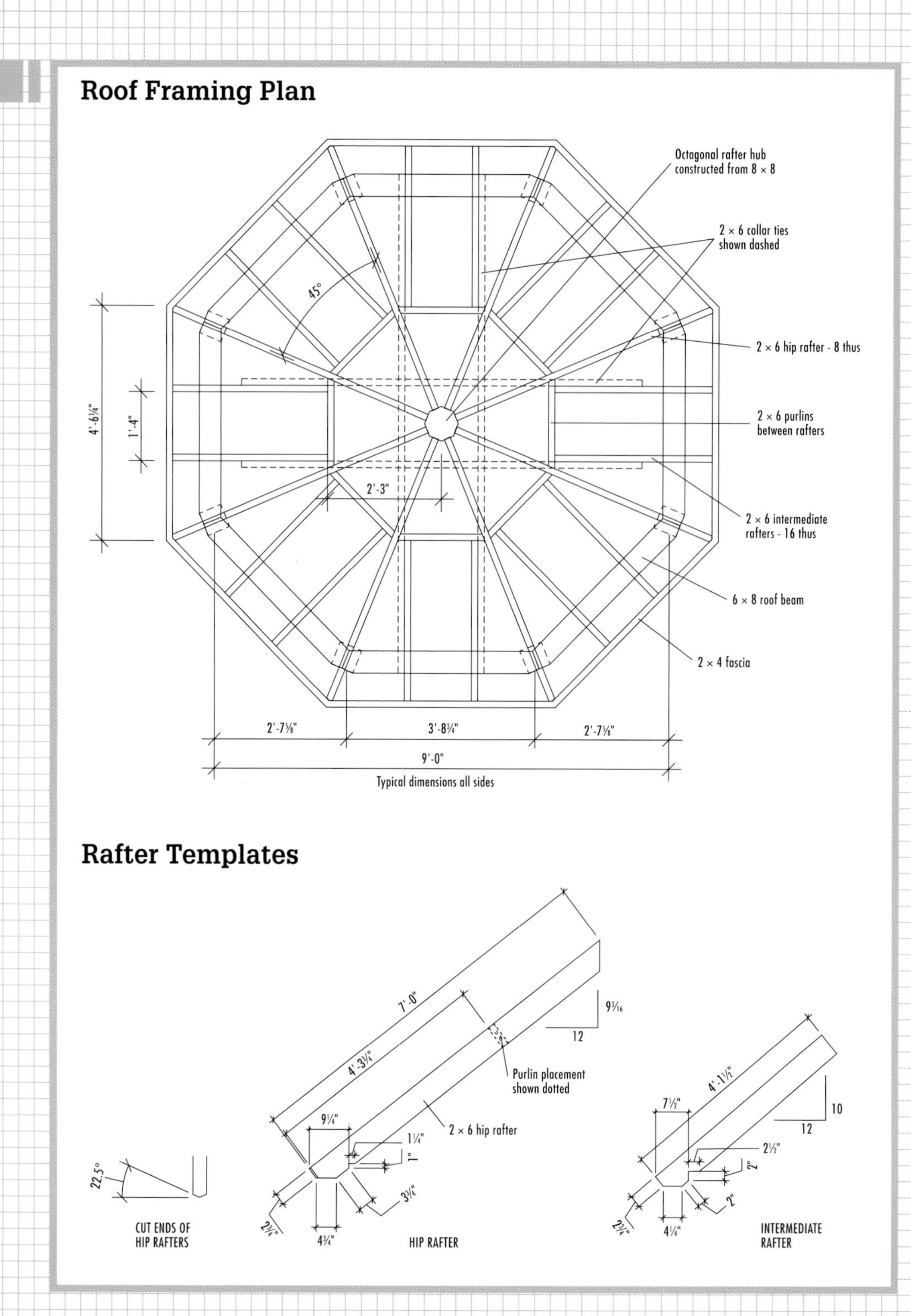

Roof Framing Plan
Octagonal rafter hub constructed from 8 × 8
2 × 6 collar ties shown dashed
2 × 6 hip rafter - 8 thus
2 × 6 purlins between rafters
2 × 6 intermediate rafters - 16 thus
6 × 8 roof beam
2 × 4 fascia
45°
4'-6 1/4"
1'-4"
2'-3"
2'-7 5/8"
3'-8 3/4"
2'-7 5/8"
9'-0"
Typical dimensions all sides
Rafter Templates
7'-0"
4'-3 3/4"
9 3/16
12
Purlin placement shown dotted
2 × 6 hip rafter
9 1/4"
1 1/4"
1"
3 3/4"
2 3/4"
4 3/4"
22.5°
CUT ENDS OF HIP RAFTERS
HIP RAFTER
4'-1 1/2"
7 1/2"
10
12
2 1/2"
2"
2"
2 3/4"
4 1/4"
INTERMEDIATE RAFTER

Decking Plan/Floor Beam Support Detail

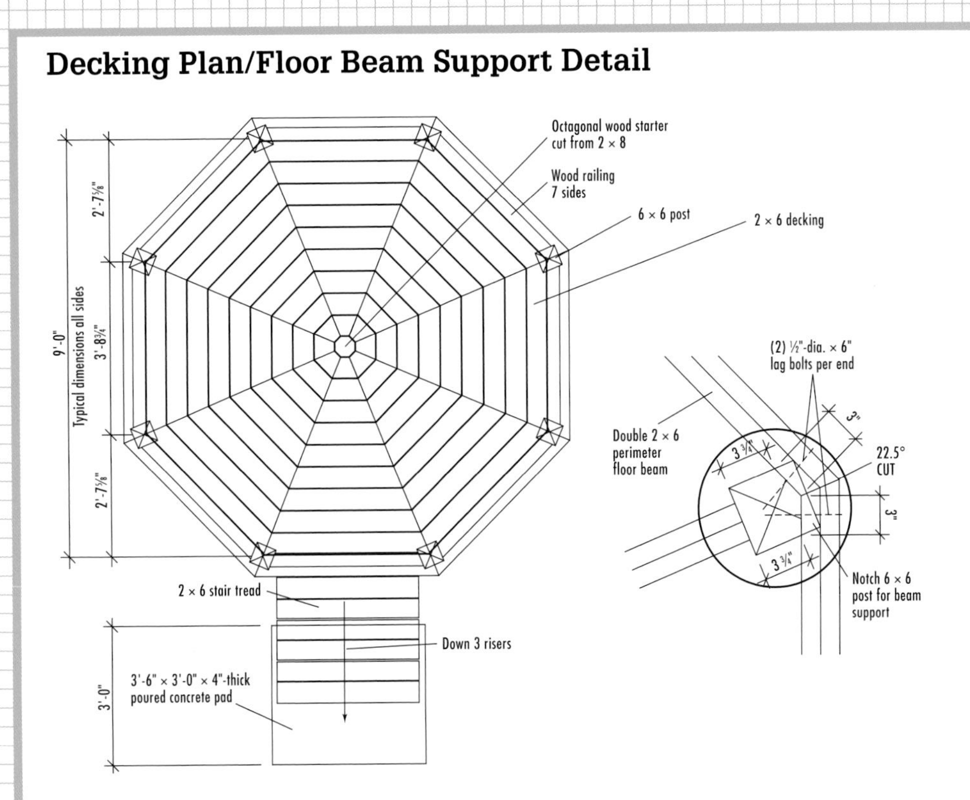

Stringer Template/ Stair Detail

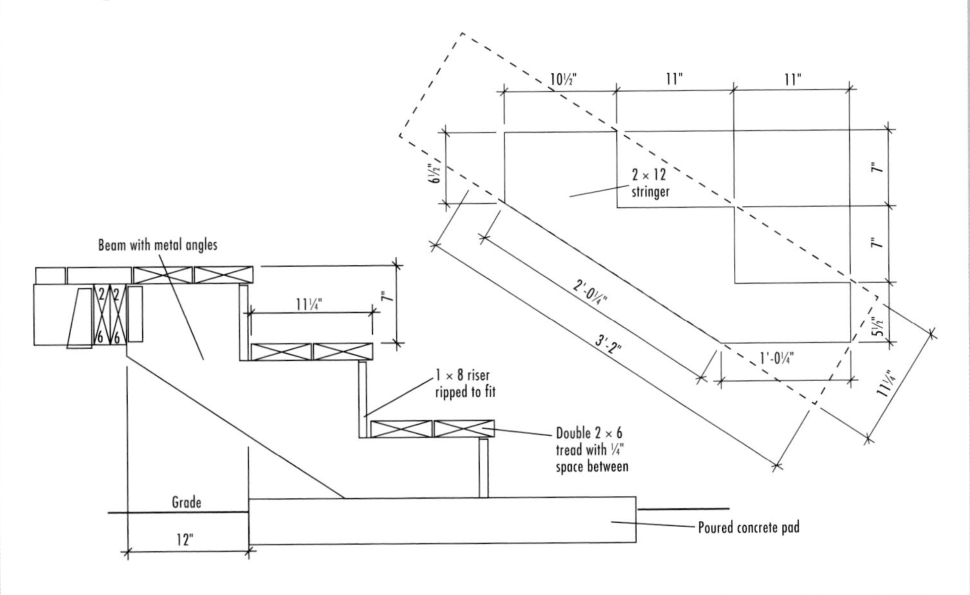

Detail at Deck Edge

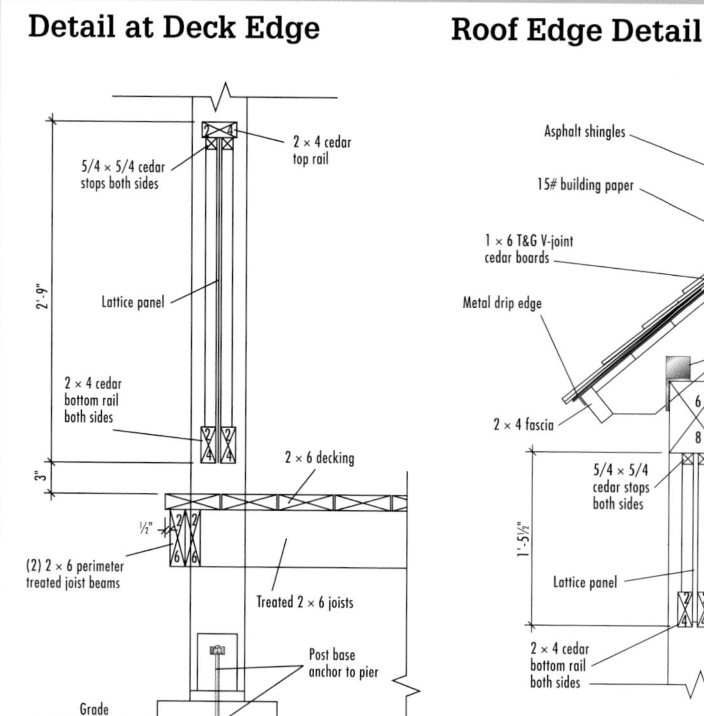

Roof Edge Detail

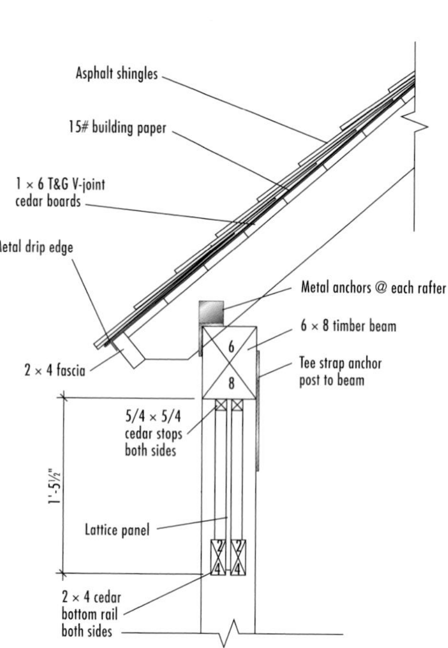

Rafter Hub Detail

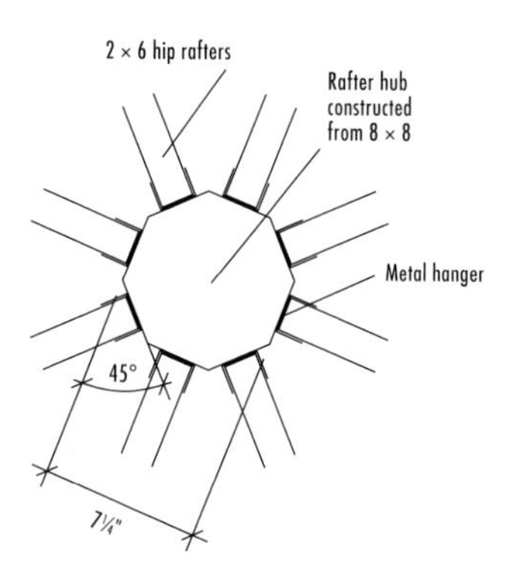

Corner Detail at Roof Beam Line

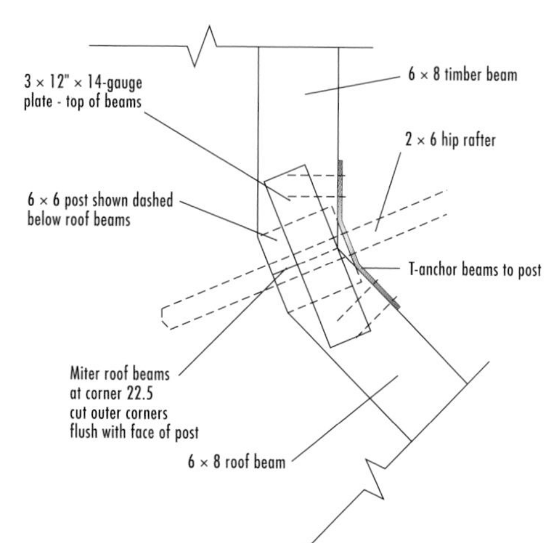

How to Build a Classic 8-sided Gazebo

STEP 1: POUR THE CONCRETE PIER FOOTINGS

Note: See pages 434 to 437 for instructions on laying out and pouring concrete pier footings. Use 12"-dia. cardboard tube forms for the eight outer piers and a 16"-dia. form for the center pier.

A. Set up batter boards in a square pattern, and attach tight mason's lines to form a 9 × 9-ft. square. Take diagonal measurements to make sure the lines are square to one another. Attach two more lines that run diagonally from the corners and cross in the center of the square—this intersection represents the center of the center footing.

B. Measure 31⅝" in both directions from each corner and make a mark on a piece of tape attached to the line.

C. At each of the nine points, use a plumb bob to transfer the point to the ground, and mark the point with a stake. Remove the mason's lines.

D. Dig holes for the forms and add a 4" layer of gravel to each hole. Set the forms so the tops of the outer forms are 2" above grade and the center form is 11" above grade. Level the forms and secure them with packed soil. Restring the mason's lines and confirm that the forms are centered under the nine points.

E. Fill each form with concrete, and screed the tops. Insert a ⅝ × 12" J-bolt in the center of each form. Use a plumb bob to align the J-bolt with the point on the line layout. On the outer footings, set the bolts so they protrude ¾" to 1" from the concrete. On the center footing, set the bolt to protrude 5". Let the concrete cure completely.

STEP 2: SET THE POSTS

A. Use a straight board to mark reference lines for squaring the post anchors. Set the board on top of one of the outer footings and on the center footing. Holding the board against the same side of the J-bolts, draw a pencil line along the board across the tops of the footings. Do the same for the remaining footings.

B. Place a metal post anchor on each perimeter footing and center it over the J-bolt. Use a framing square to position the anchor so it's square to the reference line. Secure the anchor with washers and a nut.

C. Set each post in an anchor, tack it in place with a nail, then brace it with temporary crossbraces so that it's perfectly plumb. Secure the post to the anchor using the fasteners recommended by the manufacturer. *Note: You will cut the posts to length during the construction of the roof frame.*

STEP 3: INSTALL THE PERIMETER FLOOR BEAMS

A. Starting at one of the posts that will be nearest to the stairs, measure from the ground and mark the post at 20½". Draw a level line at this mark around all four sides of the post. Transfer this height mark to the other posts using a mason's line and a line level.

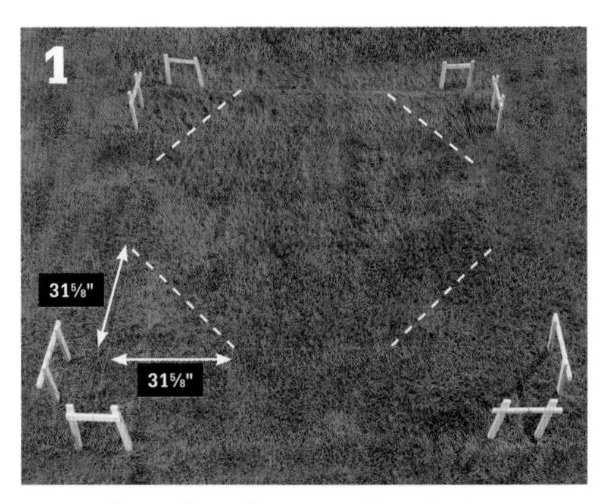

Measure in 31⅝" from the corners of the string layout to mark the centers of the outside piers.

Set a board across the center and each outer footing and mark a line across the top of the outer footing.

(continued)

These marks represent the tops of the 2 × 6 perimeter beams and the double joists of the floor frame.

B. Measure down 5½" from the post marks and make a second mark on all sides of each post. Notch the outer posts to accept the inner member of the perimeter floor beams, as shown in the FLOOR BEAM SUPPORT DETAIL on page 327 using a handsaw or circular saw and a chisel.

C. Cut the inner members of the perimeter floor beams to extend between the centers of the notches of adjacent posts, angling the ends at 22½°. Set the members into the notches and tack them to the posts with two 16d galvanized common nails.

D. Cut the outer members of the perimeter beams to fit around the inner members, angling the ends at 22½° so they fit together at tight miter joints (you may have to adjust the angles a little). Anchor the perimeter beams to the posts with two ½ × 6" lag screws at each end, as shown in the FLOOR BEAM SUPPORT DETAIL. Fasten the inner and outer beams together with pairs of 10d galvanized common nails driven every 12".

STEP 4: INSTALL THE DOUBLE JOISTS

A. Fasten metal hangers to the inside centers of the posts so the tops of the joists will be flush with the upper line drawn in Step 3 (also, see the FLOOR FRAMING PLAN on page 324).

B. Cut two 2 × 6 joists to span between two opposing posts, as shown in the FLOOR FRAMING PLAN. Nail the joists together with pairs of 10d galvanized common nails spaced every 12".

C. Set the double joist into the hangers and leave it in place while you build and fit the wood pad that supports the joists at the center pier (see the CENTER PIER DETAIL on page 323).

D. Cut two 2 × 12 plates—one from two of the boards you'll use for the stair stringers—and cut a shim at 11¼". Use treated plywood or treated lumber for the shim (if necessary, sand a lumber shim to the correct thickness with a belt sander.) Test-fit the pad, then remove the joist.

E. Fasten together the plates and shim with 16d galvanized nails. Drill a counterbored hole for the anchor nut and washer into the top plate, then drill a ⅝" hole through the center of the plates and shim. Secure the pad to the pier with construction adhesive, anchor nut, and washer.

F. Install the double joist, fastening it to the hangers with the recommended nails and toenailing it to the center pad with 10d galvanized nails.

G. Cut and assemble two double joists that run perpendicular to the full-length double joist. Install the joists at the midpoint of the full-length joist, toenailing them to the joist and pad.

H. Cut the remaining four double joists so their inside ends taper together at 45°. Install the joists following the FLOOR FRAMING PLAN.

STEP 5: INSTALL THE ANGLED FLOOR JOISTS

A. Mark the perimeter beam 11" from the post sides to represent the outside faces of the sixteen floor joists. Then, measure from the inside face of each post

Cut the post notches by making horizontal cuts with a handsaw or circular saw, and then remove the remaining material with a chisel.

Miter the ends of four of the double joists so they meet flush with the full-length joist and those perpendicular to it.

toward the center and mark both sides of the double joists at 25"—this mark represents the end of the angled joist.

B. Install metal joist hangers on the perimeter beams and skewable (adjustable) hangers on the double joists using the recommended fasteners.

C. Cut and install the 2 × 6 angled floor joists, following the hanger manufacturer's instructions.

STEP 6: POUR THE STAIR PAD

A. Using stakes or mason's line, mark a rectangular area that is 39 × 49", positioning its long side 10½" from the perimeter beam. Center the rectangle between the two nearest posts.

B. Excavate within the area to a depth of 7". Add 4" of compactable gravel and tamp it thoroughly.

C. Build a form from 2 × 4 lumber that is 36 × 42" (inner dimensions). Set the form with stakes so that the inside face of its long side is 12" from the perimeter beam and the form is centered between the nearest posts. Make sure the top of the form is level and is 19½" from the top of the perimeter beam.

D. Fill the form with concrete and screed the top flat with a 2 × 4. Float the concrete, if desired (see pages 32 to 33), and add a broomed or other textured finish for a slip-resistant surface. Round over the edges of the pad with a concrete edger. Let the concrete cure, and then remove the form and backfill around the pad with soil or gravel.

STEP 7: BUILD THE STAIRS

Note: The STRINGER TEMPLATE, on page 327, is designed for a gazebo that measures 21" from the stair pad to the top of the floor deck. If your gazebo is at a different height, adjust the riser dimension of the steps to match your project: divide the floor height (including the decking) by three to find the riser height for each step.

A. Use a framing square to lay out the first 2 × 12 stair stringer, following the STRINGER TEMPLATE: Starting at one end of the board, position the framing square along the top edge of the board. Align the 11" mark on the square's blade (long part) and the 7" mark on the tongue (short part) with the edge of the board. Trace along the outer edges of the blade and tongue, then use the square to extend the blade marking to the other edge of the board. The tongue mark represents the first riser.

B. Measure 1½" from the bottom mark and draw another line that is parallel to it—this is the cutting line for the bottom of the stringer (the 1½" is an allowance for the thickness of the treads of the first step).

C. Continue the step layout, starting at the point where the first riser mark intersects the top edge of the board. Draw lines for the tread of the first step and the riser of the second step. Repeat this process to draw one more step and a top cutting line.

D. Measure 10½" from the top riser and make a mark on the top cutting line. Draw a perpendicular line from the cutting line to the opposite edge of the board—this line represents the top end cut.

E. Cut the stringer and test-fit it against the stair pad and perimeter beam. Make any necessary adjustments. Using the stringer as a pattern, trace the layout onto the two remaining stringer boards, and then cut the stringers.

Fasten the angled floor joists to the sides of the double joists with skewable metal anchors.

Fill the 2 × 4 form for the stair pad with concrete, then screed the top with a straight piece of 2 × 4.

(continued)

F. Attach the stringers to the perimeter floor beam with metal angles, following the layout shown in the FLOOR FRAMING PLAN.

G. From scrap pressure-treated 2 × 4 lumber, cut kicker blocks to fit between the bottom ends of the stair stringers. Fasten the blocks to the concrete pad with construction adhesive and masonry screws or nails, then nail through the sides of the stringers into the kickers with 16d galvanized common nails.

STEP 8: INSTALL THE DECKING

A. Cut an octagonal starter piece from a cedar 2 × 8: Draw two lines across the board to make a 7¼ × 7¼" square. Make a mark 2⅛" in from each corner, and then connect the marks to form an octagon. Cut the starter piece and position it in the center of the floor frame with each point centered on a double joist. Drill pilot holes and attach the piece with 3" deck screws.

B. Cut the 2 × 6 deck boards for each row one at a time. The end cuts for each boards should be 22½°, but you may have to adjust the angles occasionally to make tight joints. Gap the boards, if desired, but make sure the gaps are consistent—use scrap wood or nails as spacers. Drill pilot holes and drive two screws wherever a board meets a framing member. Periodically measure to make sure the boards are parallel to the perimeter beams. Overhang the perimeter beams by ½" with the outer row of decking.

C. Install the 2 × 6 treads and 1 × 8 riser boards on the stairs following the STAIR DETAIL, on page 327.

STEP 9: SET THE ROOF BEAMS

A. Measure up from the floor deck and mark one of the posts at 101½". Transfer that mark to the remaining posts using a mason's line and a line level. Mark a level cutting line around all sides of each post, then cut the posts with a reciprocating saw or handsaw (see page 369).

B. On the top of each post, draw a line down the middle that points toward the center of the structure. Cut each of the four 6 × 8 roof beams in half so you have eight 5-ft.-long beams.

C. Set each roof beam on top of two neighboring posts so its outside face is flush with the outside corners of the posts. Mark the inside face of the roof beam where it meets the post centerlines—these marks represent cuts at each end (see the CORNER DETAIL AT ROOF BEAM LINE, on page 328). Also mark the underside of the beam by tracing along the outside faces of the posts—these lines show you where to trim the beams so they will be flush with the outside post faces. Use a square to extend the marks down around the post sides to help keep your cuts straight.

D. Starting from the end-cut marks, cut the beam ends at 22½°. Trim off the corners at the underside

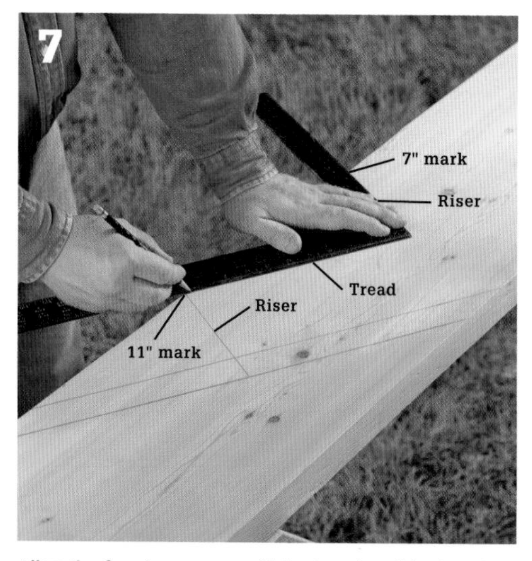

Align the framing square with the top edge of the board. Make the 11" tread mark by tracing along the square's blade, the riser mark along the tongue.

Install the decking by completing one row at a time.

marks. Mark and cut the remaining beams, test-fitting the angles as you go.

E. Install the beams, securing them to the posts with metal T-anchors. Bend the side flanges of the anchors, as shown in the CORNER DETAIL AT ROOF BEAM LINE, and fasten the anchors with the recommended fasteners. Tie the beams together with galvanized metal plates fastened with 16d galvanized box nails.

STEP 10: INSTALL THE HIP RAFTERS

A. Cut the roof hub from an 8 × 8 post, following the RAFTER HUB DETAIL on page 328. You can have the hub cut for you at a woodworking shop or cut it yourself using a table saw or circular saw. Cut the post at 16", then mark an octagon on each end: make a mark 2⅛" in from each corner and then join the marks. The cuts are at 45°. If you use a circular saw, extend the cutting lines down the sides of the post to ensure straight cuts.

B. Draw a line around the perimeter of the hub, 3½" from the bottom end. Center a metal anchor on each hub side, with its bottom flush to the line, and fasten it to the hub using the recommended nails.

C. Cut two pattern 2 × 6 hip rafters, following the RAFTER TEMPLATES on page 326. Tack the rafters to opposing sides of the hub and test-fit the rafters on the roof beams. The bottom rafter ends should fall over the post centers. Make any necessary adjustments to the rafter cuts.

D. Use a pattern rafter to mark and cut the six remaining hip rafters. Install the rafters, toenailing the bottom ends to the roof beams with one 16d common nail on each side. Fasten the top ends to the hangers with 1½" galvanized joist hanger nails, then install metal hangers at the bottom rafter ends.

STEP 11: INSTALL PURLINS & INTERMEDIATE RAFTERS

A. On each side of each hip rafter, measure up from the cut edge at the lower rafter end and make a mark at 51¾"—these marks represent the lower faces of the purlins (see the ROOF FRAMING PLAN on page 326; the BUILDING SECTION on page 325; and the RAFTER TEMPLATES).

B. Cut the 2 × 6 purlins, beveling the ends at 22½°. Position them between the rafters so their top edges are flush with the top edges of the rafters. Endnail or toenail each purlin to a rafter with 16d common nails.

C. Mark the layout for the intermediate rafters onto the tops of the roof beams, following the ROOF FRAMING PLAN.

D. Cut a pattern intermediate rafter, following the RAFTER TEMPLATES. Test-fit the rafter and make any necessary adjustments. Use the pattern rafter to mark and cut the fifteen remaining rafters.

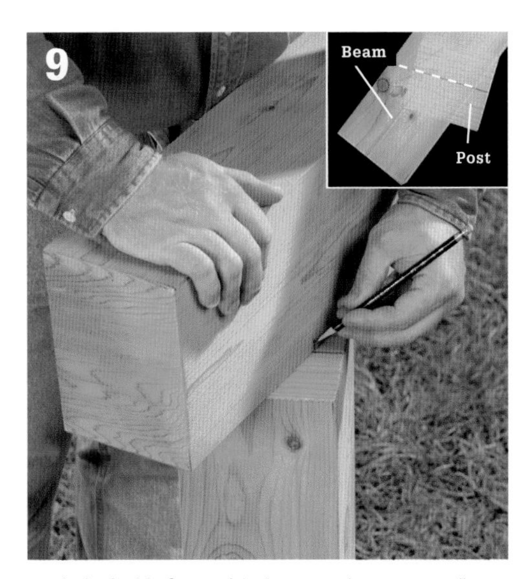

Mark the inside faces of the beams at the post centerlines. Mark the beam undersides along the outside post faces.

Attach the rafters to the hangers on the roof hub, driving the nails at a slight angle, if necessary.

(continued)

E. Install the rafters, endnailing their top ends to the purlins and toenailing their bottom ends to the roof beams with 16d nails. Install metal anchors to secure the bottom rafter ends to the roof beams.

STEP 12: INSTALL THE COLLAR TIES

A. Cut two 2 × 6 collar ties to span between the outsides of the roof beams. Clip the top corners of the collar ties so they don't project above the top edges of the intermediate rafters.

B. Install the ties to the outside faces of neighboring intermediate rafters, as shown in the ROOF FRAMING PLAN—it doesn't matter which rafters you use, as long as the basic configuration matches the plan. Fasten the collar ties with 10d nails.

C. Set two uncut 2 × 6 collar ties on top of, and perpendicular to, the installed collar ties so both ends extend beyond the intermediate rafters on opposing sides of the roof. Mark the ends of the ties by tracing along the top rafter edges.

D. Cut the marked collar ties, and then clip the top corners. Fasten the collar ties to the outside faces of the intermediate rafters with 10d nails.

STEP 13: ADD THE FASCIA & ROOF SHEATHING

A. Cut the 2 × 4 fascia, mitering the ends at 22½°. Install the fascia with its top edges ¾" above the rafters so it will be flush with the roof sheathing. Use 16d galvanized casing nails.

B. Install the 1 × 6 tongue-and-groove roof sheathing, starting at the lower edge of the roof. Angle-cut the ends of the boards at 22½°, cutting them to length so their ends break on the centers of the hip rafters. Fit the tongue-and-groove joints together, and facenail the sheathing to the hip and intermediate rafters with 8d galvanized box nails.

STEP 14: INSTALL THE ROOFING

A. Install metal drip edge along the bottom edges of the roof, angle-cutting the ends.

B. Lay 15# building paper over the sheathing and drip edge. Overlap the paper at each hip by 6".

C. Install the asphalt shingles on one section of the roof at a time. Trim the shingles flush with the hip ridges.

D. Cover the hip ridges with manufactured cap shingles or caps you cut from standard shingles.

E. Piece in metal flashing around the roof hub, and seal all flashing seams and cover all exposed nail heads with roofing cement.

F. Install the wood sphere on the center of the roof hub using a large dowel screw.

STEP 15: BUILD THE OVERHEAD LATTICE SCREENS

A. On the side faces of each post, mark the center of the post width. Then measure over, toward the

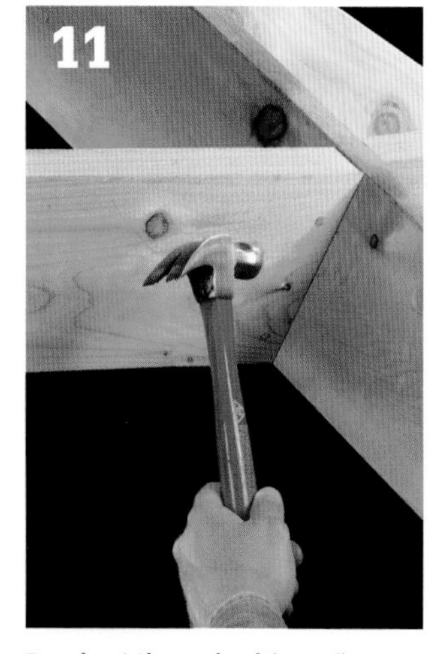

Bevel-cut the ends of the purlins so they meet flush with the rafter faces, and install them between the hip rafters.

Install the collar ties so that the upper pair rest on top of, and are perpendicular to, the lower pair.

Miter the ends of the sheathing boards and make sure the tongue-and-groove joints are tight before nailing.

gazebo center, one-half the thickness of the lattice panels and make a second mark. Use a level to draw a plumb line, starting from the second mark and extending down 17½" from the roof beam. Draw a level line across the post face at the end of the vertical line (at the 17½" mark). Also, snap a chalk line between the vertical lines on the underside of the beams. These will guide the placement of the top inner stops.

B. Cut a cedar 2 × 4 rail to span between each set of posts so the bottom rail edge is on the level line and the side face is on the plumb line. Bevel the ends at 22½°. Fasten the rails to the posts with 3" deck screws.

C. Cut 5/4 × 5/4 (about 1⅛ × 1⅛" actual dimensions) cedar inner stops to span between posts underneath the roof beams. Bevel the ends at 22½° and fasten the stops to the beams with 8d galvanized finish nails so their side faces are flush to the chalk lines.

D. The vertical stops of the overhead screens and the screens below the railings (Step 16) are 5/4 × 5/4s that have one edge beveled at 22½°. It will save time to rip all of them at once using a table saw, if available. You'll need about 110 linear feet.

E. Cut and install the inner vertical stops with their sides flush to the plumb lines drawn on the posts.

F. Cut eight lattice panels at 16 × 39⅝". Set the panels against the inner stops and rails, and fasten them with 3d galvanized finish nails.

G. Cut and install the outer rails and stops to complete the screens. Fasten the rails with 3" deck screws driven through the inner rails, and fasten the stops with 8d galvanized finish nails driven into the posts and beams.

STEP 16: BUILD THE RAILINGS & LOWER LATTICE

A. Measure up from the deck and mark the side faces of each post at 3" and 36". Draw level lines across the faces at these marks. Draw a plumb line between the level marks by finding the post center and moving inward one-half the thickness of the lattice, as you did in Step 15.

B. Cut the 2 × 4 cedar top rails to fit between seven pairs of posts (skipping the two posts flanking the stairs), as shown in the DETAIL AT DECK EDGE on page 328. Miter the rail ends at 22½° and install them with 3" deck screws so they are centered on the posts and their top faces are on the upper level lines.

C. Cut and install the 2 × 4 inner bottom rails and 5/4 × 5/4 stops, following the procedure in Step 15.

D. Cut the lattice panels at 31 × 39⅝". Fasten the panels against the stops and lower rails with 3d galvanized finish nails.

E. Cut and install the outer bottom rails, securing them with screws, then cut and install the outer horizontal and vertical stops.

Shingle the roof sides individually, then cover the hip ridges with caps, overlapping the shingles equally on both sides.

Install the inner stops and rails on the layout lines. The vertical stops are beveled at 22½° (inset).

Set the lattice panels against the inner stops and rails, and fasten them with 3d galvanized finish nails.

Pool Pavilion

A pavilion adds casual grandeur to your backyard. Defined by a stately roof, the project featured here has four open sides that invite entry from all directions and offer open views from the shaded interior. This structure also invites customization. For example, you could add curtains or shades to the beams to increase your privacy and create an elegant atmosphere.

Of course, a backyard pool is not required. The structure's simple design makes it suitable for a variety of uses. As shown in the plans, its four support posts are buried in the ground with concrete, but you can adapt the project for an existing patio by using post bases and adding structural supports to the post-and-beam frame for lateral strength (consult a professional for design modifications).

In addition to hanging curtains or shades from the beams, you might consider installing a set of weather-resistant cabinets and a stone or metal countertop to create an outdoor kitchen or bar.

A pool pavilion is the perfect place to get out of the sun, set out a summer meal, or gather with friends—with or without a pool!

Tools, Materials & Cutting List

Clamshell digger or power auger (available for rent)
Tape measure
Shovel
Level
Mason's trowel
Mason's string
Circular saw, reciprocating saw, or power miter saw
Line level
Clamps
Drill with bits
Ratchet wrench
Hammer
Table saw
Chalk line
Metal flashing
Caulk gun
Eye, ear, & glove protection

Description (No. finished pieces)	Quantity/Size	Material*
Posts (4)	4 @ field measure (114" minimum)	4 × 8
Post Piers		
Concrete tube forms	4—field measure for length	12"-diameter cardboard forms
Gravel	Field measure	Compactable gravel
Concrete	Field measure	3,000 PSI concrete
Beams		
Main beams (2)	2 @ 12'	4 × 12
Notched brackets (4)	1 @ 10'	4 × 4
Inner brackets (4)	1 @ 8'	2 × 4
Roof Frame		
Rim beams—2 × 6 (2)	2 @ 10'	2 × 6
Rim beams—2 × 8 (2)	2 @ 10'	2 × 8
Roof hub (1)	1 @ 2'	4 × 4
Hip rafters (4)	4 @ 8'	2 × 6
Intermediate rafters (4)	2 @ 12'	2 × 6
Roofing		
Sheathing	4 @ 4 × 8'	¾" exterior-grade plywood
Shingles and 15# building paper	150 square feet, plus ridge caps	
Hardware & Fasteners		
½ × 7" galvanized carriage bolts	16, with washers and nuts	
16d galvanized common nails		
3½" galvanized wood screws		
8d galvanized box nails		
Heavy-duty staples		
Roofing nails		
Metal roof flashing		
Roofing cement		

Note: All exposed lumber parts must be pressure-treated or a rot-resistant grade of lumber.

Elevation

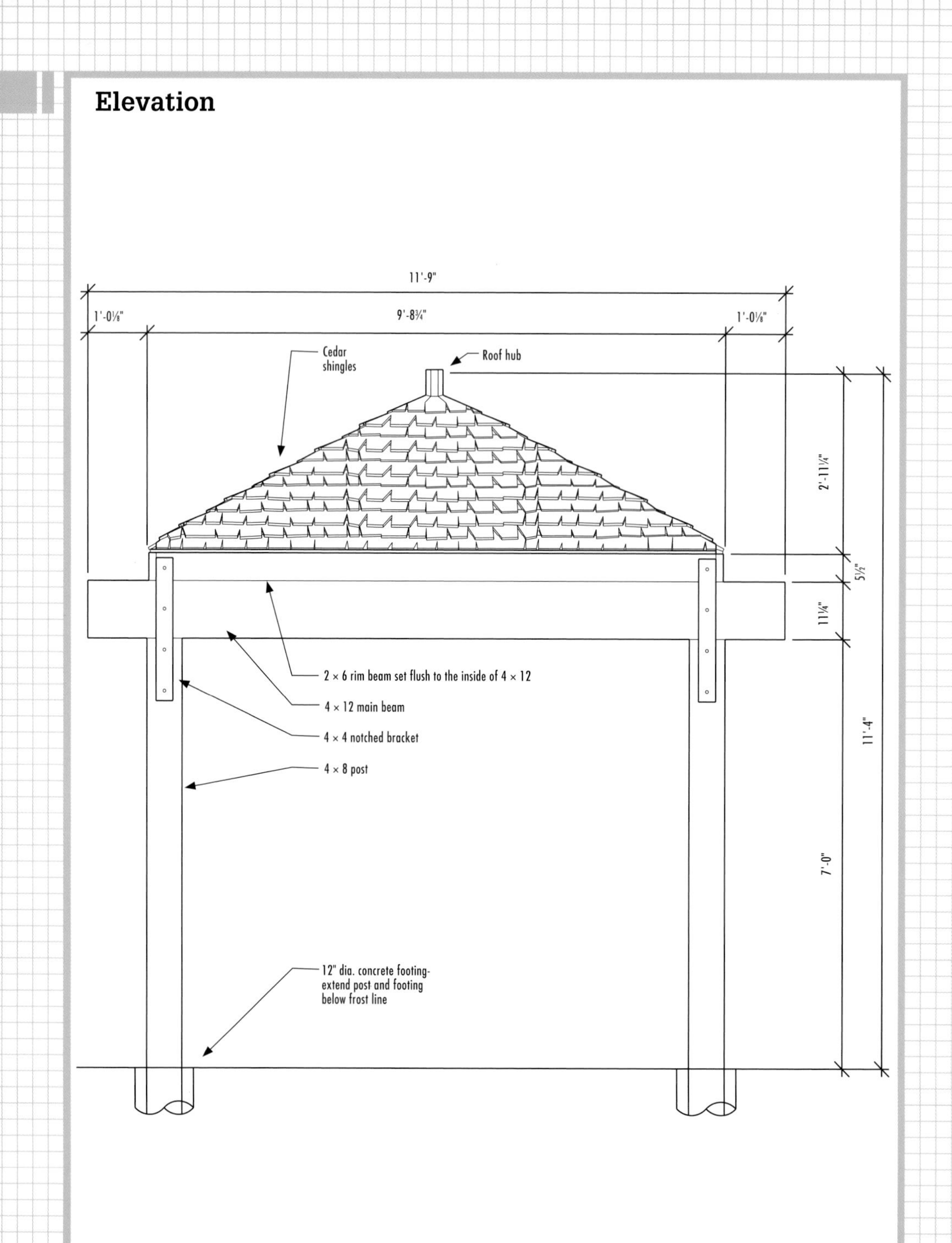

11'-9"
1'-0⅛"
9'-8¾"
1'-0⅛"
Cedar shingles
Roof hub
2'-11¼"
5½"
11¼"
11'-4"
7'-0"
2 × 6 rim beam set flush to the inside of 4 × 12
4 × 12 main beam
4 × 4 notched bracket
4 × 8 post
12" dia. concrete footing-extend post and footing below frost line

Building Section

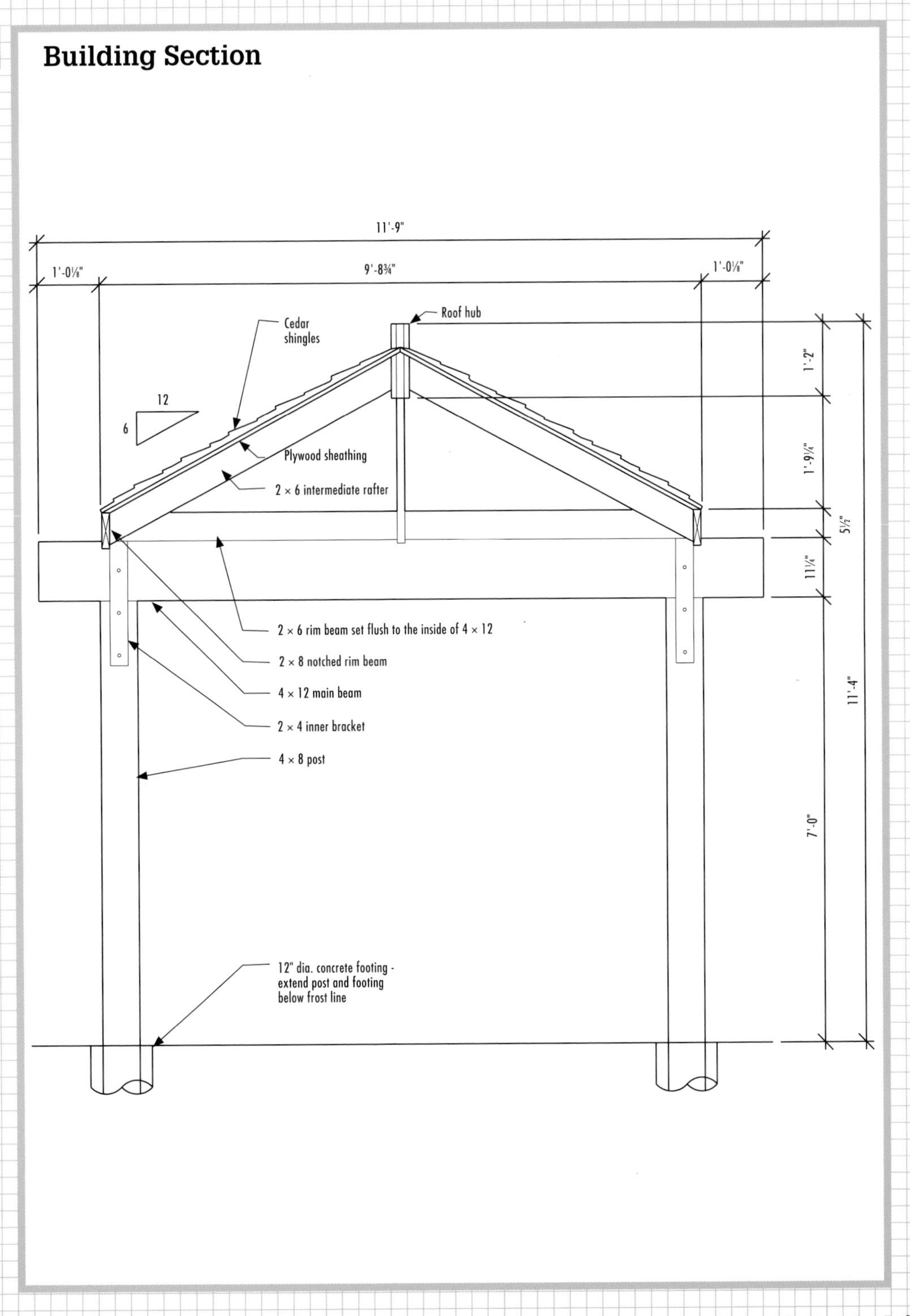

Floor Plan

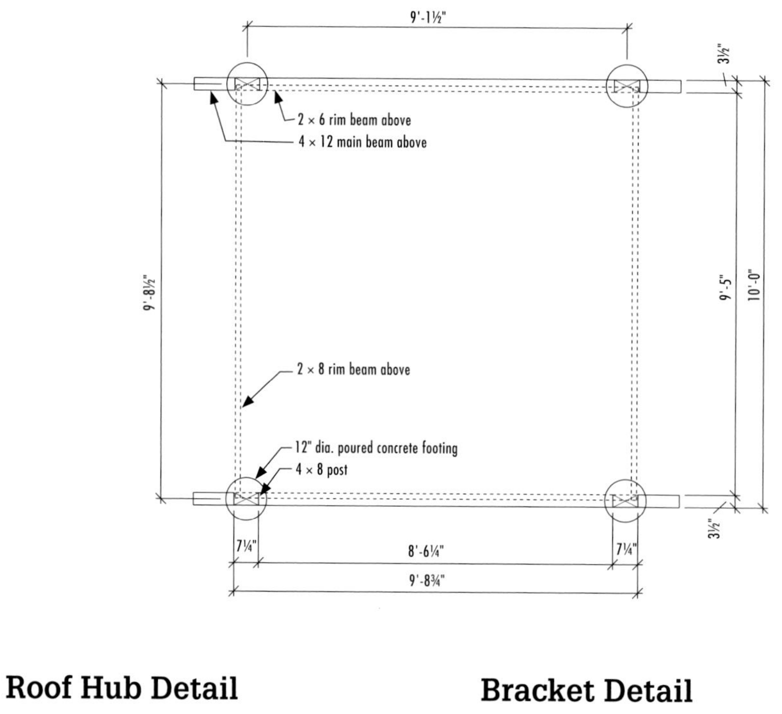

Roof Hub Detail

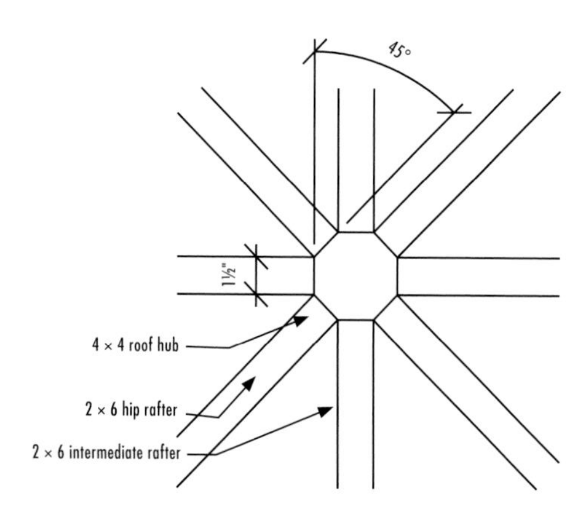

Bracket Detail

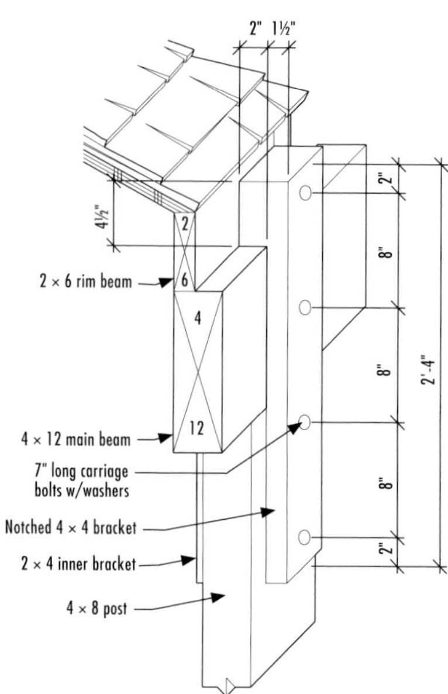

Roof Framing Plan

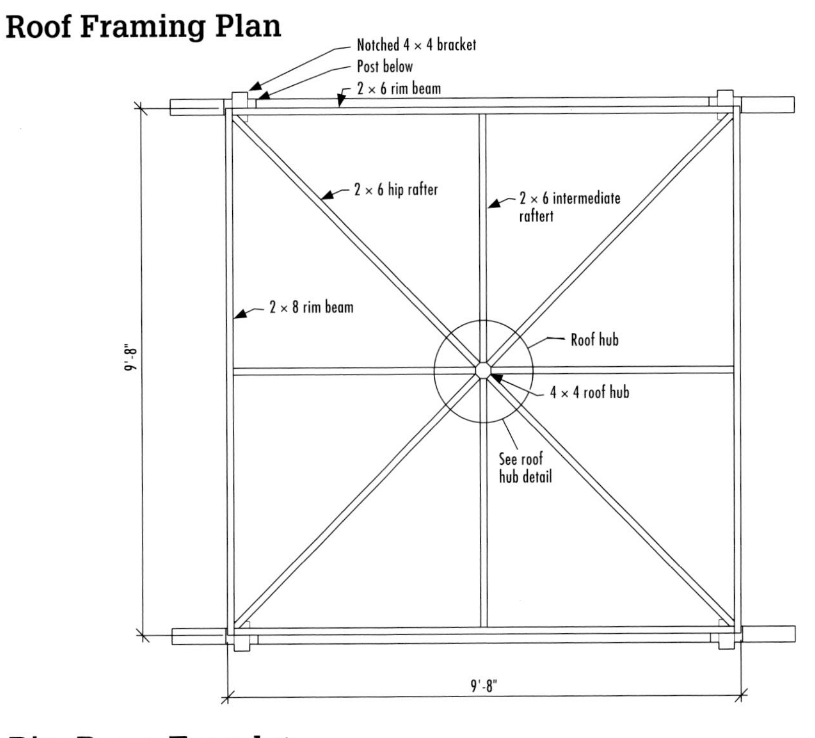

Rim Beam Template

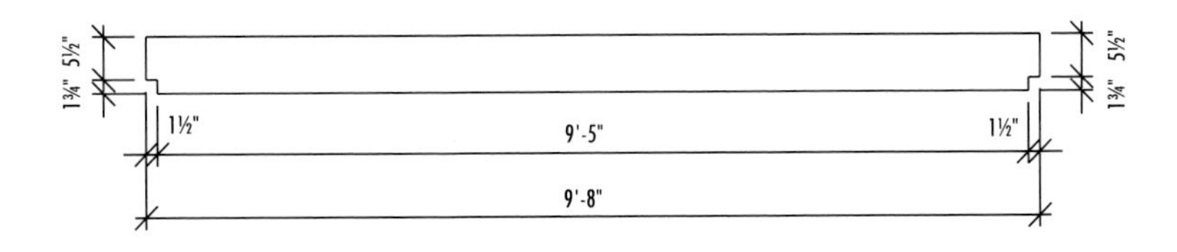

Intermediate Rafter Template

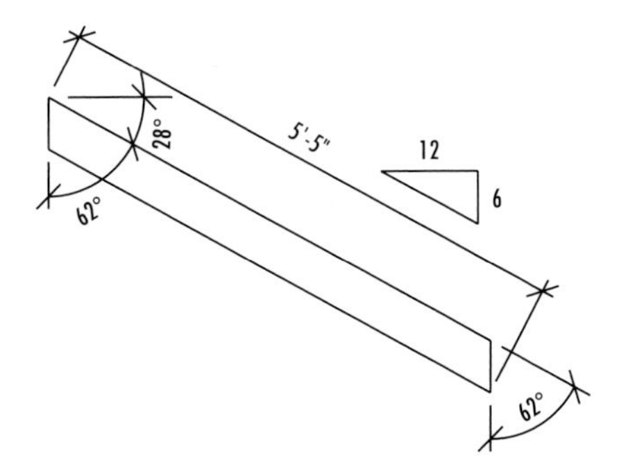

Hip Rafter Template

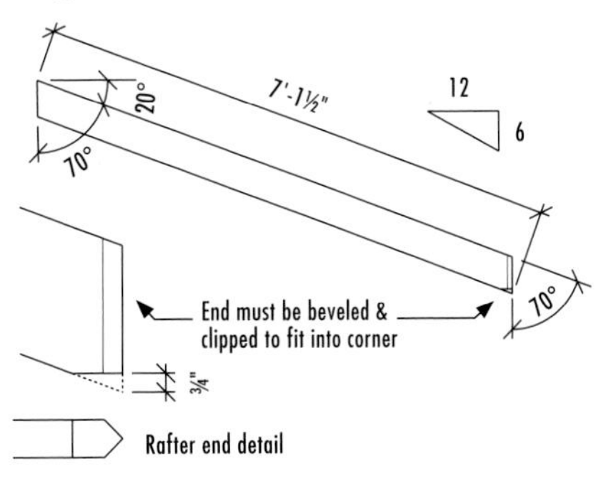

How to Build a Pool Pavilion

STEP 1: SET THE POSTS

The four 4 × 8 posts are buried in concrete piers made with 12"-diameter cardboard tube forms. The post depth must extend below the frost line and meet the requirements of your local Building Code. Treat the bottom ends of the posts for rot resistance (see page 438) before setting them.

A. Lay out the centers of the four post locations onto the ground; follow the FLOOR PLAN drawing on page 340. Diagonally measure between the posts to check for squareness: the layout is perfectly square when the diagonal measurements are equal.

B. Dig 15"-diameter holes for the post piers at the pier depth required for your area, plus 4". Fill the holes with a 4" layer of compactable gravel.

C. Cut the tube forms to length so they will slightly extend above the ground level. Set the tubes in the holes, hold them plumb, and then tightly pack around the outside with soil.

D. Place a post into each form and brace it in position with 2 × 4 cross bracing. Make sure the posts are perfectly plumb and the layout is square.

E. Fill the forms with concrete, smooth the tops of the piers, and let the concrete dry.

STEP 2: SET THE MAIN BEAMS

A. Mark one of the posts 84" above the ground. Using mason's string and a line level, transfer this height mark to the remaining posts. Cut the posts at the height mark (see page 369).

B. Cut four 4 × 4 notched bracket pieces to length at 28". Notch the brackets as shown in the BRACKET DETAIL on page 340.

C. Cut four 2 × 4 inner brackets to length at 23½".

D. Cut the two 4 × 12 main beams to length at 141". Set each beam on a pair of posts so it overhangs the posts equally at both ends.

E. Position a notched bracket on the outer surface over the beam and post so it's centered on the post. Position an inner bracket on the opposite side, centered on the post with its top end flush with the top of the beam. Clamp the pieces together and drill pilot holes for ½ × 7" carriage bolts; follow the layout shown in the BRACKET DETAIL. Anchor the parts with carriage bolts in the three lower positions only.

STEP 3: INSTALL THE RIM BEAMS

A. Cut the two 2 × 6 rim beams to length at 113".

B. Cut the two 2 × 8 rim beams to length at 116". Check all of the beams for crowning; always install them crown up. Notch the bottom corners of the 2 × 8 beams; follow the 2 × 8 RIM BEAM TEMPLATE on page 341.

C. Fit the 2 × 8 rim beams over the main beams so their inside faces are flush with the brackets.

D. Set the 2 × 6 rim beams between the 2 × 8s, flush with the inside edges of the main beams. Make sure the rim beams are flush at the tops and fasten them together with 16d common nails driven through the 2 × 8s and into the ends of the 2 × 6s.

E. Anchor the 2 × 6 beams to the notched brackets using the top carriage bolt pilot hole. Countersink the washer and nut on the inside face of the 2 × 6, then cut off the bolt flush with the beam so it won't interfere with the hip rafters.

Set the posts in concrete. Cut off the exposed portion of the cardboard tube form after the concrete cures.

Anchor the main beams to the posts with two brackets secured with carriage bolts.

Endnail through the notched beam into the 2 × 6 rim beams, (not seen), and then anchor the 2 × 6s with carriage bolts.

(continued)

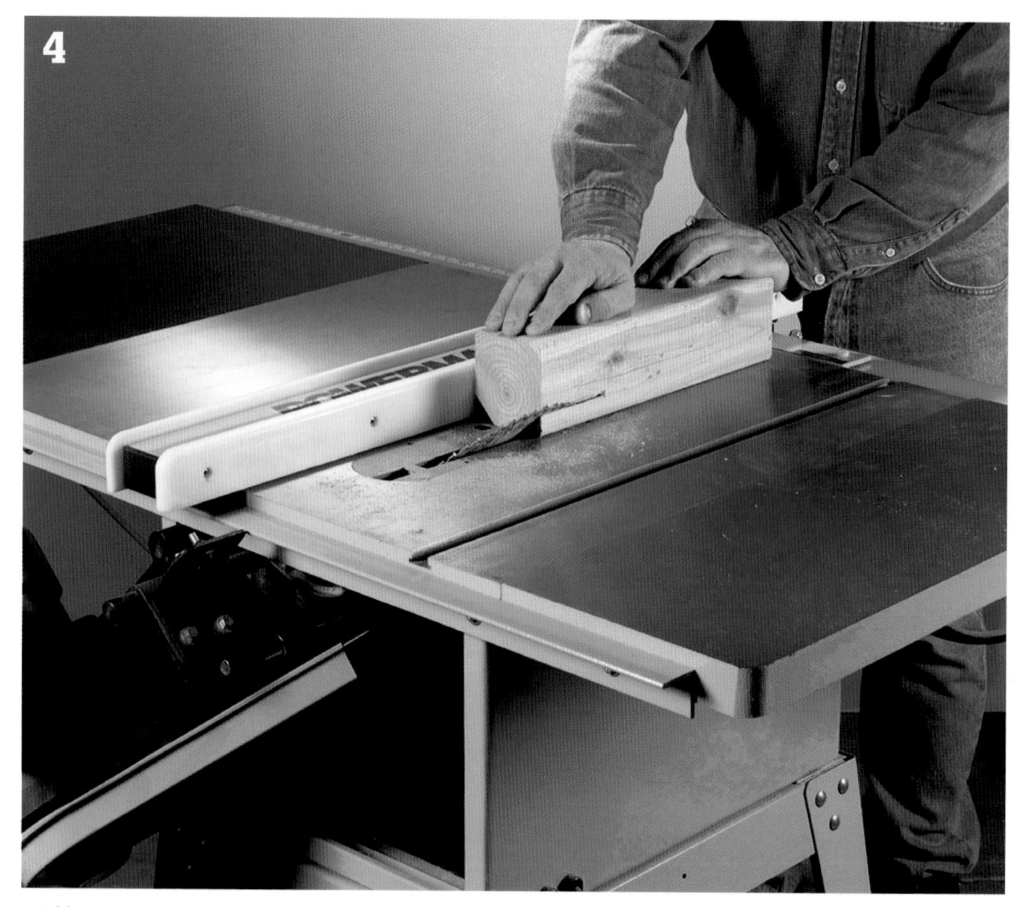

A table saw is the best tool for making the four cuts to create the octagonal roof hub.

STEP 4: CUT THE ROOF HUB

With the blade of a table saw or circular saw set at a 45° angle, four cuts turns a 4 × 4 into an octagon.

Tip: Use a scrap piece of 4 × 4 to set up the cuts, or use one end of the workpiece that you can cut off to create the finished hub.

A. If you're using a table saw, set the saw fence about 2½" from the blade. If you're using a circular saw, draw cutting lines down two adjacent sides about 1" from the corner edges.

B. Make a couple of test cuts, then measure the facets. Adjust the fence or cutting lines as needed so that all eight facets will be equal.

C. Make the final cuts down the full length of the hub. Cut the hub to length at 14".

Note: If the facets are slightly smaller than the thickness of the rafters, plane the rafter ends or taper them with a chisel to match.

STEP 5: FRAME THE ROOF

A. Select two straight 2 × 6s for the pattern hip rafters, and check the boards for crowning.

B. Cut the rafters following the HIP RAFTER TEMPLATE on page 341. The roof slope is 6-in-12. In addition to having a plumb cut on its bottom end, the hip rafter gets 45° bevel cuts so the end fits into the corner of the beams, as shown in the ROOF FRAMING PLAN on page 341. The bottom end also gets clipped to clear the 2 × 4 inner bracket.

C. Use the roof hub to test-fit the hip rafters against the rim beams. The tops of the rafters should be flush with the inside top edges of the rim beams. Make any necessary adjustments for a good fit.

D. Use one of the pattern rafters to mark the remaining two hip rafters, then make the cuts.

E. Install the hip rafters by screwing through the top and sides of each rafter into the roof hub using 3½" wood screws. The hub should extend about 6½" above the tops of the rafters. Fasten the bottom ends of the rafters to the beams with 16d common nails. Alternate between opposing sides as you work to ensure the hub remains centered.

F. Repeat the cutting and test-fitting process to cut four 2 × 6 intermediate rafters; follow the INTERMEDIATE RAFTER TEMPLATE on page 341. Install the intermediate rafters so they're flush with the hip rafters at the hub and flush with the tops of the rim beams at the bottom ends. Also make sure the bottom ends are centered along the lengths of the rim beams.

Drive angled screws to fasten the rafters to the roof hub.

STEP 6: SHEATH & SHINGLE THE ROOF

As shown in the plans, the pavilion roof is sheathed with ¾" exterior-grade plywood. For a more attractive ceiling surface—and one that is less likely to show nails coming through—you can use 5/4 cedar decking instead of plywood.

A. Install plywood sheathing from the bottom up. To lay out the plywood cuts, measure from the intermediate rafter center to the hip rafter center at the bottom ends. Mark the intermediate rafter 4 feet from the rim beam, and measure straight across to the center of the hip rafter. Transfer the two dimensions to a sheet of plywood, measuring from the square factory edge.

B. Snap a chalk line between the marks. Cut the plywood along the line.

C. Position the plywood so its bottom edge is butted against the rim beam and its side edges break on the centers of the rafters. Fasten the sheathing to the rafters with 8d box nails spaced every 6".

D. Use the angled edge of the leftover plywood and make a square cut for the next piece of sheathing. When the bottom row of sheathing is finished, use the cutoffs to fill in the top row up to the hub.

E. Install 15# building paper and cedar shingles, following the steps on pages 440-441. Install the shingles so their top ends run long over the ridges of the roof, then trim them off with a saw. Cap the ridges with premade shingle caps or with 1× cedar boards. Use metal flashing and roofing cement to seal around the roof hub.

Align premade or site-built shingle caps over ridges, working from the bottom up.

Patio Shelter

If you have the perfect patio adjoining your home, but you'd like some protection from the rain and strong winds, this stylish, contemporary patio shelter may be just what you're looking for. Designed as a cross between an open-air arbor or pergola and an enclosed three-season porch, this structure has clear glazing panels on its roof and sides, allowing plenty of sunlight through while buffering the elements and even blocking harmful UV rays.

The roof of the patio shelter is framed with closely spaced 2 × 4 rafters to create the same light-filtering effects of a slatted arbor roof. The rafters are supported by a doubled-up 2 × 10 beam and 4 × 6 timber posts. Because the shelter is attached to the house, the posts are set on top of concrete foundation piers, or footings, that extend below the frost line. This prevents any shifting of the structure in areas where the ground freezes in winter.

The patio shelter's side panels cut down on wind while providing a degree of privacy screening. Their simple construction means you can easily alter the dimensions or locations of the panels to suit your own plans. In the project shown, each side has two glazing panels with a 3½" space in between, for airflow. If desired, you can use a single sheet of glazing across the entire side section. The glazing is held in place with wood strips and screws so they can be removed for seasonal cleaning.

Slats of white oak sandwich clear polycarbonate panels to create walls that block the wind without blocking light and views.

Building against a solid wall makes the space inside this contemporary shelter much more usable. The corrugated roof panels (see Resources, page 442) made of clear polycarbonate allow light to enter while keeping the elements out.

Patio Shelter

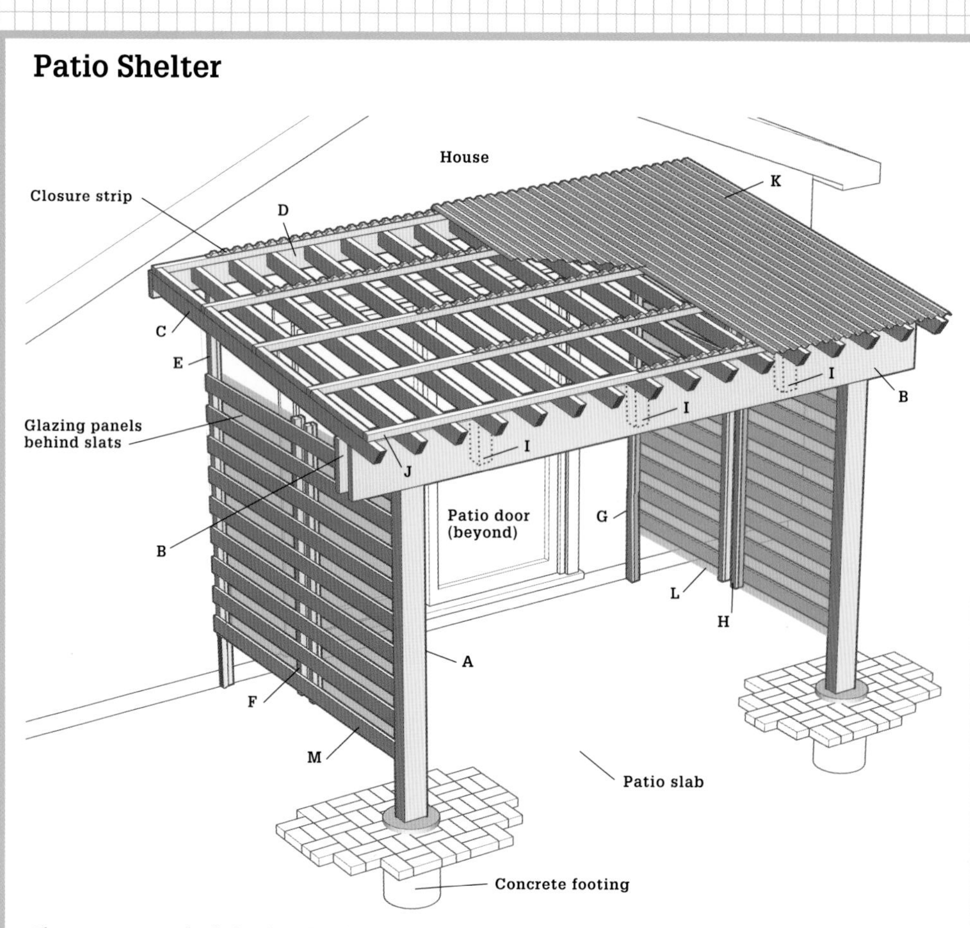

Plan your own patio shelter based on the requirements set by your local Building Code. Your city's building department or a qualified building professional can help you with the critical structural specifications, such as the size and depth of the concrete post footings, the sizing of beam members, and the overall roof construction. The building department will help make sure your shelter is suitable for the local weather conditions (particularly wind and snow loads).

Cutting List

Key	Part	No.	Size	Material
A	Post	2	$3\frac{1}{2} \times 5\frac{1}{2} \times 144$"	4 × 6 treated pine
B	Beam member	2	$1\frac{1}{2} \times 9\frac{1}{4} \times 120$"*	2 × 10 treated pine
C	Rafter	16	$1\frac{1}{2} \times 3\frac{1}{2} \times 120$"*	2 × 4 pine
D	Ledger	1	$1\frac{1}{2} \times 5\frac{1}{2} \times 144$"	2 × 6 treated pine
E	Back post	2	$1\frac{1}{2} \times 1\frac{1}{2} \times 96$"*	2 × 2 treated pine
F	Slat cleat	6	$1\frac{1}{2} \times 1\frac{1}{2} \times 62\frac{1}{2}$"	2 × 2 treated pine
G	Back post cap	2	$\frac{3}{4} \times 1\frac{1}{2} \times 96$"*	1 × 2 treated pine

Key	Part	No.	Size	Material
H	Slat cleat cap	6	$\frac{3}{4} \times 1\frac{1}{2} \times 62\frac{1}{2}$"	1 × 2 treated pine
I	Beam blocks	3	$3\frac{1}{2} \times 3\frac{1}{2} \times 8$"	4 × 4 treated pine
J	Purlin	7	$1\frac{1}{2} \times 1\frac{1}{2} \times 120$"	2 × 2 treated pine
K	Roof panel	6	26 × 96"	Corrugated polycarbonate
L	Side panel	4	$\frac{1}{4} \times 36 \times 58$"	Clear polycarbonate
M	Slat	18	$\frac{3}{4} \times 3\frac{1}{2} \times 80$"*	White oak

**Size listed is prior to final trimming*

Tools & Materials

Chalk line
4-ft. level
Plumb bob
Mason's string
Digging tools
Concrete mixing tools
Circular saw
Ratchet wrench
Line level
Reciprocating saw or handsaw
Table saw
Drill with bits
Finish application tools
Gravel
12"-dia. concrete tube forms
Concrete mix
⅝"-dia. J-bolts
⅜ × 4" corrosion-resistant lag screws
Flashing
Silicone caulk
Corrosion-resistant metal post bases and hardware
Scrap lumber
Lumber (see Cutting List, page 348)
Galvanized 16d and 8d common nails
½"-dia. galvanized carriage bolts and washers
Exterior wood glue or construction adhesive
Corrosion-resistant framing anchors (for rafters)
1½" and 3" deck screws
Corrugated polycarbonate roofing panels
Clear polycarbonate wall panels
Closure strips
Roofing screws with EPDM washers
Roofing adhesive/sealant
Wood finishing materials
Neoprene weatherstripping
Dark exterior wood stain
Stakes
Maul
Hammer
Eye and ear protection

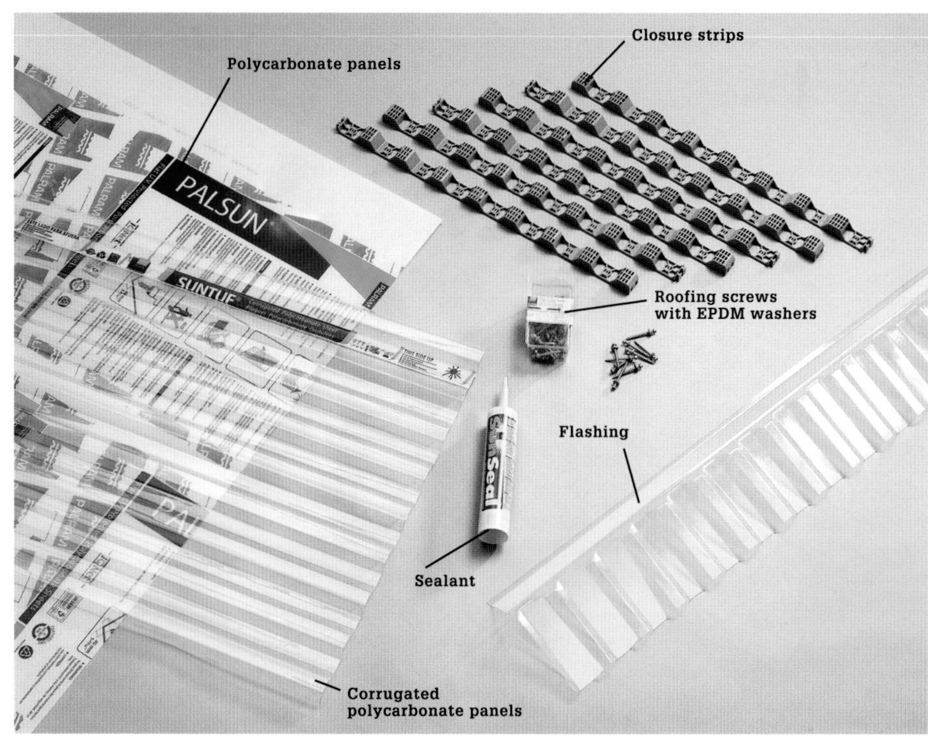

The roofing and side glazing panels of the patio shelter are made with tough polycarbonate materials. The corrugated roofing panels allow up to 90 percent light transmission while blocking virtually 100 percent of harmful UV rays. The flat side panels offer the transparency of glass but are lighter and much stronger than glass. Also shown are: wall flashing designed to be tucked under siding; closure strips that fit between the 2 × 2 purlins and the corrugated roof panels; self-sealing screws and polycarbonate sealant. For more information on these products, see Resources, page 442.

How to Build a Patio Shelter

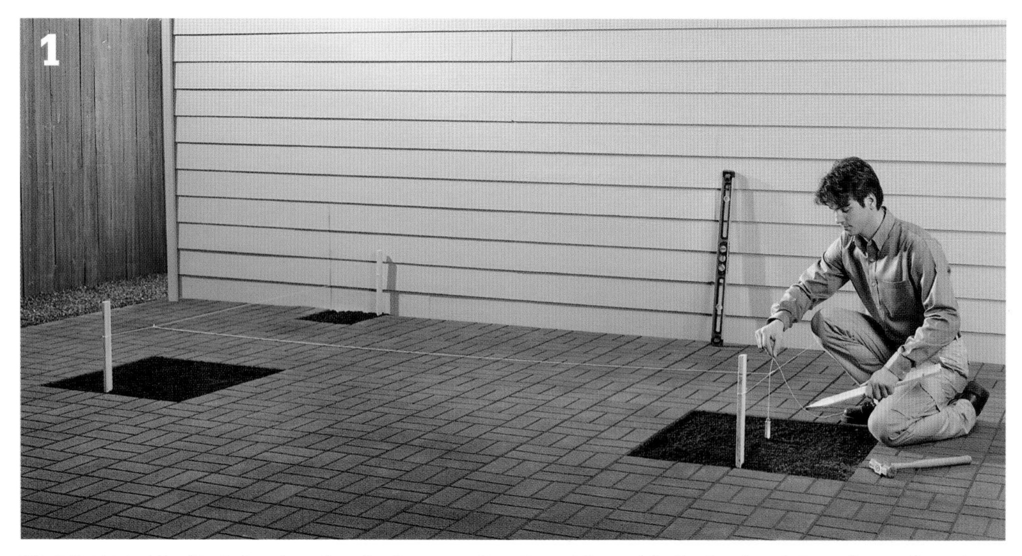

Mark the layout for the ledger board on the house wall, and lay out the post footing locations in the patio area. To mark the cutout for the ledger board, include the width of the ledger board, plus the height of the roofing, plus 1½" for the flashing. The length of the cutout should be 1" longer than the length of the ledger board (12 ft. as shown). Plumb down from the ends of the ledger, then measure in to mark the locations of the post centers. At each of these points, run a perpendicular string line from the house out to about 2 ft. beyond the post locations. Set up a third string line, perpendicular to the first two, to mark the centers of the posts. Plumb down from the string line intersections and mark the post centers on the ground with stakes.

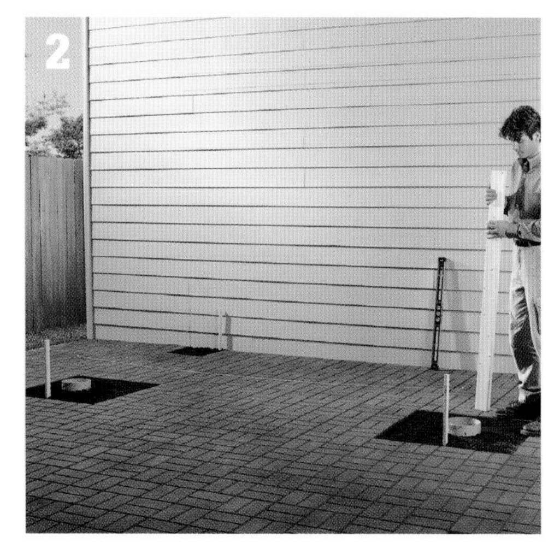

Dig a hole for a concrete tube form at each post location following the local Building Code for the footing depth (plus 6" for gravel). Add 6" of gravel and tamp it down. Position the tube forms so they are plumb and extend at least 2" above the ground. Backfill around them with soil and compact thoroughly.

Fill the tube forms with concrete and screed it level with the tops of the forms. At each post-center location, embed a J-bolt into the wet concrete so it extends the recommended distance above the top of the form. Let the concrete cure.

Cut out the house siding for the ledger board using a circular saw. Cut only through the siding, leaving the wall sheathing intact. *(Note: If the sheathing is fiberboard instead of plywood, you may have to remove the fiberboard; consult your local building department.)* Replace any damaged building paper covering the sheathing.

Stain the wood parts before you begin installing the shelter closure strips and panels. We used a black, semitransparent deck and siding stain.

Apply a protective finish to the wood slats as desired. We used a semitransparent deck stain.

(continued)

Install the ledger. First, slip corrugated roof flashing or metal roof flashing behind the siding above the ledger cutout so the vertical flange extends at least 3" above the bottom of the siding. Cut the ledger board to length. Fasten the ledger to the wall using ⅜ × 4" lag screws driven through counterbored pilot holes at each wall-stud location. Seal over the screw heads and counterbores with silicone caulk.

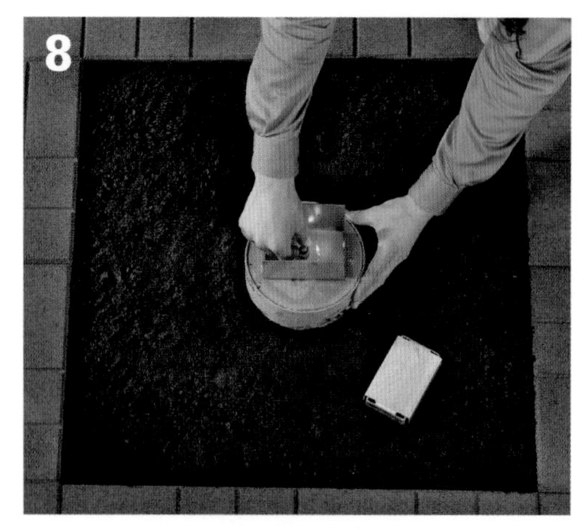

Anchor the post bases to the concrete footing, securing them with the base manufacturer's recommended hardware. Make sure the bases are aligned with each other and are perpendicular to the house wall.

Cut off the bottom ends of the posts so they are perfectly square. Set each post in its base and hold it plumb. Fasten the post to the base using the manufacturer's recommended fasteners. Brace the posts with temporary bracing. *Note: You will cut the posts to length in a later step.*

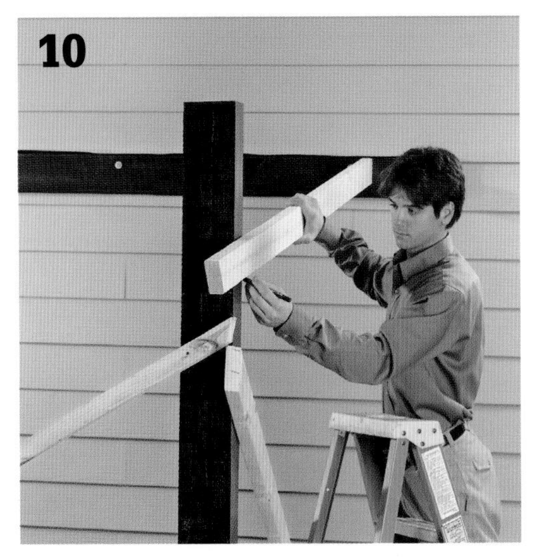

Cut a pattern rafter from 2 × 4 lumber using the desired roof slope to find the angle cut for the top end. Angle the bottom end as desired for decorative effect. Set the rafter in position so its top end is even with the top of the ledger and its bottom end passes along the side of a post. Mark along the bottom edge of the rafter onto the post. Repeat to mark the other post. Use a string and line level to make sure the post marks are level with each other.

Cut the inner beam member to length from 2 × 10 lumber, then bevel the top edge to follow the roof slope. Position the board so its top edge is on the post markings and it overhangs the posts equally at both ends (12" of overhang is shown). Tack the board in place with 16d nails.

Cut the outer beam member to length from 2 × 10 lumber. Bevel the top edge following the roof slope, and remove enough material so that the bottom edges of the two beam members will be level with each other. Tack the member in place with nails.

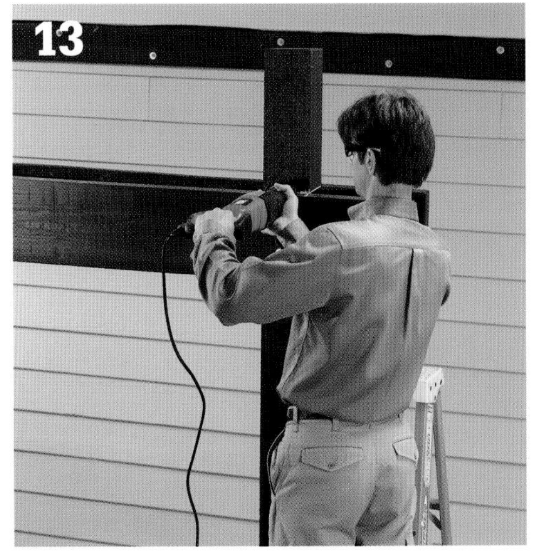

Anchor the beam members together and to the posts with pairs of ½"-dia. carriage (or machine) bolts and washers. Cut the posts off flush with the tops of the beam members using a handsaw or reciprocating saw.

(continued)

Trim the cutoff post pieces to length and use them as blocking between the beam members. Position the blocks evenly spaced between the posts and fasten them to both beam members with glue and 16d nails. *Note: Diagonal bracing between the posts and beam may be recommended or required in some areas; consult your local building department.*

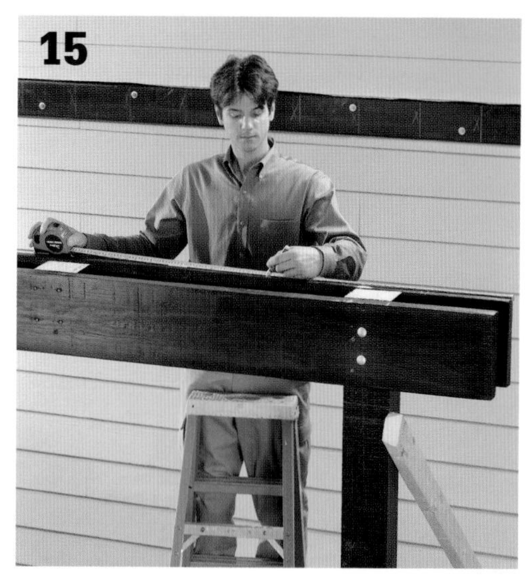

Mark the rafter layout onto the ledger and beam. As shown here, the rafters are spaced 9½" apart on center. The two outer rafters should be flush with the ends of the ledger and beam.

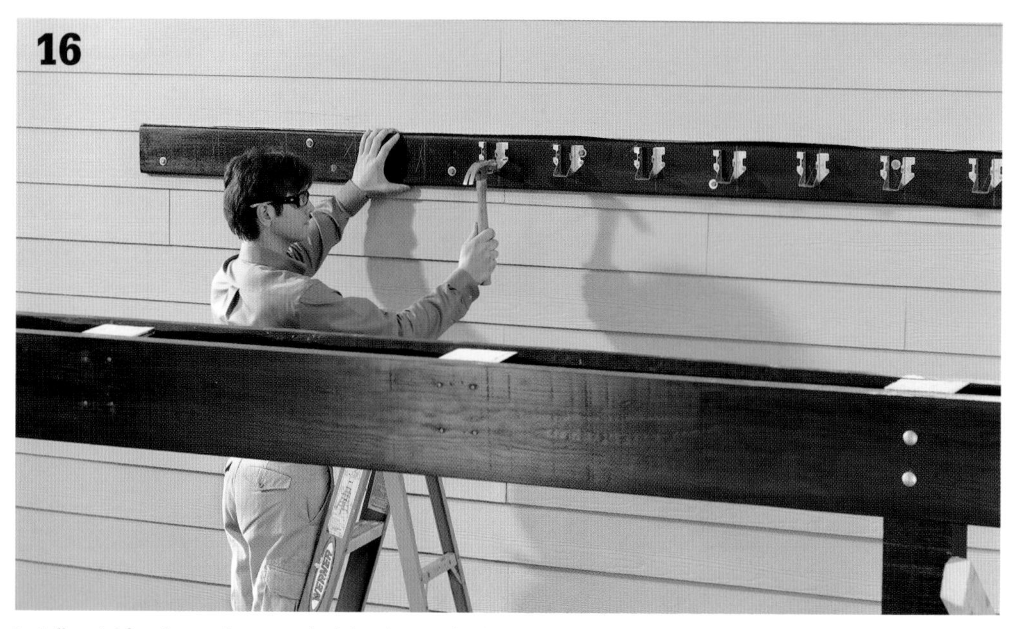

Install metal framing anchors onto the ledger for securing the top rafter ends using the anchor manufacturer's recommended fasteners. Use the pattern rafter or a block to position the anchors so the rafters will be flush with the top of the ledger.

Use the pattern rafter to mark the remaining rafters, and then cut them. Install the rafters one at a time. Fasten the top ends to the metal anchor using the recommended fasteners. Fasten the bottom ends to both beam members by toenailing one 8d nail through each rafter side and into the beam member.

Install the 2 × 2 purlins perpendicular to the rafters using 3" deck screws. Position the first purlin a few inches from the bottom ends of the rafters. Space the remaining purlins 24" on center. The ends of the purlins should be flush with the outside faces of the outer rafters.

Add 2 × 2 blocking between the purlins along the outer rafters, and fasten them with 3" deck screws. This blocking will support the vertical closure strips for the roof panels.

Starting at one side of the roof, install the roof panel closure strips over the purlins using the manufacturer's recommended fasteners. Begin every run of strips from the same side of the roof, so the ridges in the strips will be aligned.

(continued)

Add vertical closure strips over the 2 × 2 purlin blocking to fill in between the horizontal strips.

Position the first roofing panel along one side edge of the roof. The inside edge of the panel should fall over a rafter. If necessary, trim the panel to length or width following the manufacturer's recommendations.

Drill pilot holes, and fasten the first panel to the closure strips with the recommended type of screw and rubber washer. Fasten the panel at the peak (top) of every other corrugation. Drive the screws down carefully, stopping when the washer contacts the panel but is not compressed. This allows for thermal expansion of the panel.

Apply a bead of the recommended adhesive/sealant (usually supplied by the panel manufacturer) along the last trough of the roofing panel. Set the second panel into place, overlapping the last troughs on both panels. Fasten the second panel. Install the remaining panels using the same procedure. Caulk the seam between the roof panels and the roof flashing.

Closure Strips ▸

If you do not have wall flashing designed to work with the roof profile, place closure strips upside down onto the roof panels and run another bead of adhesive/sealant over the tops of the strips. Work the flashing down and embed it into the sealant. Seal along all exposed edges of the ledger with silicone caulk.

To create channels for the side glazing panels, mill a rabbet into each of the six 2 × 2 slat cleats. Also rabbet the panel portion of each 2 × 2 back post. Consult the glazing manufacturer for the recommended channel size, making sure to provide space for thermal expansion of the panels. Mill the rabbets using a table saw, router, or circular saw. Stop the rabbets so the bottom edges of the panels will be even with, or slightly above, the bottom edge of the lowest side slat.

Position a cleat on each post at the desired height, with the cleat centered from side to side on the post. The rabbeted corner should face inside the shelter. Fasten the cleats to the posts with 3" deck screws. Fasten the back posts to the house wall so they are aligned with the post cleats.

Cut the side slats to length to fit between the posts and the house wall. Mark the slat layouts onto the outside faces of the cleats and back posts, and install the slats with 1½" deck screws or exterior trim-head screws. Space the slats 3½" apart or as desired.

(continued)

Fasten the middle cleats to the slats on each side of the shelter, leaving about 3½" of space between the cleats (or as desired). The cleats should overhang the top and bottom slats by 1½" (or as desired).

White Oak ▸

Used for the decorative accent slats on this patio shelter, white oak is a traditional exterior wood that was employed for boatbuilding as well as outdoor furnishings. Although it requires no finishing, we coated the white oak with a dark, penetrating wood stain to bring out the grain.

Cut the cap strips for the glazing panels from 1 × 2 material. Position each cap over each cleat and back post and drill evenly spaced pilot holes through the cap and into the cleat. Make sure the holes go into the solid (non-rabbeted) portion of the cleat. Drill counterbores, too (below), to help conceal the screw heads.

Option: Add a 2 × 4 decorative cap on the outside face of each post. Center the cap side-to-side on the post and fasten it with 16d casing nails.

Trim the side glazing panels to size following the manufacturer's directions. Apply neoprene or EPDM stripping or packing to the side edges of the panels. Fit each panel into its cleat frame, cover the glazing edges with the 1 × 2 caps, and secure the caps with 1½" deck screws. *Note: If the glazing comes with a protective film, remove the film during this step as appropriate and make sure the panel is oriented for full UV protection.*

Arbor Retreat

The airy, sun-filtered space under an arbor always makes you want to stay awhile—making it a perfect place for built-in seating. This project has plenty of room for lounging or visiting, but it's designed to do much more. Viewed from the front, it is an elegant passageway. The bench seating is obscured by latticework, and your eyes are drawn toward the central opening and striking horizontal beams. This makes the structure a grand garden entrance or a perfect landscape focal point. Or, if you prefer seclusion, tuck this arbor behind some foliage.

Sitting inside the retreat you can enjoy privacy and shade behind the lattice screens. The side roof sections over the seats are lowered to follow a more human scale and create a cozier sense of enclosure. Each bench comfortably fits three people, and the two sides face each other at a range that's ideal for conversation.

An arbor with plenty of seating space is the perfect place to appreciate the natural beauty of your property—whether that may be stunning terrain or the verdant beauty of your own flower garden.

Tools, Materials & Cutting List

Tools and materials for setting posts (page 438)
Mason's string
4- or 6-ft. level
Eye and ear protection
Speed or combination square
Circular saw, reciprocating saw, or power miter saw
Router and roundover bit
Drill with bits
Ratchet wrench
Protractor (optional)
Cardboard
Jigsaw or bandsaw
Sandpaper
Compass
Tape
Tape measure

Description (No. finished pieces)	Quantity/Size	Material
Posts		
Inner posts (4)	4 @ field measure	4 × 4
Outer posts (4)	4 @ field measure	4 × 4
Concrete	Field measure	3,000 PSI concrete
Gravel	Field measure	Compactable gravel
Roof		
Beams (6 main, 4 cross)	8 @ 8'	4 × 4
Roof slats (10 lower, 11 upper)	21 @ 8'	2 × 2
Seats		
Seat supports, spacers, slats (6 horizontal supports, 6 vertical supports, 4 spacers, 16 slats)	16 @ 8'	2 × 6
Aprons (2)	2 @ 6'	1 × 8
Lattice Screens		
Arches (4)	1 @ 8'	2 × 8
Slats—arched sides (20 horizontal, 8 vertical)	12 @ 8'	2 × 2
Slats—back (8)	8 @ 8'	2 × 2
Hardware & Fasteners		
3/8 × 7" galvanized lag screws	12, with washers	
3" deck screws		
3½" deck screws		
2½" deck screws		
¼ × 3" galvanized lag screws	16, with washers	

Front Elevation

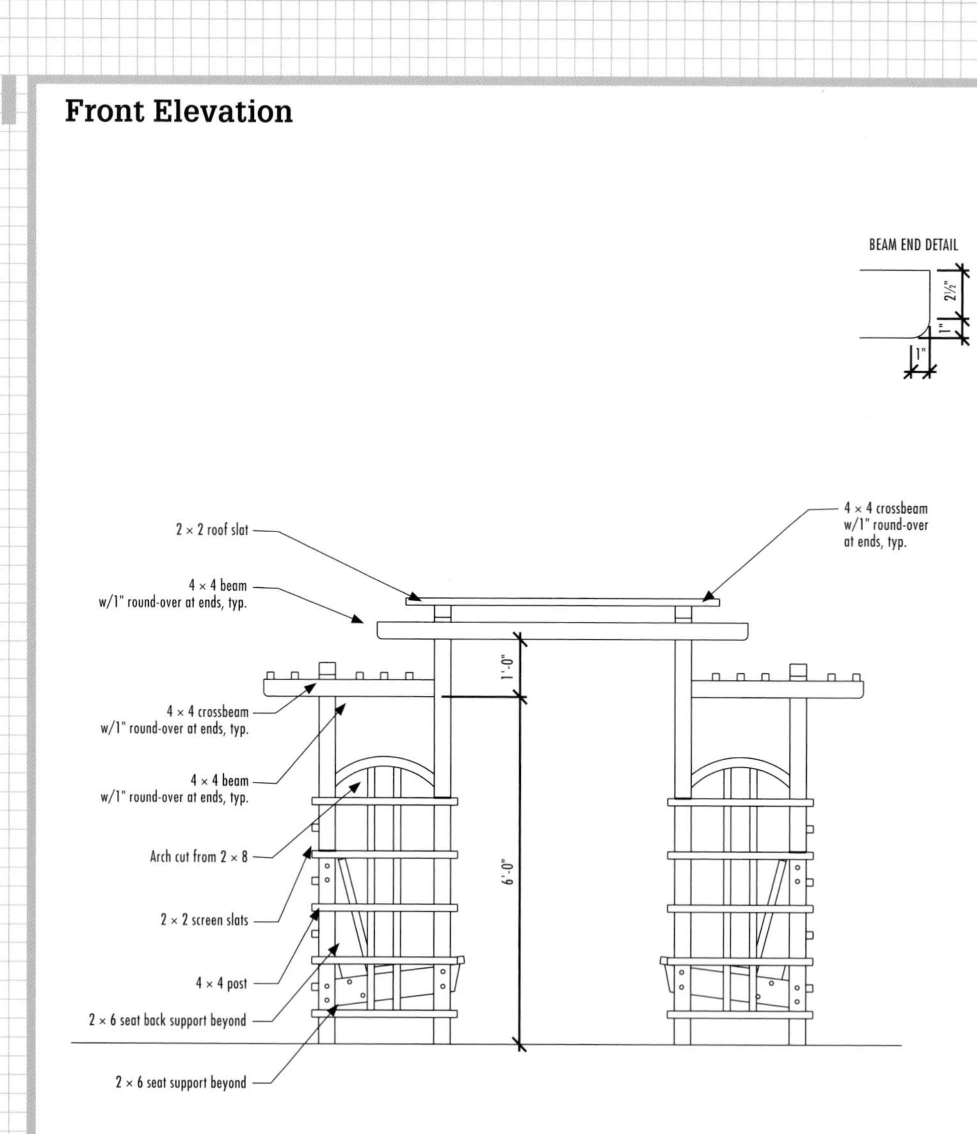

Side Elevation

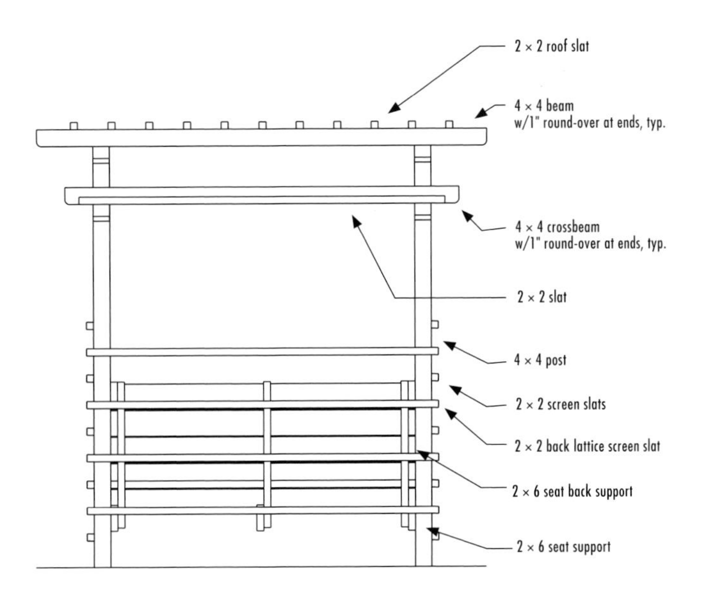

Post Layout

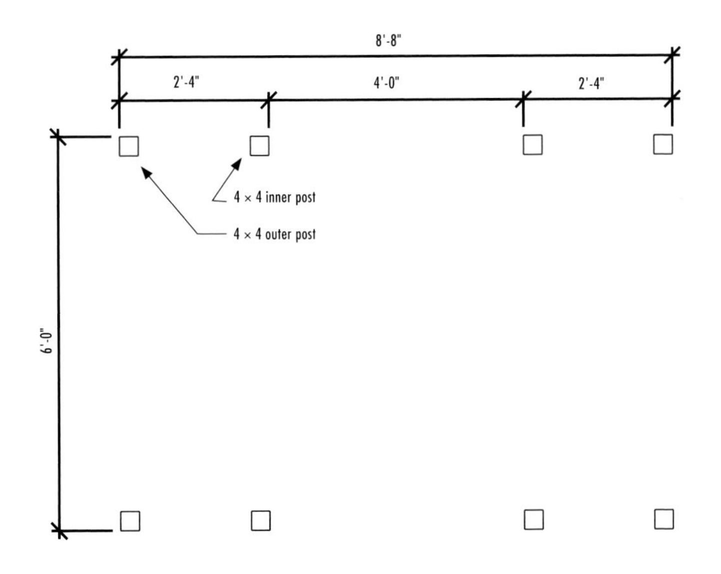

Upper Level Roof Framing

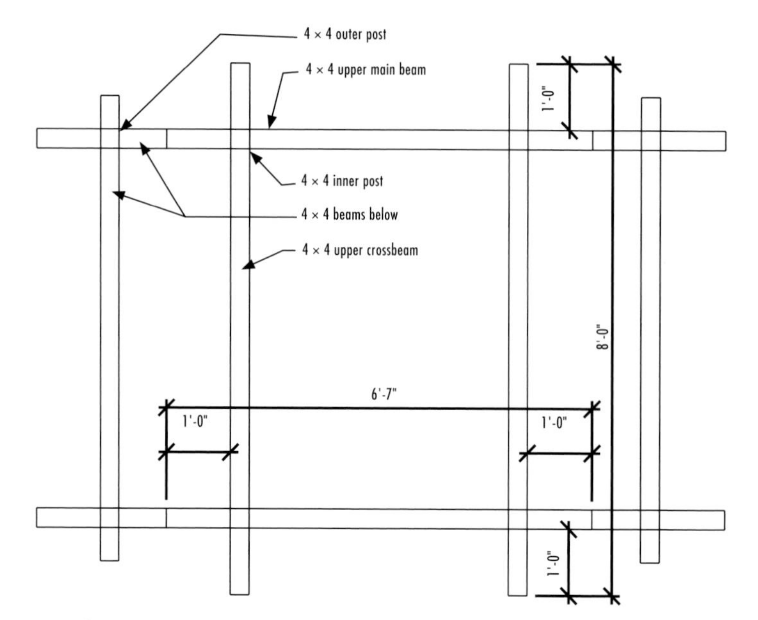

Seat Framing

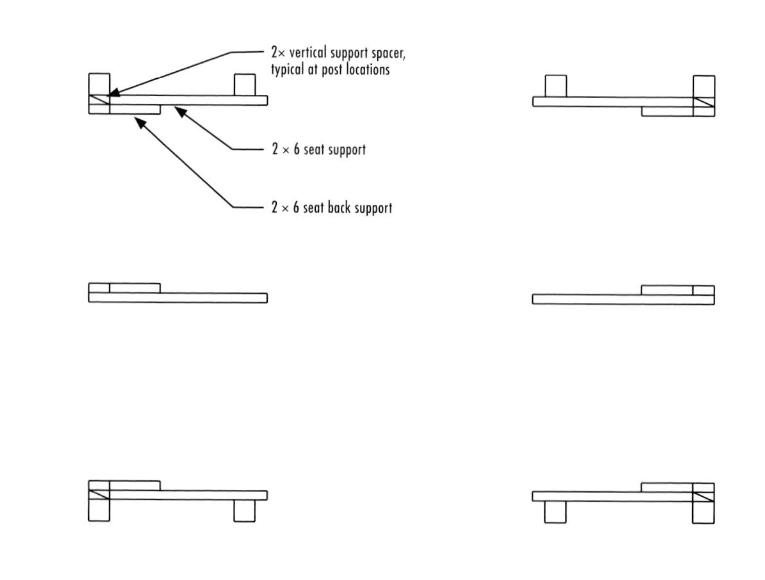

Roof/Slat Plan

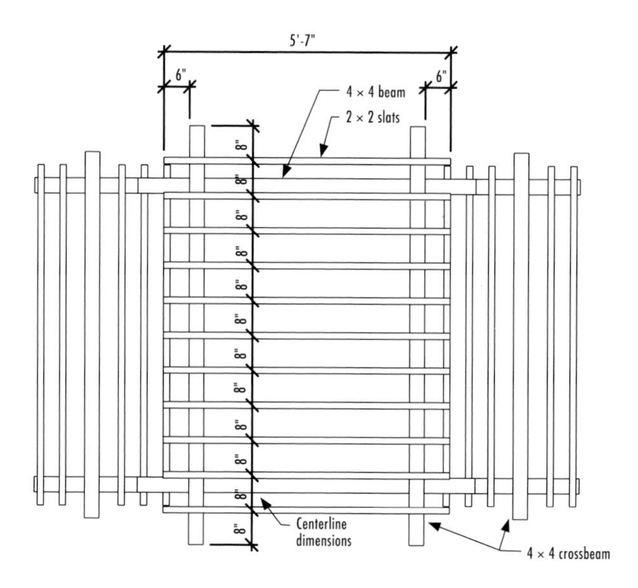

Slat Plan at Seating

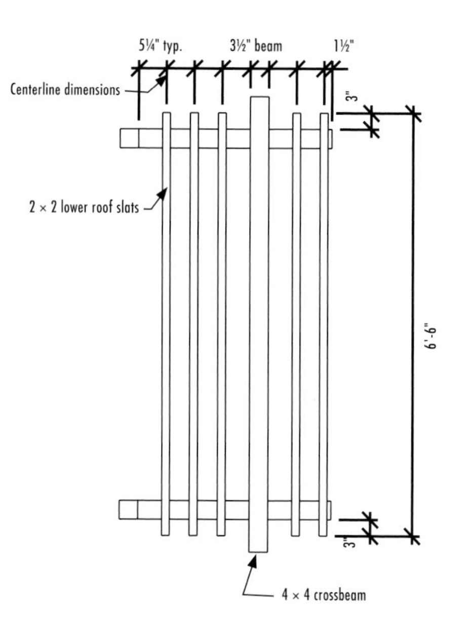

Seat Section

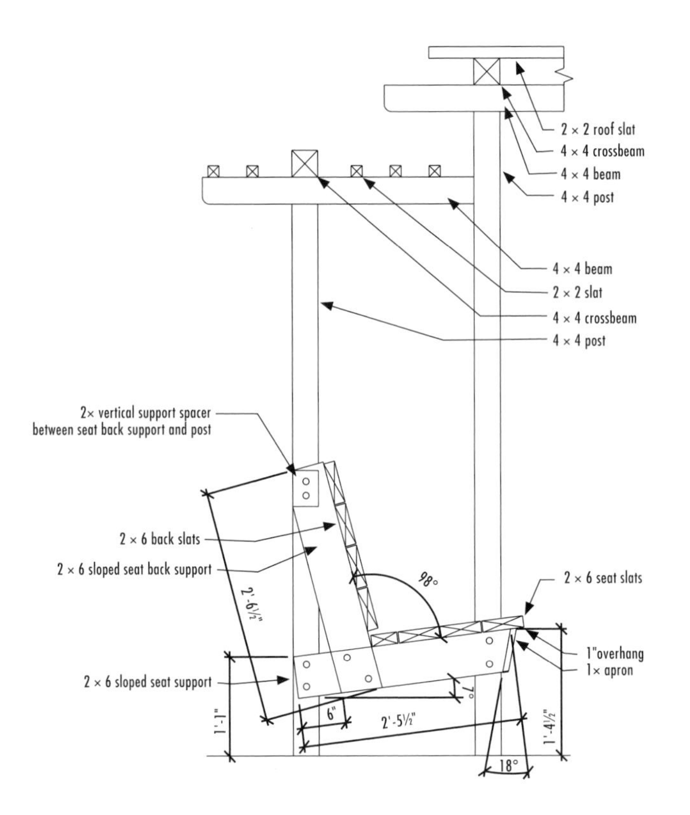

Seat Level Roof Framing

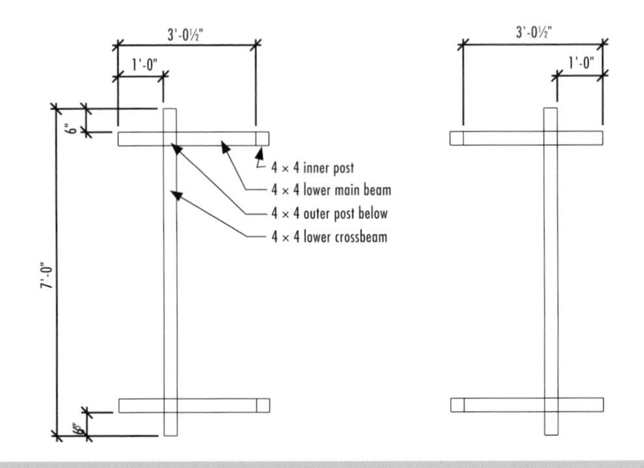

Seat Slat Layout

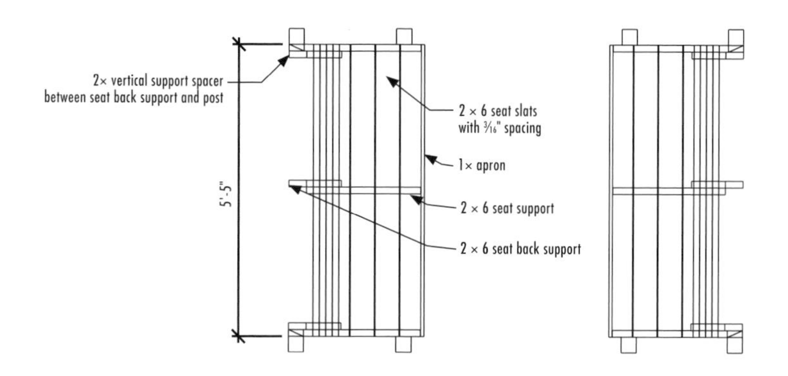

Arch Detail/Screen Layout

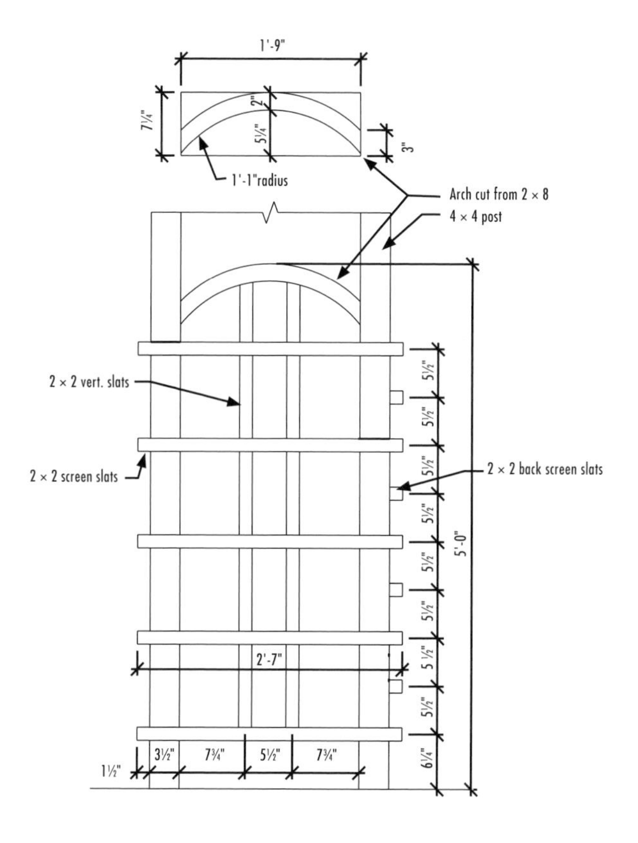

How to Build an Arbor Retreat

STEP 1: SET THE POSTS

A. Treat the bottoms of the posts for rot resistance (see page 438).

B. Lay out the eight post locations on the ground, following the POST LAYOUT on page 363.

C. Follow the basic procedure shown on pages 438 to 439 to set the posts in concrete. Make sure the size and depth of the postholes conform to your local Building Code. The post heights don't have to be exact at this stage; however, the four inner posts must stand at least 84" above the ground. The four outer posts must be at least 72" above the ground. Set up mason's lines to make sure the posts are perfectly aligned and the layout is square.

D. Pour the concrete and let it dry completely.

STEP 2: CUT THE POSTS TO LENGTH

A. You need a long leveling tool to mark the posts for cutting. If you don't own a 6-ft. level, you can make one using a standard 4-ft. level and a straight 2 × 4. Tape the level to a straight edge of a 7-ft.- or 8-ft.-long 2 × 4 so the level is roughly centered along the board's length.

B. Measure up from the ground and mark one of the inner posts at 84". Using the long level, transfer the height mark to the remaining inner posts.

C. Mark one of the outer posts 72" from the ground, and then transfer that mark to the other outer posts.

D. Cut the posts to length (see Cutting Lumber Posts, page 369). Cut carefully so the post tops are flat and level.

STEP 3: CUT & SHAPE THE BEAMS

The Arbor Retreat has two levels of roof beams. The lower level has four short main beams running perpendicular to the seats and two crossbeams running parallel to the seats (see SEAT LEVEL ROOF FRAMING on page 366.) The upper level has two main beams and two crossbeams (see UPPER LEVEL ROOF FRAMING on page 364).

All of the beams are 4 × 4s and have one or two ends rounded over at the bottom corners (see BEAM END DETAIL on page 362). This is an optional decorative detail that gives the project a finished look.

A. Cut the seat level main beams to length at 36½". Cut the seat level crossbeams at 84".

B. Cut the upper level main beams to length at 79". Cut the upper level crossbeams at 96".

C. Shape the beam ends, if desired using a router with a roundover bit. Or, you can simply make a 45° bevel cut with a saw. The upper level main beams and all four crossbeams are shaped at both ends. The short, lower level main beams are shaped only at the outside ends.

Align the posts with mason's lines, and use crossbracing to keep the posts plumb while the concrete sets.

Use a 6-ft. level or a standard level and 2 × 4 to mark each set of posts at a uniform height.

Cutting Lumber Posts ▸

Large posts can be tricky to cut, especially when the post is already standing. Here are some tips for making accurate cuts. If the combination of power saws and ladders is uncomfortable to you, a sharp handsaw can always do the trick, and often with greater accuracy. Start the cut carefully, and watch your lines as you work. Remember: standard handsaws cut only on the push stroke; don't waste energy by applying pressure on the pull stroke.

Here are some more tips for cutting posts:

Extend your cutting line all the way around the post. Use a speed or combination square. This helps you keep your saw on track as you cut from different sides of the post.

When cutting with a circular saw, set the saw blade to maximum depth. Cut all the way around the post, moving from one side to the next. Be careful to stay on your cutting lines, so the cut surface will be flat. On a 6 × 6, you can cut all four sides, then finish off the center with a handsaw. Because a handsaw blade is thinner than a circular saw blade, keep the handsaw flat against the wood as you cut.

A reciprocating saw with a long, woodcutting blade makes it easy to cut through posts in a single pass. Be careful to keep the saw steady and straight to ensure a good cut.

Power miter saws are the easiest tool for cutting 4 × 4s on a workbench, and some saws can cut a 4 × 4 in a single pass, so a continuous cutting line is not needed. To cut a 6 × 6—or a 4 × 4 with multiple cuts—set up a stop block. This allows you to evenly rotate the post, while cutting from all sides.

(continued)

Round over the bottom corners of the beams using a router and a roundover bit.

Test-fit the lower main beams to the posts, then drill counterbored pilot holes and fasten the beams with lag screws.

STEP 4: INSTALL THE LOWER MAIN BEAMS

A. For each of the lower level main beams, set the beam on top of an outer post and butt its unshaped end against the corresponding inner post. Hold the beam level, and mark where the top face of the beam meets the inner post. Set the beam aside.

B. On the opposite (inside) face of the inner post, mark a point for drilling a pilot hole so the hole will be centered on the end of the beam.

C. At each pilot hole mark, drill a counterbored hole just deep enough to completely recess the washer and head of a ⅜ × 7" lag screw.

D. Reposition each beam so its top face is on the post reference line. Holding the beam in place, drill a pilot hole for the lag screw through the inner post and into the end of the beam. Fasten each main beam with a ⅜ × 7" lag screw.

E. Drill a counterbored pilot hole down through the top of each lower level main beam and into the end of its outer post. Fasten the beam to the post with a ⅜ × 7" lag screw. Make sure the head of the screw is flush or slightly recessed into the beam.

STEP 5: INSTALL THE LOWER CROSSBEAMS & ROOF SLATS

A. Position the lower crossbeams on top of the lower main beams so they are centered over the outer posts and overhang the main beams by 6" at both ends (see SEAT LEVEL ROOF FRAMING).

B. Drill angled pilot holes through the sides of the crossbeams and into the main beams, about ¾" in from the sides of the main beams (to avoid hitting the lag screws). Drill two holes total on each side of the crossbeam at each joint.

C. Fasten the crossbeams to the main beams with 3½" deck screws (eight screws total per crossbeam).

D. Cut the 10 lower roof slats to length at 78".

E. Mark the roof slat layout onto the tops of the lower main beams, following the SLAT PLAN AT SEATING drawing on page 365.

F. Position the slats on the layout so they overhang the main beams by 3" at both ends. Drill pilot holes, and fasten the slats to the main beams with 2½" deck screws.

STEP 6: INSTALL THE UPPER MAIN BEAMS, CROSSBEAMS & ROOF SLATS

A. Position the upper main beams on top of the inner posts so they overhang the posts by 12" at both ends (see UPPER LEVEL ROOF FRAMING). Check the fit of the joints, and make adjustments as needed for a good fit.

B. At each post location, drill a counterbored pilot hole and secure the beam to the post with a ⅜ × 7" lag screw with washer, just as you did to fasten the lower main beams to the outer posts.

C. Position the upper crossbeams over the upper main beams so they are centered over the inner posts and overhang the main beams by 12" at each end.

D. Drill pilot holes. Fasten the crossbeams to the main beams with 3½" deck screws, just as you did with the lower crossbeams.

E. Cut the 11 upper roof slats to length at 67".

F. Mark the slat layout onto the upper crossbeams, following the ROOF/SLAT PLAN (page 365).

G. Position the slats so they overlap the crossbeams by 6" at both ends. Drill pilot holes, and fasten the slats with 2½" deck screws.

STEP 7: CUT THE SEAT SUPPORTS

Each seat has three horizontal seat supports and three vertical seat back supports, plus two vertical support spacers (see SEAT SECTION, page 366, and SEAT FRAMING, page 364). The sets of supports and spacers are identical, so once you mark and cut each type, you can use it as a pattern to mark the duplicate pieces.

A. Cut one horizontal seat support and one vertical seat back support, following the SEAT SECTION.

Tip: Cut each of the supports from a different 8-ft. 2 × 6, and save all of the cutoffs for seat slats. Also cut a vertical support spacer from a full 2 × 6.

B. Test-fit the pieces on the arbor posts. Make any necessary adjustments or re-cuts so all of the angles fit properly, as shown in the SEAT SECTION.

C. Use the cut pieces to mark the remaining supports, and then make the cuts. For the two center support assemblies, cut the rear end of the horizontal seat support so it will be flush with the rear edge of the vertical support.

STEP 8: INSTALL THE OUTER SEAT SUPPORTS & APRONS

A. On each side of the structure, measure up from the ground and mark the inner posts at 16½" and the outer posts at 13". These marks represent the top edges of the horizontal seat supports.

B. Position the horizontal seat supports on the marks so their back ends are flush with the outsides of the outer posts. Fasten the supports to the posts with pairs of ¼ × 3" lag screws driven through counterbored pilot holes.

Fasten the crossbeams to the main beams with screws set at an angle. Countersink the screws for best appearance.

Install the upper beams and slats using the same procedure for securing the lower beams and slats.

(continued)

C. Position each vertical seat back support and spacer as shown in the SEAT SECTION, and mark the location of the support spacer onto the post. Fasten the spacer to the post with 3" deck screws driven through pilot holes. Then, fasten the vertical seat back support to the spacer and horizontal seat support with 2½" deck screws; use three or four screws at each end.

D. Measure between the outside faces of the side seat supports, and then cut the 1 × 8 aprons to length at those dimensions.

E. Bevel the top edge of the aprons at 7°. Position them against the seat supports, as shown in the SEAT SECTION. Mark the bottom edges of the aprons for cutting. Bevel the bottom edges at 7°. Fasten the aprons to the ends of the seat supports with 2½" deck screws.

STEP 9: INSTALL THE SEAT SLATS & CENTER SUPPORTS

A. Measure between the inner posts to determine the length of the seat slats. Using this dimension cut eight slats to length for each side.

B. Position a slat on top of the horizontal seat supports so its front edge overhangs the supports by about 1". Fasten the slat to the supports with pairs of 3" deck screws.

C. Install the next three slats on each side, leaving a 3⁄16" gap between the slats. Rip the final seat slat to fit the remaining space.

D. Install the vertical back seat slats from the top down. Position the top slat so its highest edge is flush with, or just below, the tops of the vertical seat supports. Gap the remaining slats by 3⁄16".

E. Using 2½" deck screws, assemble the two center seat supports so they match the outer supports. Install the center supports at the midpoints of the slats by screwing through the slats and into the supports using 3" deck screws.

STEP 10: BUILD THE ARCHED LATTICE SCREENS

A. Mark the layout of the horizontal lattice pieces onto the posts, following the SCREEN LAYOUT on page 367; mark along one post, then use a level to transfer the marks to each neighboring post.

B. Cut 20 2 × 2 lattice slats to length at 31". Position the slats on the layout so they overhang the posts by 1½" at both ends. Fasten the slats to the posts with 2½" deck screws driven through pilot holes.

C. To make the arches, make a cardboard template, following the arch detail on page 367. Using the template, trace one arch onto a 2 × 8. Cut out the arch with a jigsaw or bandsaw. Test-fit the arch between the post pairs, and make any necessary adjustments for a good fit. Cut the remaining arches. Sand the cut edges smooth.

D. Position each arch on its layout marks so its outside edges are flush with the outside faces of the

Mark the end cuts on the seat supports using a speed square, or you can use a protractor to find the angles.

Fasten the horizontal seat supports to the posts with lag screws. Attach the vertical seat back supports with deck screws.

posts. At each end of the arch, drill an angled pilot hole upward through the bottom of the arch and into the post. Fasten the arch with 2½" deck screws.

E. On the top and bottom horizontal slats, make a mark 7" in from each post. These represent the outside edges of the vertical lattice slats.

F. Cut the eight vertical slats to a rough length of 54". Mark the top ends of the slats to match the arches by holding each slat on its reference marks. Use a compass to transfer the arch curve to the end of the slat. Cut the curved ends and test-fit the slats.

G. Hold each vertical slat in place against the arch and mark the bottom end for length so it will be flush with the bottom edge of the lowest horizontal slat. Cut the vertical slats to length.

H. Install the vertical slats with 3" deck screws driven down through the arches and 2½" deck screws driven through the lowest horizontal slats.

STEP 11: BUILD THE BACK LATTICE SCREENS

A. Mark the layout for the back lattice slats onto the outer posts, following the SCREEN LAYOUT.

B. Cut the eight 2 × 2 back lattice slats to length at 75". This gives you 1½" overhang at each end.

C. Position the slats on their layout marks so they overhang the posts by 1½" at both ends. Drill pilot holes, and fasten the slats to the posts with 2½" deck screws.

Install the center support so it's centered along the length of the seat slats.

Use a compass to mark a curved trim line on the end of the vertical slat where it fits into the bottom curve of the arch.

Fasten the back slats to the outer posts with 2½" deck screws.

Classical Pergola

Tall and stately, the columned pergola is perhaps the grandest of garden structures. Its minimal design defines an area without enclosing it and makes it easy to place anywhere—from out in the open yard to right up against your house. Vines and flowers clinging to the stout framework create an eye-catching statement of strength and beauty.

In this project, Tuscan-style columns supporting shaped beams mimic the column-and-entablature construction used throughout classical architecture. Painting the columns white or adding faux marbling enhances the classical styling. The columns used here are made of structural fiberglass designed for outdoor use. They even adhere to the ancient practice of tapering the top ⅔ of the shaft (see The Timeless Column, on page 380).

Structural fiberglass columns, like the ones used in this project, are available from architectural products dealers. You can order them over the phone and have them shipped to your door. This type of column is weather-resistant, but most manufacturers recommend painting them for appearance and longevity. Whatever columns you use, be sure to follow the manufacturer's instructions for all installation and maintenance.

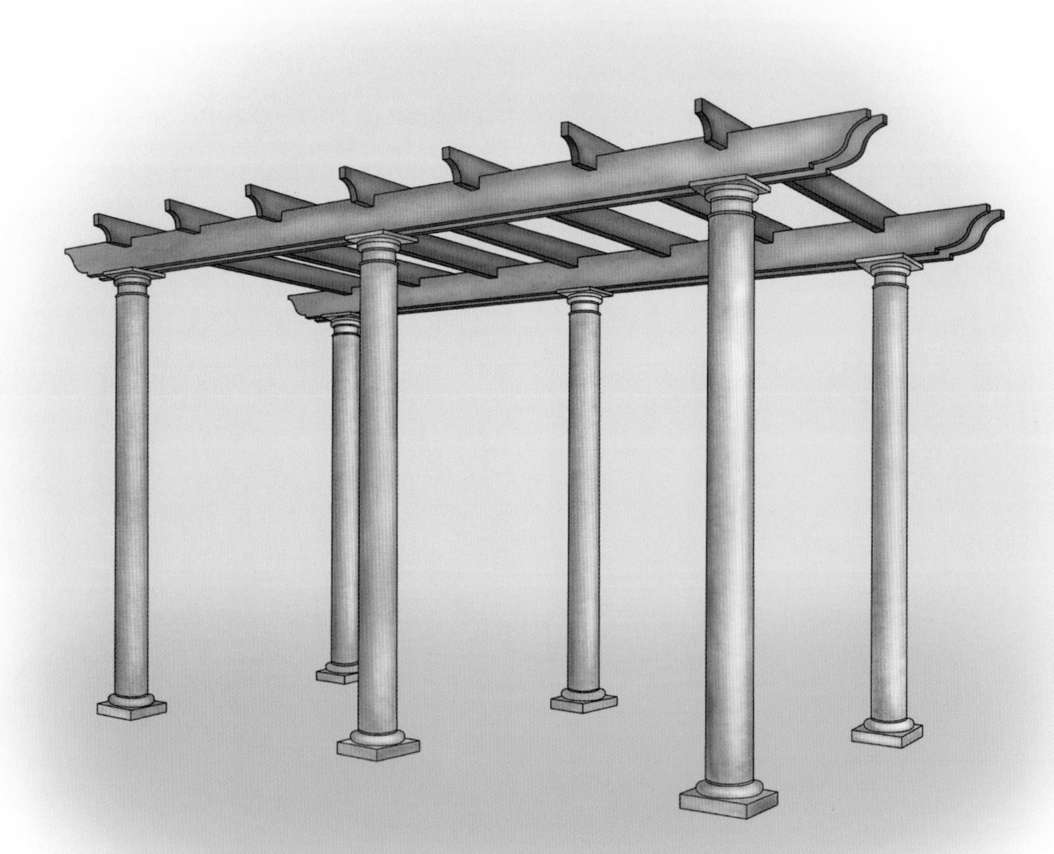

The freestanding pergola is perhaps the most adept of all outdoor structures at delineating outdoor spaces without creating a real barrier between them. The classical columns of this structure can add elegance and grandeur to any landscape.

Tools, Materials & Cutting List

Batter boards
Mason's strings
Clamshell digger or power auger (available for rent)
Shovel
2-ft. level
Line level
Concrete float
Hammer drill with masonry bit
Nylon brush
Concrete anchor adhesive
Caulk gun
Tape measure
Air compressor with trigger-type nozzle
Cardboard
Jigsaw or bandsaw
Sandpaper
Wood preservative
Construction adhesive or waterproof wood glue
Clamps
Rasp
Handsaw
Circular saw
Ratchet wrench
Chisel
Paintable caulk
Finishing materials
Eye and ear protection
Hammer

Description (No. finished pieces)	Quantity/Size	Material
Columns	6 @ 8"-dia. × 8'	Structural fiberglass column
Concrete Piers		
Concrete tube forms	6—field measure for length	16"-diameter cardboard forms
Gravel	Field measure	Compactable gravel
Concrete	Field measure	3,000 PSI concrete
Beams		
Main beams (4)	4 @ 16'	2 × 8
Crossbeams (7)	7 @ 8'	2 × 6
Blocks (6)	1 @ 50"	4 × 4 pressure-treated
Hardware & Fasteners		
½ × 6" J-bolts	6	
Threaded rod (for concrete slab foundation only)	½"-dia. × 4' corrosion-resistant threaded rod	
Concrete anchoring adhesive—concrete patio installation only		Simpson Acrylic-Tie® adhesive, or similar approved product
Construction adhesive or waterproof wood glue		
16d galvanized common nails		
½"-diameter corrosion-resistant threaded metal rod	6 @ 99"	
½" corrosion-resistant coupler nuts	6	
Corrosion-resistant bearing plates and nuts	6 each	Simpson BP1/2-3 or similar approved bearing plate. Recommended nut for ½" threaded rod
Corrosion-resistant masonry screws		
2½" deck screws		
Paintable caulk		

Front Elevation

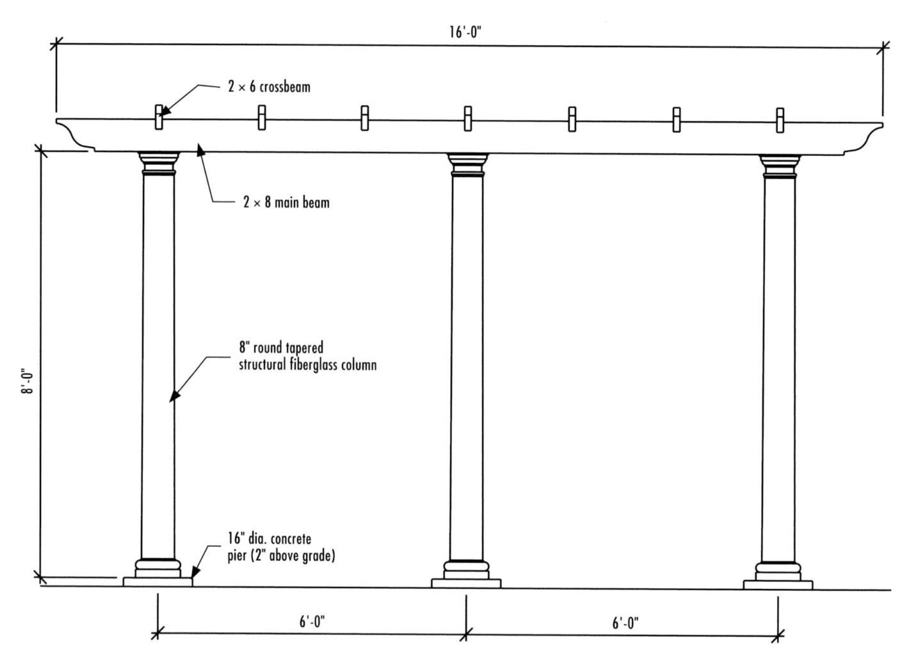

Side Elevation

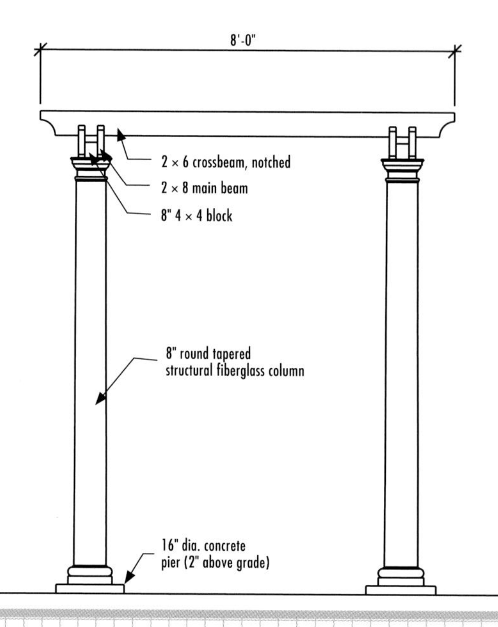

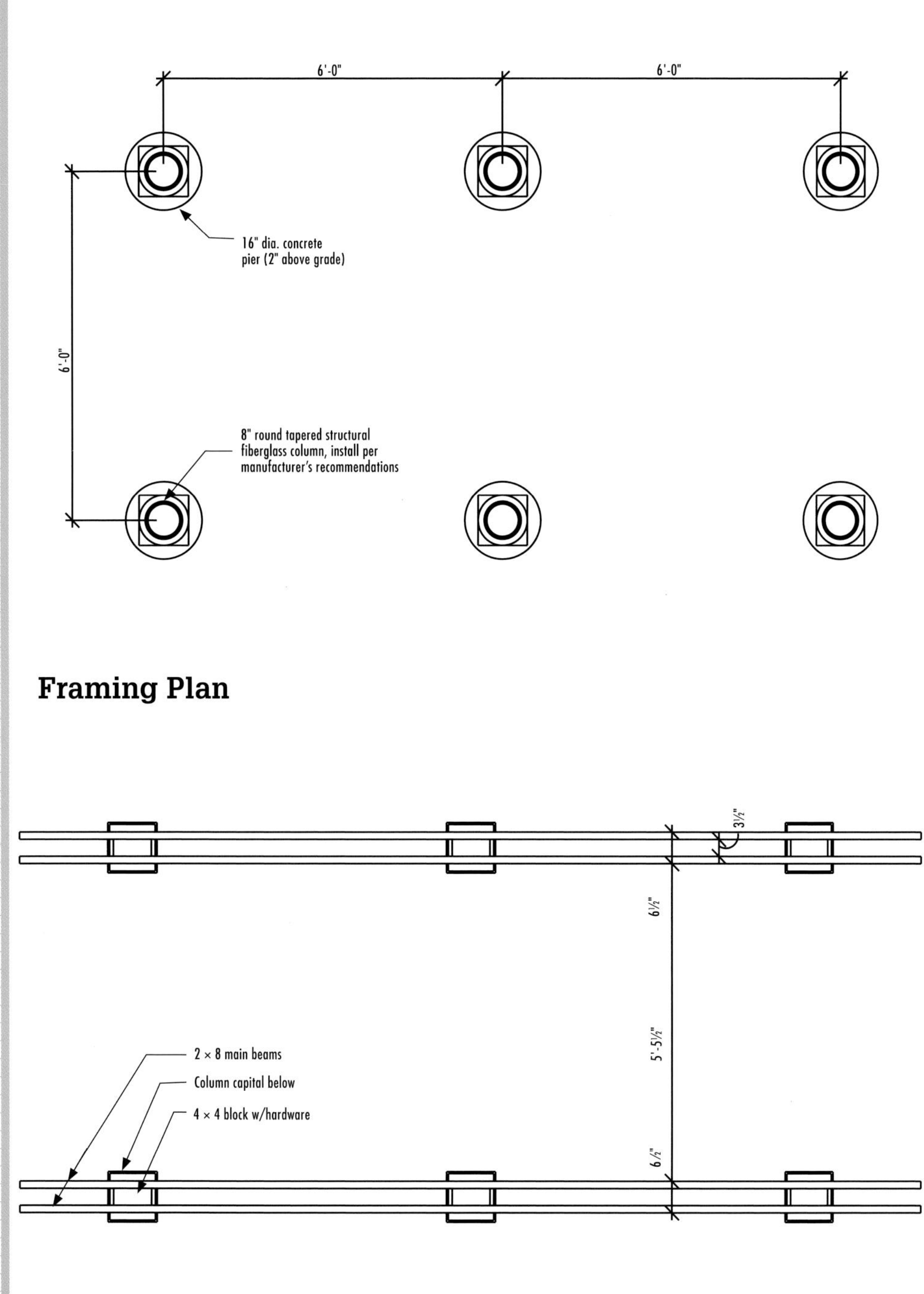
Foundation Plan
6'-0"
6'-0"
16" dia. concrete pier (2" above grade)
6'-0"
8" round tapered structural fiberglass column, install per manufacturer's recommendations
Framing Plan
3½"
6½"
5'-5½"
6½"
2 × 8 main beams
Column capital below
4 × 4 block w/hardware

Roof Framing Plan

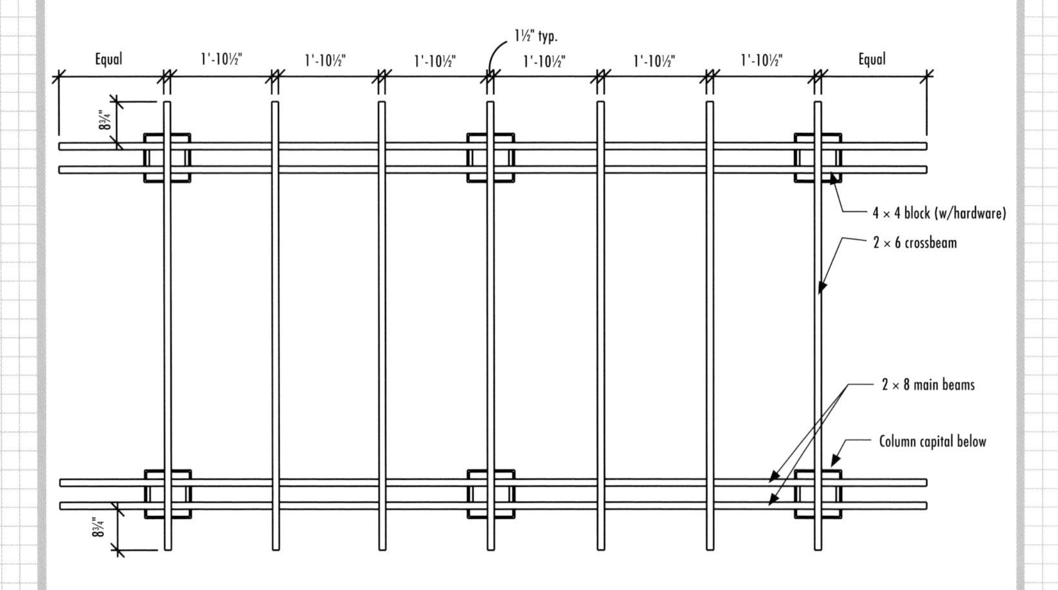

Beam End Templates

1" × 1" grid shown

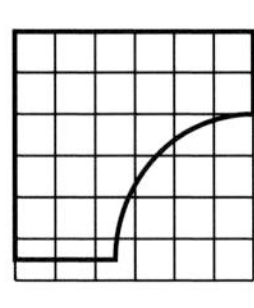

2 × 6 crossbeams

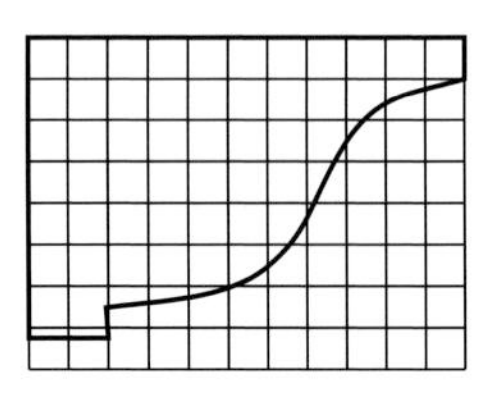

2 × 8 main beams

Column Connection

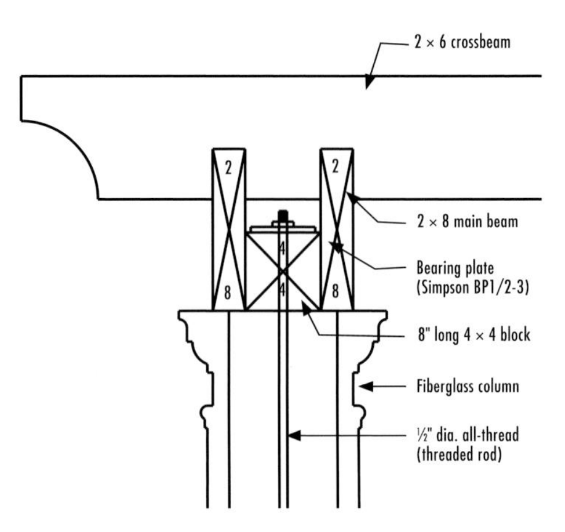

How to Build a Classical Pergola

If you're building on poured concrete piers in your yard, complete Step 1, below. To build on an existing concrete patio slab, skip ahead to Option: Existing Concrete Slab or Pier Foundation.

STEP 1: POUR THE CONCRETE PIERS

A. See pages 434 to 437 for detailed instructions on laying out and pouring concrete piers using cardboard forms. Set up batter boards and mason's lines to lay out the pergola columns following the FOUNDATION PLAN on page 377.

B. Dig the six holes for the concrete forms. Add a layer of gravel, then set and brace the forms. Make sure the pier depth and gravel layer meet the requirements of your local Building Code. For this project, the piers are 16" in diameter and extend at least 2" above the ground. You may have to adjust the height of some piers so that all of them are in the same level plane; measure against your level mason's lines to compensate for any unevenness in the ground.

C. Pour the concrete for each form, and set a ½ × 6" J-bolt in the center of the wet concrete. Make sure the bolt is perfectly plumb and extends 1¾ to 2" above the surface of the concrete.

D. Following the concrete manufacturer's instructions, finish the tops of the piers to create a smooth, attractive surface. When painted, the piers become part of the finished project.

As the concrete sets, finish the tops of the piers using a concrete float.

OPTION: EXISTING CONCRETE SLAB OR PIER FOUNDATION

Note: Follow the manufacturer's specifications and instructions for installing the anchor rods in this step.

A1. On the patio surface, mark the layout for the column centers; follow the FOUNDATION PLAN on page 377. The centers must be at least 6" from any edge of the slab. This ensures the column base (plinth) doesn't hang over the edge of the slab and gives you a little bit of wiggle room for adjustments.

Drill a ⅝"-diameter hole straight down into the concrete; use a hammer drill and ⅝" masonry bit.

Use a nylon brush to dislodge loose material.

Fill the anchor hole ½ to ⅔ full with concrete anchor adhesive.

Insert a rod into the hole, turning the rod slowly until it contacts the bottom of the hole.

(continued)

At each column centerpoint, drill a ⅝"-diameter hole straight down into the concrete using a hammer drill and ⅝" masonry bit. Make the hole at least 4¼" deep.

A2. Spray out the holes to remove all dust and debris using an air compressor with a trigger-type nozzle. Make sure the air is completely oil-free. If necessary, use a clean nylon brush to dislodge any loose material, then spray again with compressed air to completely remove all dust.

A3. Cut six pieces of ½"-diameter corrosion-resistant threaded rod to length at 8". Make sure the rods are clean and oil-free. Fill each anchor hole ½ to ⅔ full with concrete anchor adhesive (sold at most building centers). Fill the hole starting from the bottom and working up to prevent air pockets. Keep the nozzle of the adhesive dispenser above the adhesive as the hole fills.

A4. Insert a rod into each hole, turning the rod slowly until it contacts the bottom of the hole. Position the rod plumb. Leave the rod undisturbed until the adhesive has fully cured.

STEP 2: CUT & SHAPE THE BEAMS

A. Cut the four main beams to length at 192". Cut the seven crossbeams to length at 96".

B. Check all of the beams for crowning—a slight arching shape that's apparent when the board is set on edge. Hold each board flat and sight along its narrow edges. If the board arches, mark the top (convex) side of the arch. This is the crowned edge and should always be installed facing up.

C. Make cardboard patterns for shaping the ends of the main beams and crossbeams; follow the BEAM END TEMPLATES on page 378. Use the patterns to mark the shapes onto the beam ends.

D. Shape the beam ends using a jigsaw or bandsaw, and then sand the cuts smooth.

The Timeless Column ▸

The ancient Greeks and Romans used columns everywhere, and they designed them to exact specifications. A column just wasn't respectable if it didn't have the right shape and proportion. Many of those same rules are still followed today.

According to the ancients, a good column must have a tapered shaft. This is because a perfectly straight shaft appears to be smaller in the center, thus conveying a sense of weakness (not a popular trait among Romans). Some columns are straight along the lowest ⅓ of the shaft and taper inward along the top ⅔; others slightly bulge out in the center (called entasis).

All but the earliest forms of columns had a base and a capital, the style of which largely determined the "order" or type of column it was. Remember art history class? The three Greek orders are Doric, Ionic, and Corinthian. The Roman orders are Tuscan and Composite. Doric and Tuscan have simply ornamented capitals, while Ionic are the ones with the scrolls. Corinthian and Composite follow an anything-goes style and might be decorated with leaves, scrolls, cherubs, goat heads… you name it.

Today, column suppliers offer a variety of capitals and bases. Although it usually costs extra, you can swap out the standard capital or base with one that better suits your style.

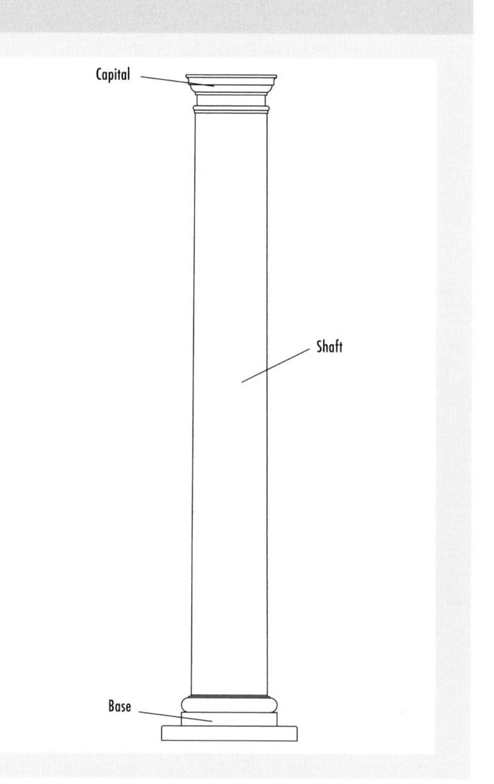

STEP 3: CONSTRUCT THE MAIN BEAM ASSEMBLIES

A. Cut six 4 × 4 blocks at 8".

B. Lay each block flat, and drill a 9/16"-diameter hole through the center of one side.

C. Coat the ends of the blocks and the insides of the holes with wood preservative, following the manufacturer's instructions. The blocks are the main structural connecting points for the pergola, and the preservative helps prevent rot from moisture over the years.

Sight along both narrow edges of the beams. If a beam is arched, mark the beam on the convex side of the arch.

D. Make a mark 20" in from the end of each main beam. These marks represent the outside ends of the blocks.

E. Construct the main beam assemblies by applying construction adhesive or waterproof wood glue to the side faces of the blocks and sandwiching the beams over the blocks. Make sure the blocks are flush with the bottoms of the beams and their ends are on the reference marks. The holes must be face up (vertical). Clamp the assembly, and then fasten the beams to the blocks with 16d common nails. Drive four nails on each side, making sure to avoid the center hole in the blocks. Let the glue dry completely.

F. Mark the crossbeam layout on to the top edges of the main beams, following the ROOF FRAMING PLAN on page 378.

STEP 4: PREPARE & SET THE COLUMNS

You'll need at least two helpers for this step and the following step. Once you set the columns for one side, continue to the next step to install the main beam. Then, repeat the two steps for the other side of the pergola.

A. Cut the threaded rods to length at 99".

B. Add a corrosion-resistant coupler nut to each J-bolt (threaded anchor rod for patio installation).

C. Lay the columns down next to their respective piers. Slip the base and capital over the ends of the column shafts; these will stay loose so you can slide them out of the way until you secure them in Step 7.

Sandwich the blocks between the main beams, and fasten the assemblies with glue and nails.

Have one person lift up the column while another tightens the coupler nut to the J-bolt and threaded rod.

(continued)

D. Run the threaded rod through the center of each column.

E. Tip up the column and center it on top of a pier. Check the joint where the column meets the pier; it should make even contact all the way around the column. If necessary, use a rasp to shave the end of the column to ensure even contact.

F. While one person lifts and holds the column out of the way, thread the rod into the coupling nut. Adjust the nut so the rod and J-bolt have equal penetration into the nut, and tighten the nut following the manufacturer's instructions. Temporarily brace the column if necessary, or have a helper hold it upright. Repeat steps D–F to set the remaining two columns.

STEP 5: SET THE MAIN BEAMS

A. Using stepladders set up next to the columns, place one of the main beams onto the columns, inserting the rod ends through the blocks. Check for even contact of the beam on all three columns. If necessary, you can trim a column: Cut from the bottom end only using a sharp handsaw.

Center the column at both ends, then tighten the nut over the bearing plate to secure the entire assembly.

Set a circular saw to cut just above the notch seat; clean up the notch with a chisel.

Note: If there's a slight gap above the center column due to a crowning beam, it will most likely be gone once the beam is anchored.

B. Add bearing plates and nuts to the end of each threaded rod, and loosely thread the nuts.

C. Working on one column at a time, make sure the column shaft is centered on the pier and is centered under the beam block at the top end. Place a 2-ft. level along the bottom, untapered section of the column shaft and check the column for plumb. Hold the column plumb while a helper tightens the nut on the rod. Repeat to adjust and secure the remaining columns.

D. Repeat the procedure to install the columns and beam on the other side of the pergola.

STEP 6: NOTCH & INSTALL THE CROSSBEAMS

A. Place each crossbeam onto the layout marks on top of the main beams so the crossbeam overhangs equally at both ends. Mark each edge where the main beam pieces meet the crossbeam. This ensures the notches will be accurate for each crossbeam. Number the crossbeams so you can install them in the same order. On your workbench, mark the notches for cutting at 2½" deep.

B. To cut the notches, you can save time by clamping two beams together and cutting both at once. Using a circular saw or handsaw, first cut the outside edges of the notches. Next, make a series of interior cuts at ⅛" intervals. Use a chisel to remove the waste and smooth the seats of the notches.

C. Set the crossbeams onto the main beams following the marked layout. Drill angled pilot holes through the sides of the crossbeams and into the main beams; drill one hole on each side, offsetting the holes so the screws won't hit each other. Fasten the crossbeams with 2½" deck screws.

STEP 7: FINISH THE COLUMNS

A. Fit each column base against its pier. Secure the base to the pier with corrosion-resistant masonry screws: First, drill pilot holes slightly larger than the screws through the base. Using a masonry bit, drill pilot holes into the pier. Fasten the base with the screws.

B. Fit each capital against the main beam, drill pilot holes, and fasten the capital with deck screws.

C. Caulk the joints around the capital and base with high-quality, paintable caulk.

D. Paint the columns—and beams, if desired—using a primer and paint recommended by the column manufacturer.

After fastening the base and capital, caulk all of the joints to hide any gaps and create a watertight seal.

Design Variation ▸

Add an extra layer of overhead slats for additional shade and to further substantiate the structure.

Corner Lounge

Decks and patios extend your home into the open air and they're the best places for all kinds of activities—parties, evening meals, afternoon naps, and sunbathing. This project is designed to make the most of all the ways you use your deck or patio.

This project combines the sheltering and light-filtering qualities of an arbor roof with the convenience of built-in bench seating. And it fits into the corner, so it won't take up a lot of floor space on your deck. You can add as much or as little lattice screening as you like for just the right amount of shade or privacy. An optional roof design lets you extend the roof over an 11 × 11-ft. area—perfect for adding a table that takes advantage of the bench seating.

You can build this project on most traditional decks and concrete patios. The location will dictate how you install the posts; steps are given here for elevated wood decks, as well as ground-level decks and concrete patios. You can also locate this structure anywhere in your garden by setting the posts in concrete (see pages 438-439).

Adding a corner lounge to your deck or patio provides some shade and helps to buffer the wind, as well as providing built-in seating for outdoor gatherings.

Tools, Materials & Cutting List

Drill with bits
Jigsaw
Tape measure
Circular saw or handsaw
Level
Hammer
Clamps
Ratchet wrench
Hammer drill with bits
Chisel
Drill with bits
Sandpaper
Waterproof wood glue
Eye, ear, & glove protection

Description (No. finished pieces)	Quantity/Size	Material
Posts		
Full-height posts* (7)	7 @ field measure	4 × 4
*Add 1 post for optional full roof		
Seat support post (1)	1 @ field measure	4 × 4
Post blocking	2 blocks for each post; field measure	2× pressure-treated lumber; size to match existing deck joists
Roof Frame		
Beams (8)	8 @ 12'	2 × 8
Roof slats (20)	10 @ 8'	2 × 2
Optional Full Roof		
Beams (6)	6 @ 12'	2 × 8
Roof slats (12 long, 12 short)	18 @ 8'	2 × 2
Seats		
Seat supports (6 sides, 6 ends)	5 @ 8'	2 × 6
Seat slats (27)	9 @ 8'	2 × 6
Lattice Screens		
Lattice slats and backer strips	Field measure	1 × 1 (¾ × ¾" actual dimensions)
Hardware & Fasteners		
16d galvanized common nails or 3½" deck screws		
Post bases (for concrete patios or ground-level decks only)	8, with recommended anchors and fasteners	Simpson AB44 or similar approved base
⅜ × 8" galvanized carriage bolts	16, with washers and nuts	
¼ × 6¼" galvanized carriage bolts	20, with washers and nuts	
2½" deck screws		
3½" deck screws		
3½" galvanized lag screws	22, with washers	
Galvanized metal angle	1	
3d galvanized finish nails		
Waterproof glue		
6d galvanized finish nails		

Elevation

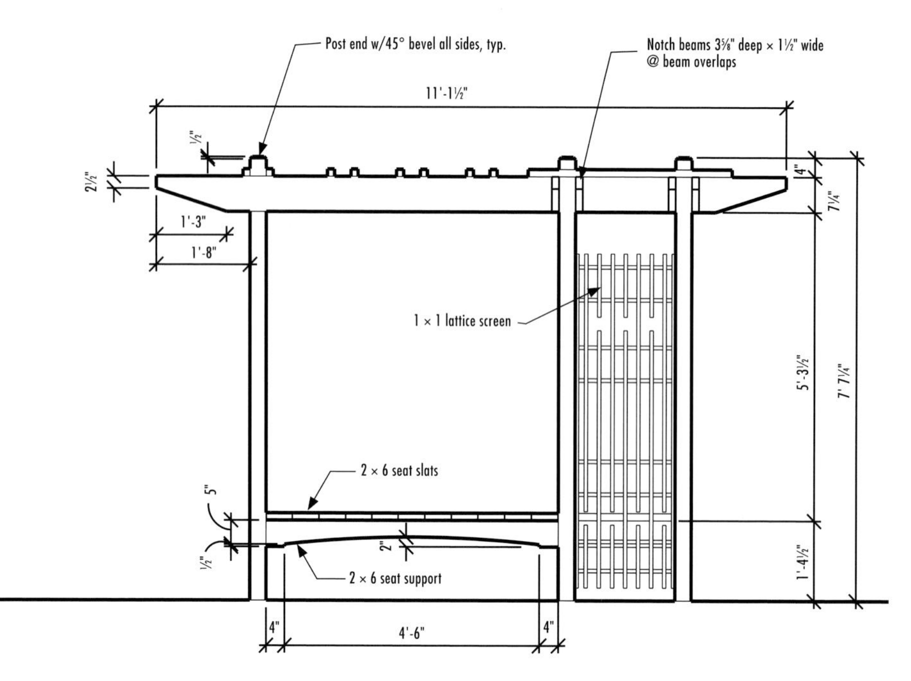

Post Layout Plan

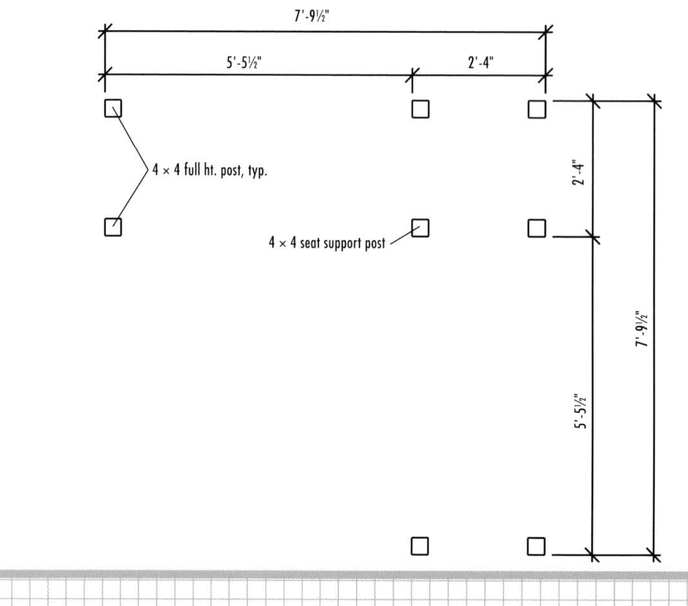

Roof Slat Plan

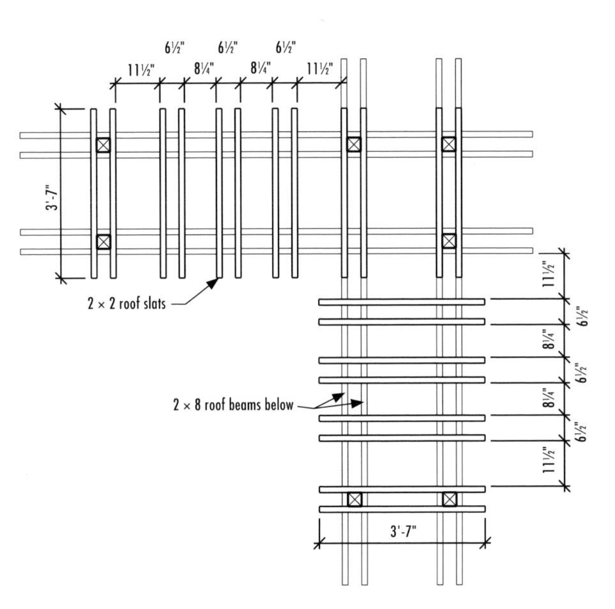

Screen Layout

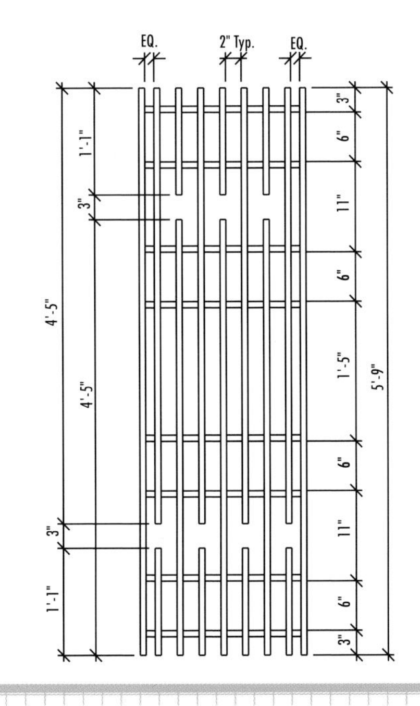

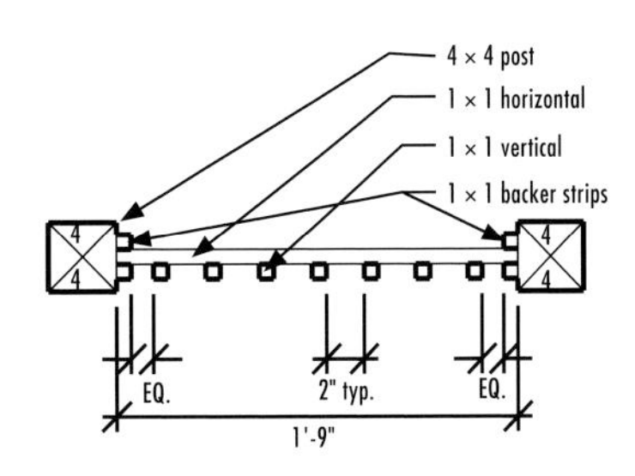

Roof Framing Plan

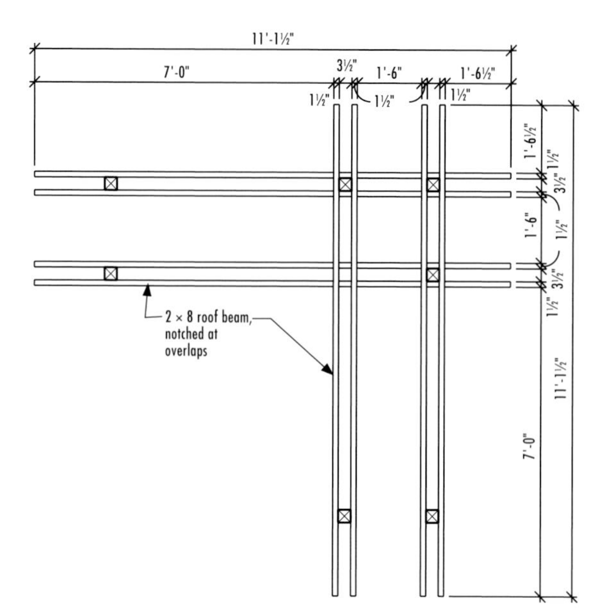

Optional Fullum Roof Frame

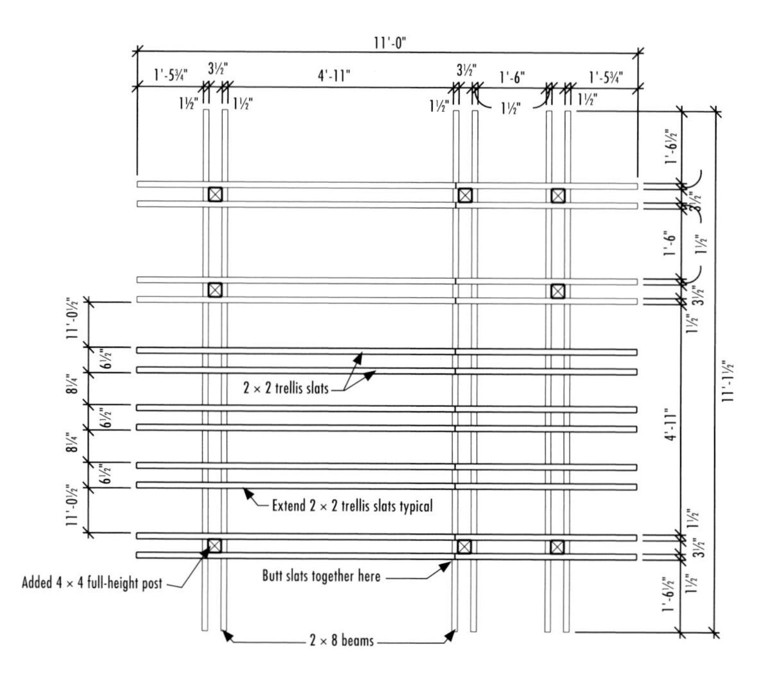

Seat Slat Layout

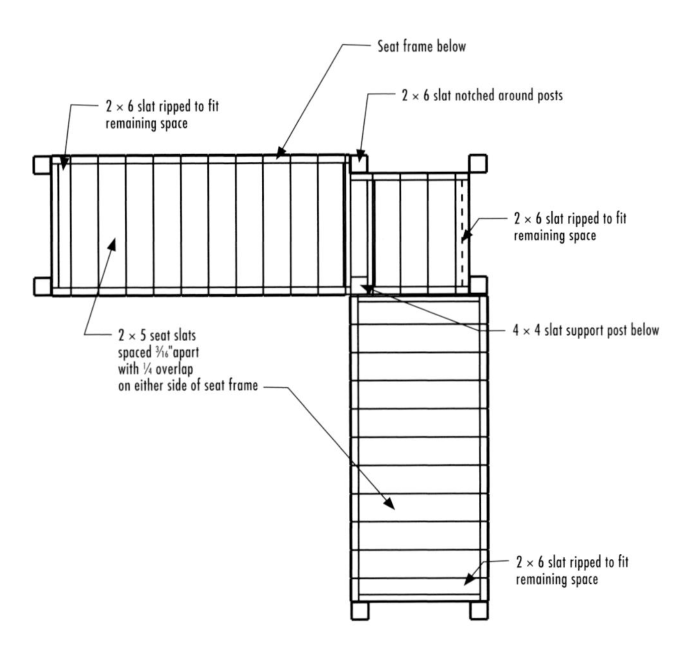

Seat Framing Plan

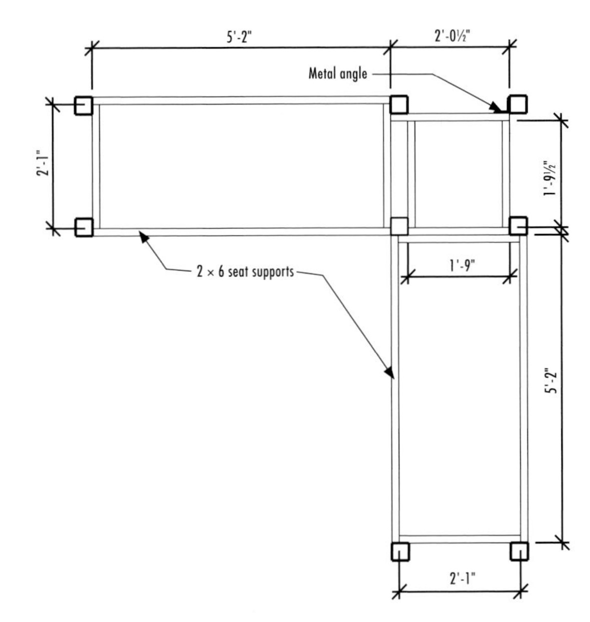

How to Build a Corner Lounge

The proper method for installing the posts depends on your situation. If you have an elevated wood deck, complete the following steps, but skip the Alternative Post Installation. For a concrete patio or ground-level deck, skip ahead to the Alternative Post Installation (page 391). Or, if installing on the ground, follow the procedure on pages 438 to 439 to bury the posts in concrete. *Note: This requires longer posts to compensate for the buried portion.*

STEP 1: CUT THE POSTHOLES

A. Lay out the post locations on your deck, following the POST LAYOUT PLAN, on page 386.

Note: If a series of posts falls over a deck joist, move the layout just enough so that the posts will be flush against the side of the joist. You can either move the entire structure or move only the affected posts. If you choose the latter, you can cut-to-fit the affected pieces to complete the project—just be aware that the plan measurements might not always apply.

B. Mark cutout holes for the posts onto the decking boards. Measure each post to find its exact dimensions (they often vary slightly), then mark the cutout onto the decking boards.

C. Drill starter holes inside each cutout marking, then use a jigsaw with a down-cutting blade to make the cutouts.

STEP 2: INSTALL THE POST BLOCKING

A. Underneath the deck, measure between the neighboring joists at each post location. Cut two pieces of blocking to fit in between the joists at each location. Use pressure-treated lumber that is the same size as the joists (e.g., 2 × 10, 2 × 12, etc.).

B. Get someone to help with this step, so one person is on top of the deck and one is below. Have the top person insert a short length of post (such as the seat support post) into a post cutout, extending it down so it's even with the bottoms of the joists. The person below sandwiches the post with blocking. While the top person uses a level to hold the post perfectly plumb, the bottom person marks the outsides of the blocking onto the neighboring joists.

C. Remove the post, then fasten the blocking to the joists with 16d common nails or 3½" deck screws. Drive three fasteners through the joists and into the ends of the blocking, making sure the blocking stays on the marks made in step B.

Drill starter holes inside the post marks, then make the cutouts with a jigsaw.

Hold the blocks on their layout marks, and fasten them to the joists at both ends.

STEP 3: CUT & INSTALL THE POSTS

A. For the full-height posts: Measure from the bottom of the blocking to the top of the deck surface. Add that dimension to 91¼" to find the total post length. For the seat support post: Add 16½" to the deck-depth measurement above to find the total length. Cut the posts to length (see page 369).

B. If desired, bevel the top ends of the full-height posts at 45°, as shown in the ELEVATION (page 386).

C. Measure from the top end of each full-height post and make a reference mark at 91¼". On the seat support post, make a reference line 16½" from the top end.

D. With one person on top of the deck and one below, set each post into its hole so the reference line is aligned with the deck surface. Use a level to hold the post perfectly plumb, then have the person below clamp the post to the blocking.

E. Drill pilot holes and anchor each post to the blocking with two ⅜ × 8" carriage bolts.

Use a clamp to help hold the post at the proper height. Anchor the post with carriage bolts.

ALTERNATIVE POST INSTALLATION: SETTING POSTS ON A CONCRETE PATIO OR A GROUND-LEVEL DECK

A. Lay out the post locations on your patio or deck; follow the POST LAYOUT PLAN (page 386).

B. For concrete patios: Use a hammer drill to drill a hole for each post base anchor. Refer to the base manufacturer for the size and type of anchor to use. Secure the anchor to the concrete, then bolt the post base to the anchor using the recommended hardware.

For decks: Fasten a post base to the deck at each post location using the fasteners recommended by the base manufacturer.

C. Cut the full-height posts to length so that they will stand 91¼" above the patio or deck surface when they're installed on the post bases. Cut the seat support post so it will stand 16½" above the surface when installed on its base.

D. If desired, bevel the top ends of the full-height posts at 45°, as shown in the ELEVATION (page 386).

E. Set each post on its base and support it with temporary braces so that it stands perfectly plumb. Fasten the post to the base using the fasteners recommended by the base manufacturer.

Alternative Post Installation: Use post bases with metal standoff plates to protect the posts from surface moisture.

STEP 4: CUT & SHAPE THE ROOF BEAMS

A. Cut the eight 2 × 8 beams to length at 133½".

B. To shape the beam ends, make a mark 2½" down from the top corner at each end. Make another

(continued)

mark 15" in from the bottom corner. Draw a line connecting the two marks. Cut along this diagonal line.

C. At the corner, the four sets of beam pairs intersect with half-lap joints. To mark the notches for the half-lap joints, measure the depth (width) and thickness of the beams. The width of the notches must match the thickness of the beams; the length of the notches must equal half the depth of the beams. Mark the layout of the notches, following the ROOF FRAMING PLAN (page 388).

D. Cut the notches. Save time by clamping two or more beams together and cutting them at once. Using a circular saw or handsaw, cut the outside edges of the notches first. Then, make a series of interior cuts at 1/8" intervals. Use a chisel to remove the waste and smooth the seats of the notches.

E. Test-fit the notches on the ground and make any necessary adjustments for a good fit.

STEP 5: INSTALL THE ROOF BEAMS

A. Mark the sides of the posts that will receive the beams 11¼" down from the top ends.

B. Starting with the beams with the top-down notches, sandwich one set of posts so the notches clear the posts on both sides and the bottom edges of the beams are on the 11¼" reference marks. Clamp the beams in place.

C. Drill two pilot holes for ¼ × 6½" carriage bolts through both beams and the post. On the less-visible beam sides, countersink the holes just enough to completely recess the washer and nut. Fasten the beams to the posts with the carriage bolts.

D. Repeat sub steps B and C to install the other set of parallel beams.

E. Install the perpendicular beams, fitting the notches together so all the beams are flush at the top and bottom edges. Clamp the beams as before, then drill pilot holes and attach the beams with carriage bolts.

Tip: Drill the pilot holes from the outsides of the beam intersections, so you have enough room for the drill bit.

STEP 6: INSTALL THE ROOF SLATS

A. Cut the 20 roof slats to length at 43".

Test-fit the beam joints and make adjustments before installing the beams.

B. Mark the slat layout on the tops of the roof beams; follow the ROOF SLAT PLAN on page 387.

C. Position each slat on its layout mark so it overhangs the outer roof beams by 6" on both sides. Fasten the slat to each intersecting beam with 2½" deck screws driven through pilot holes.

STEP 7: BUILD THE SEAT FRAMES

A. Look at the SEAT FRAMING PLAN (page 389) to understand the seat frame layout. There are three, 4-sided seat frames. You can build them on a workbench, then install them—just make sure they fit snugly between the sets of posts.

B. Measure between the posts for each seat frame. Cut the side seat supports to length so they extend from post to post. Cut the end seat supports to length so they fit between the side supports.

C. Lay out the arched cutout on one side support; follow the ELEVATION. Make the cut with a jigsaw or bandsaw, then sand the arch smooth. Use the support as a pattern to mark the three remaining long seat supports, then make the cuts.

D. Assemble the seat frames with 3½" deck screws drilled through the side supports and into the

Counterbores help hide the bolt hardware. Locate them in the least conspicuous areas.

Drill pilot holes for the roof slats, and fasten them to the roof beams with 2½" deck screws.

(continued)

end supports. Make sure the pieces are flush along their top edges.

E. Measure up from the deck surface, and mark the inside faces of the posts at 16½". Install the seat frames as shown in the SEAT FRAMING PLAN, so their top edges are on the reference marks; fasten through the end seat supports and into the posts with two 3½" lag screws at each location. Using a metal angle and screws, fasten the frame at the outside corner of the lounge.

STEP 8: ADD THE SEAT SLATS

A. Measure between the outside faces of the seat frames to find the lengths of the seat slats. You can either make the slats flush to the frames or overhang the frames by ¼" on either side.

B. Notch the first slat to fit around the posts where the left side frame meets the corner frame; see SEAT SLAT LAYOUT (page 389). Drill pilot holes, and then fasten the slat to the seat frames with pairs of 2½" deck screws.

C. Cut and install the remaining slats, gapping them at 3⁄16". Rip the last slat in each section to fit the remaining space.

STEP 9: BUILD THE LATTICE SCREENS

How you use the lattice screening is your choice. You may want screens only on the ends of the seats, as shown in the plan drawings, or you might cover the entire back side of the project. To avoid blocking your view behind your deck, you can build short screens that match the deck rail height. The basic procedure for building screens is shown here.

A. Construct a jig for assembling the lattice screens: on a sheet of plywood, fasten two straight 2 × 4s in an "L" pattern using a framing square to set the pieces at an exact 90° angle.

B. Cut several 11"-long spacers from ¾"-thick plywood.

C. Cut the lattice pieces to length from ¾ × ¾" (actual dimensions) lumber; follow the SCREEN LAYOUT on page 387.

D. Using the jig and spacers, assemble the screens according to the drawing, or create your own pattern. Fasten the pieces with waterproof wood glue and 3d finish nails.

E. To install the screens, fasten a lattice backer strip to each post, as shown in the plan detail of the SCREEN LAYOUT using 6d finish nails. Finally, install the screens against the backer strips.

A flexible wood strip helps you make a perfect curve for the long side seat supports.

Notch the slats as needed to fit snugly around the posts. Gap the slats by 3⁄16".

Construct a right-angle jig for assembling the lattice screens. Use plywood spacers to help you keep the lattice screens square as you work.

Designing with Shade ▸

One of the most useful design modifications you can make to a slatted-roof project is to update or change the roof slats to meet your needs. To determine what kind of configuration will be best for you, ask yourself when you'll use the space the most and how much shade or sunlight you'd like in the space at your favorite times.

Then, modify your project by changing the size, orientation, number, and spacing of the overhead slats or crosspieces. For example, for more shade use more slats placed closer together. You could even slant them to increase the design effect and increase functionality.

After the primary structure of posts and beams is in place, lay the slats on top in different configurations. When you've settled on your favorite, simply fasten them down.

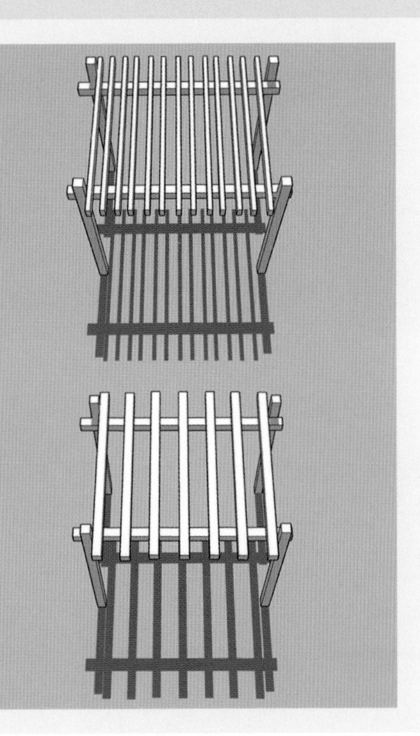

MOTORCYCLE
RACES
SUNDAY MAY 28
2318

Garages

First and foremost, a garage is a sheltered building where you can park your vehicles safely. But it can be much more than that, and it often is. A garage may also serve as an organized and climate-controlled workspace to pursue hobbies or as a utility shed for storing gardening and snow-removal equipment. It may be a workshop, a walk-in sports locker, or an overflow storage area. How can one room do it all? Truth be told, having a versatile, hard-working, well-organized garage is a very tall order—but with practical projects and the right approach it definitely can be done.

This chapter includes complete plans and materials lists for two versatile, freestanding garage designs that demonstrate different ways you could configure your garage, depending on how you intend to use it. If you're interested in a less intensive project and simply need a canopy to keep your car out of the rain, perhaps the clever carport project is the right choice for you. Or, browse through our inspirational Additional Garage Plans section, which will likely include the right garage for your needs.

In this chapter:

- Single Detached Garage
- Compact Garage
- Carport
- Additional Garage Plans

Single Detached Garage

A detached garage is a great convenience for any homeowner. The one-stall garage featured here is among the simpler designs you can find for a detached garage. The trickiest parts are cutting the rafter ends and making the cornices. One way to simplify these tasks is to replace the rafters with a truss system (see pages 50 to 51).

The plan drawings that follow on the next five pages deal primarily with the structure of the building. Finished details such as trim and siding are left somewhat open, since it is likely that you'll choose finish materials that match your own house. The garage shown here features fiber-cement siding on top, with the bottom section of each wall sided with cast veneer stone. As shown, this garage is 14 feet wide and 22 feet from front to back. If you choose to alter any of the dimensions for your project, do so with great care and make certain to update all of your part dimensions.

This efficient garage is built from the ground up using common building materials available at any building center. This plan was the basis for the Construction Techniques chapter featured earlier in this book. Materials lists and plan drawings for the garage are included on pages 399 to 403.

Materials & Cutting List

Description	Quantity/Size
Wall bottom (plate treated)	1 pc. / 2 × 4 × 14'
Wall bottom (plate treated)	5 pcs. / 2 × 4 × 12'
Precut wall studs	58 pcs. / 2 × 4 × 92⅝"
Wall top/tie plates	3 pcs. / 2 × 4 × 14'
Wall top/tie plates	6 pcs. / 2 × 4 × 12'
Wall top/tie plates	3 pcs. / 2 × 4 × 10'
Header over garage door	2 pcs. / 2 × 12 × 10'
Header blocking	2 pcs. / 2 × 4 × 10'
Header over window & door	2 pcs. / 2 × 8 × 10'
Cripple studs	6 pcs. / 2 × 4 × 8
Garage door	1 pc. / 2 × 6 × 10'
Garage door	2 pcs. / 2 × 6 × 8'
Corner brace	6 pcs. / 1 × 4 × 12'
Rafter tie	6 pcs. / 2 × 6 × 14'
Rafters & gable blocking	12 pcs. / 2 × 6 × 18'
Ridgeboard	2 pcs. / 2 × 8 × 12'
Gable studs	3 pcs. / 2 × 4 × 12'
Gable nailer (top plate)	4 pcs. / 2 × 4 × 8'
Soffit nailer	4 pcs. / 2 × 2 × 12'
Horizontal fiber-cement siding 10½ exp.	617 sq. ft. / 7/16" × 12"
Metal corners for siding	40 pcs.
Rake fascia	2 pcs. / 1 × 8 × 18'
Rake soffit	2 pcs. / 1 × 8 × 18'
Rake shingle mold	36 L.F.
Fascia & soffit	8 pcs. / 1 × 8 × 12'
Aluminum foil kraft paper	1 roll / 36" wide
C-D ext. plywood roof sheathing	13 pcs. / 4 × 8' × ½"
Roofing felt	1 roll / 15#
Asphalt shingles	4⅓ sq. / 235#
Sliding window unit	1 ea. / 4 × 3'
Exterior caulk	2 tubes / 101
Garage service door 6 panel	1 ea. / 2'8" × 6'8"
Sectional overhead garage door	1 ea. / 9 × 7'
Door jambs	42 L.F. / 1 × 4'
Brickmold casing	42 L.F.
Shingle mold stop	42 L.F.
Concrete slab foundation & floor	8 cu. yd.
Wire mesh	308 sq. ft. / 6 × 6 × #10
Reinforcing bars	144 L.F. / ½" dia.
Exterior paint	3 gal.
Nails	
16d common nails coated	20 lb.
10d common nails coated	2 lb.
8d common nails coated	2 lb.
6d common nails coated	5 lb.
8d galvanized siding nails	5 lb.
1¼" galvanized roofing nails	15 lb.
8d casing nails	2 lb.
Anchor bolts w/nuts & washers	20 ea. / ½ dia. × 12"
Key in knob cylinder lockset	1 ea.
Door butts	1 pr. / 3½ × 3½"
Optional	
Ext. plywood sheathing (corner bracing)	8 pcs. / 4 × 8' × ½"
Insulating sheathing	12 pcs. / 4 × 8' × ½"
Galvanized nails	10 lb. / 1½"
Optional for Alternate Formed Foundation	
Concrete for footing	3 cu. yd.
Concrete for walls	6 cu. yd.
Concrete for floor	4 cu. yd.

These plans have been prepared to meet professional building standards. However, due to varying construction codes and local building practices, these drawings may not be suitable for use in all locations. Results may vary according to the quality of material purchased and the skill of the builder. Always use eye, ear, and glove protection for safety.

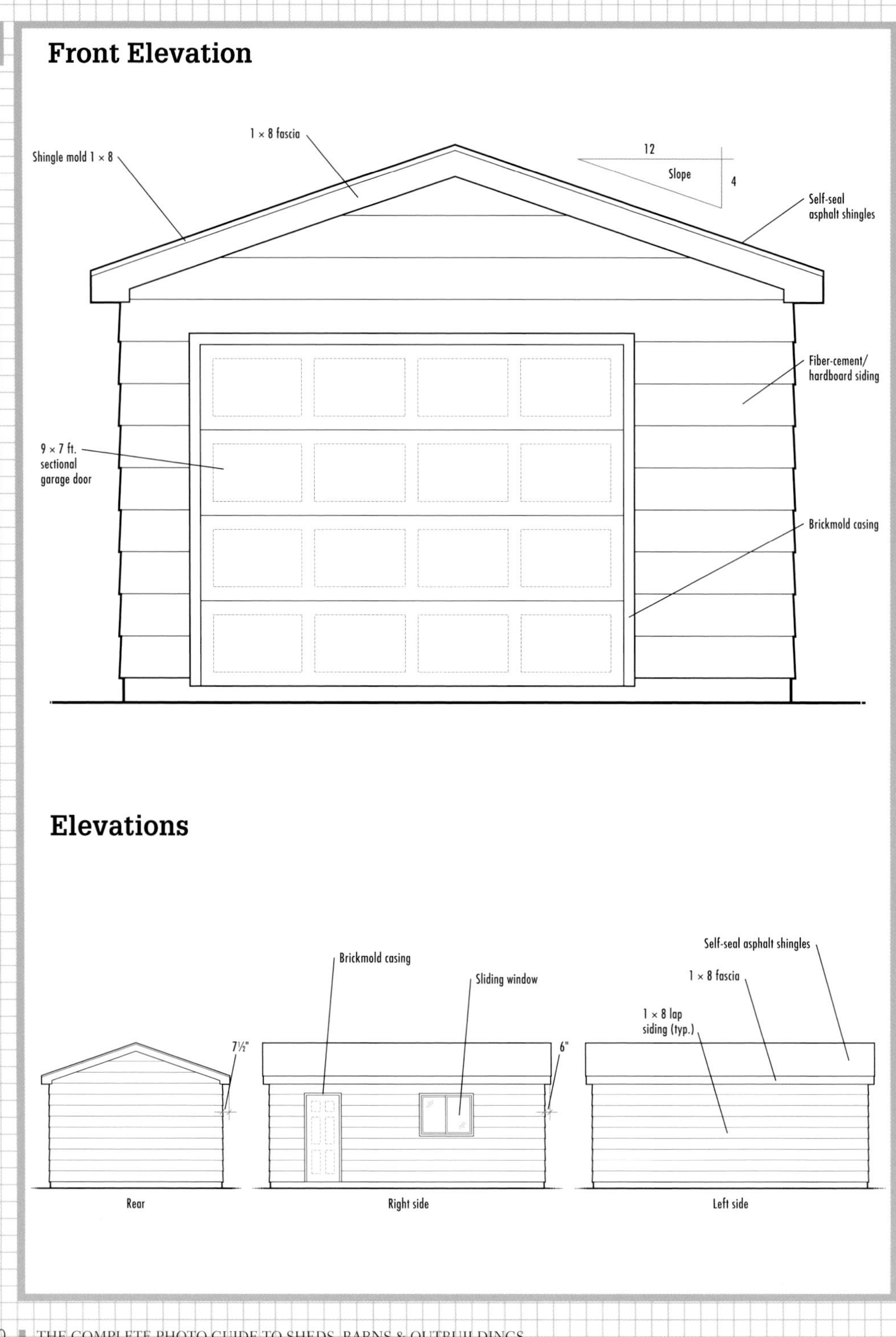
Front Elevation
1 × 8 fascia
Shingle mold 1 × 8
12
Slope
4
Self-seal asphalt shingles
Fiber-cement/ hardboard siding
9 × 7 ft. sectional garage door
Brickmold casing
Elevations
Self-seal asphalt shingles
Brickmold casing
Sliding window
1 × 8 fascia
1 × 8 lap siding (typ.)
7½"
6"
Rear
Right side
Left side

Floor Plan

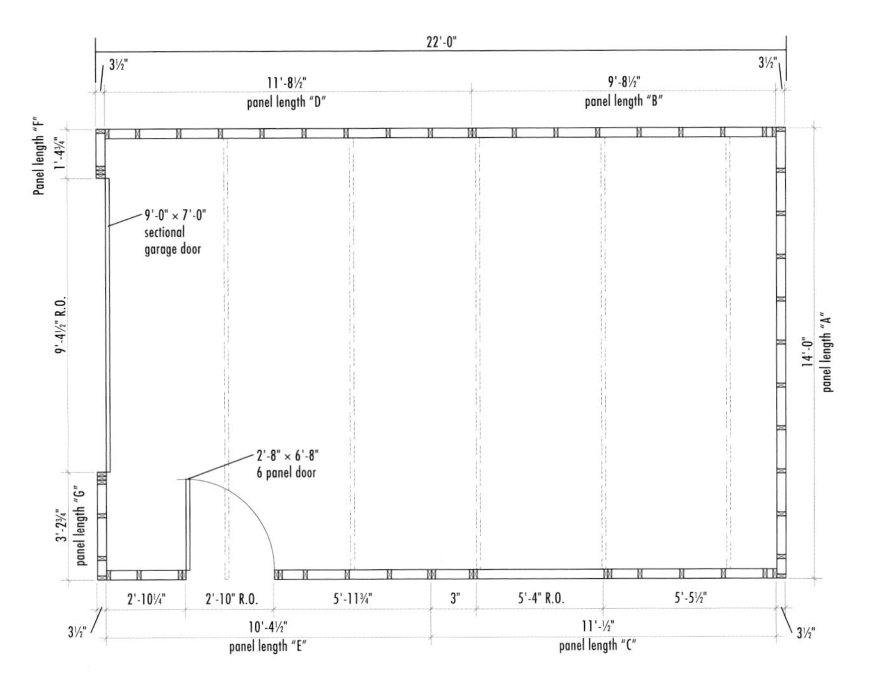

Overhead Door Jamb Detail

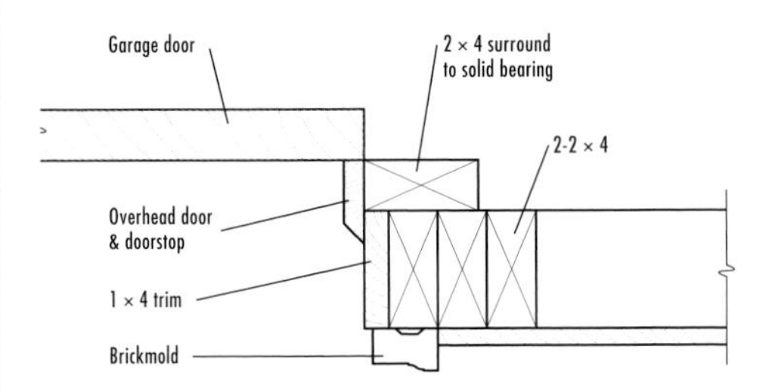

Typical Corner Detail

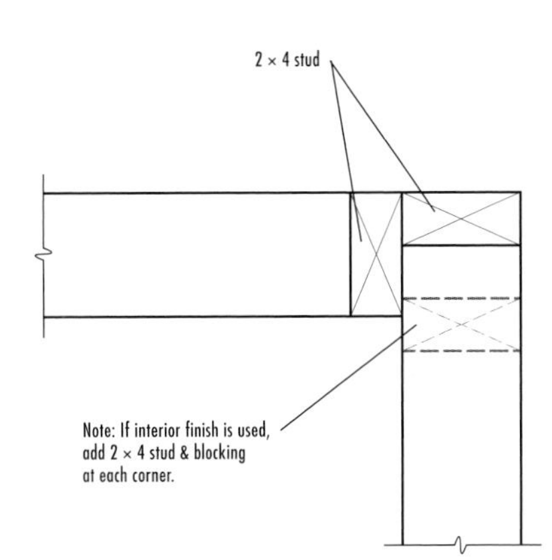

Service Door Jamb Detail

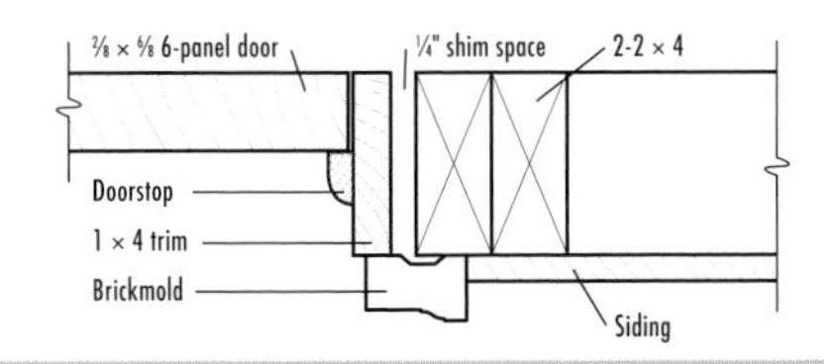

Building Section

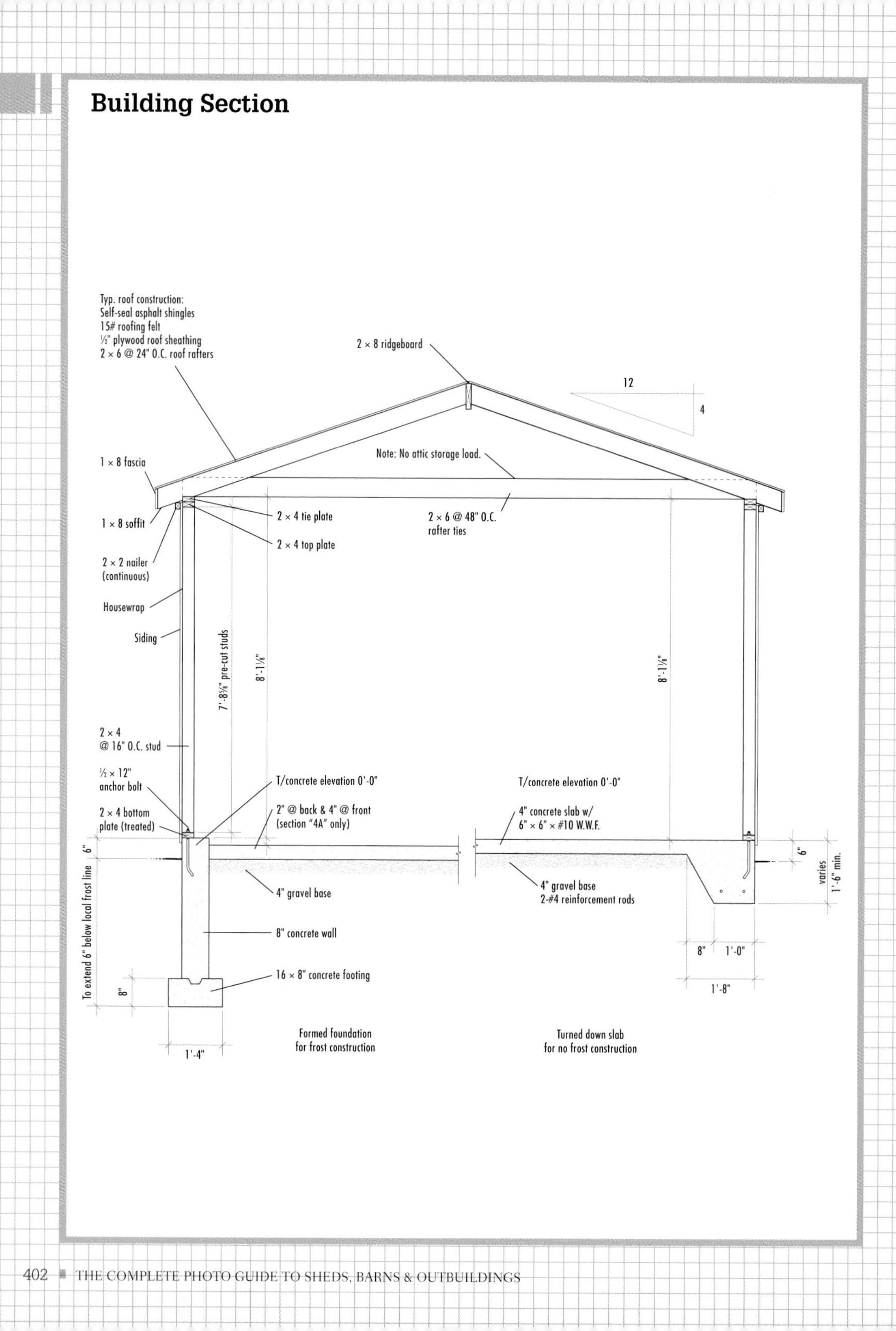

Side Wall Framing Elevation

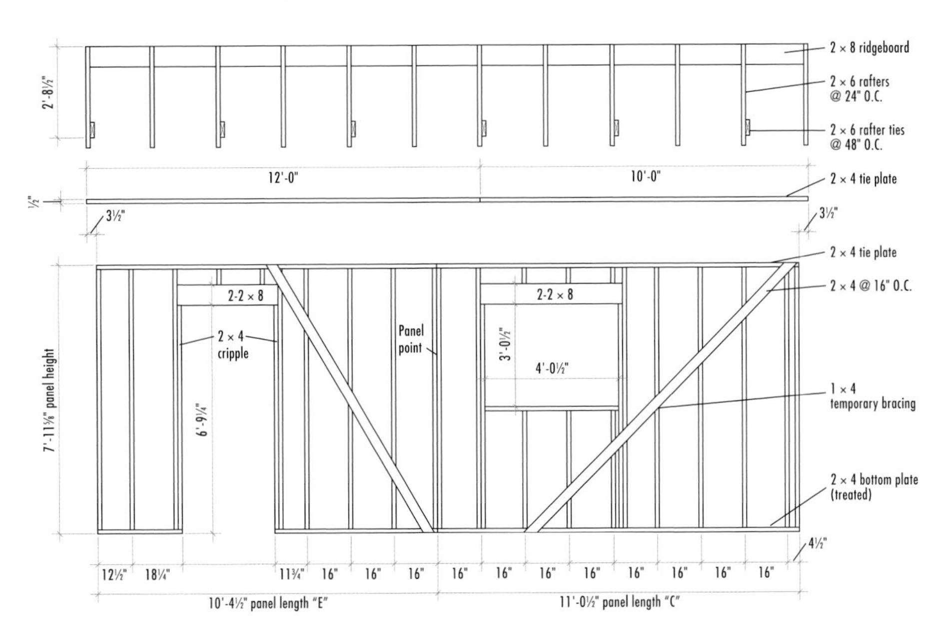

Front Framing Elevation

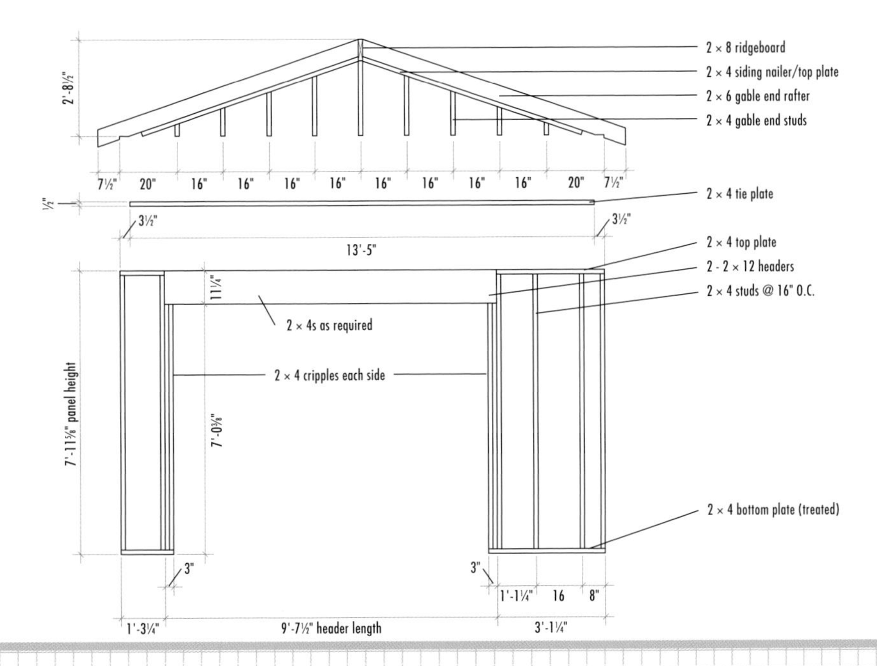

Compact Garage

The Compact Garage is designed for exceptional versatility and ample storage space. This classic gabled outbuilding has a footprint that measures 12 × 16 feet, and it includes several useful features. For starters, its 8-foot-wide overhead garage door provides easy access for large equipment, supplies, projects, or even a small automobile. The foundation and shed floor is a poured concrete slab, so it's ideal for heavy items like lawn tractors and stationary tools.

To the right of the garage door is a box bay window. This special architectural detail gives the building's façade a surprising houselike quality while filling the interior with natural light. And the bay's 33"-deep × 60"-wide sill platform is the perfect place for herb pots or an indoor flower box. The adjacent wall includes a second large window and a standard service door, making this end of the garage a pleasant, convenient space for all kinds of work or leisure.

Above the main space of the garage is a fully framed attic built with 2 × 6 joists for supporting plenty of stored goods. The steep pitch of the roof allows for over three feet of headroom under the peak. Access to the attic is provided by a drop-down staircase that folds up and out of the way, leaving the workspace clear below.

The garage door, service door, staircase, and both windows of the garage are prebuilt factory units that you install following the manufacturer's instructions. Be sure to order all of the units before starting construction. This makes it easy to adjust the framed openings, if necessary, to match the precise sizing of each unit. Also consult your local building department to learn about design requirements for the concrete foundation. You may need to extend and/or reinforce the perimeter portion of the slab or include a footing that extends below the frost line. An extended apron (as seen in the Mini Gambrel Barn, page 114) is very useful if you intend to house vehicles in the garage.

This garage's ample interior space is perfect for hobbies, a workshop, or housing vehicles and equipment. An attic with a drop-down staircase is the perfect hideout for seldom-accessed stored items.

For extra punch, paint your overhead garage door to match your building's interior, or purchase an overhead door in a color complementary to your landscape.

The service door and bay windows make this project easily accessible and full of natural light.

Cutting List

Description	Quantity/Size	Material
Foundation		
Drainage material	2.75 cu. yd.	Compactable gravel
Concrete slab	Field measure	3,000 psi concrete
Mesh	200 sq. ft.	6 × 6", W2.9 × W2.9 welded wire mesh
Reinforcing bar	As required by local code	As required by local code
Wall Framing		
Bottom plates	1 @ 16', 2 @ 12' 1 @ 10'	2 × 4 pressure treated
Top plates	2 @ 14', 4 @ 12' 4 @ 10'	2 × 4
Standard wall studs	51 @ 8'* *may use 92⅝" precut studs	2 × 4
Diagonal bracing	5 @ 12'	1 × 4 (std. lumber)
Jack studs	5 @ 14'	2 × 4
Gable end studs	5 @ 8'	2 × 4
Header, overhead door	2 @ 10'	2 × 12
Header, windows	2 @ 10'	2 × 12
Header, service door	1 @ 8'	2 × 12
Header & stud spacers		See Sheathing, right
Box Bay Framing		
Half-wall bottom plate	1 @ 8'	2 × 4 pressure-treated
Half-wall top plate & studs	3 @ 8'	2 × 4
Joists	3 @ 8'	2 × 6
Window frame	4 @ 12'	2 × 4
Sill platform & top	1 sheet @ 4 × 8'	½" plywood
Rafter blocking	1 @ 8'	2 × 8
Roof Framing		
Rafters (& lookouts, blocking)	36 @ 10'	2 × 6
Ridgeboard	1 @ 18'	2 × 8
Attic		
Floor joists	16 @ 12'	2 × 6
Floor decking	6 sheets @ 4 × 8'	½" plywood
Staircase	1 unit for 22 × 48" rough opening	Disappearing attic stair unit
Exterior Finishes		
Eave fascia	2 @ 18'	2 × 8 cedar
Gable fascia	4 @ 10'	1 × 8 cedar

Description	Quantity/Size	Material
Drip edge & gable trim	160 lin. ft.	1 × 2 cedar
Siding	15 sheets @ 4 × 8'	⅝" T 1-11 plywood siding w/ vertical grooves 8" on center (or similar)
Siding flashing	30 lin. ft.	Metal Z-flashing
Overhead door jambs	1 @ 10', 2 @ 8'	1 × 6 cedar
Overhead doorstops	3 @ 8'	Cedar door stop
Overhead door surround	1 @ 10', 2 @ 8'	2 × 6
Corner trim	8 @ 8'	1 × 4 cedar
Door & window trim	4 @ 8', 5 @ 10'	1 × 4 cedar
Box bay bottom trim	1 @ 8'	1 × 10 cedar
Roofing		
Sheathing (& header, stud spacers)	14 sheets @ 4 × 8'	½" exterior-grade plywood roof sheathing
15# building paper	2 rolls	
Shingles	4⅔ squares	Asphalt shingles—250# per sq. min.
Roof flashing	10'6"	
Doors & Windows		
Overhead garage door w/hardware	1 @ 8 × 7'	
Service door	1 unit for 38 × 72⅞" rough opening	Prehung exterior door unit
Window	2 units for 57 × 41⅜"	Casement mullion window unit—complete
Fasteners & Hardware		
J-bolts w/nuts & washers	14	½-dia. × 12"
16d galvanized common nails	3 lbs.	
16d common nails	15 lbs.	
10d common nails	2½ lbs.	
8d box nails	16 lbs.	
8d common nails	5 lbs.	
8d galvanized siding nails	10 lbs.	
1" galvanized roofing nails	10 lbs.	
8d galvanized casing nails	3 lbs.	
Entry door lockset	1	

Foundation Plan

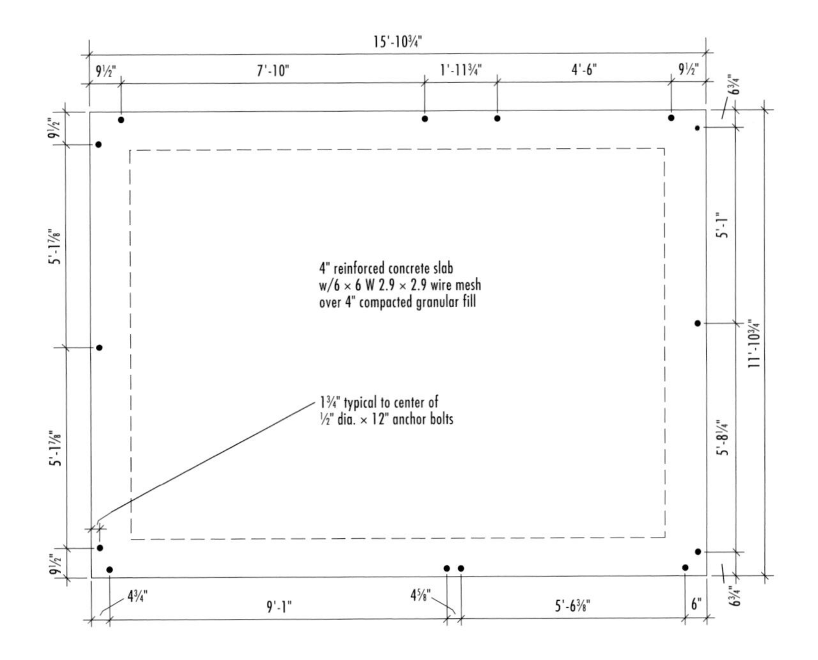

Foundation Detail

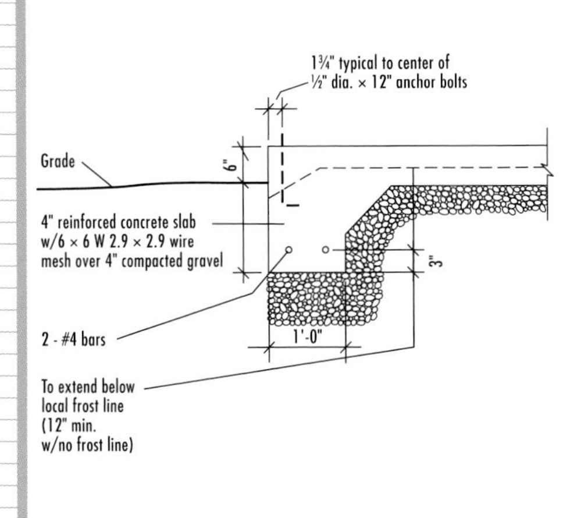

Front Elevation

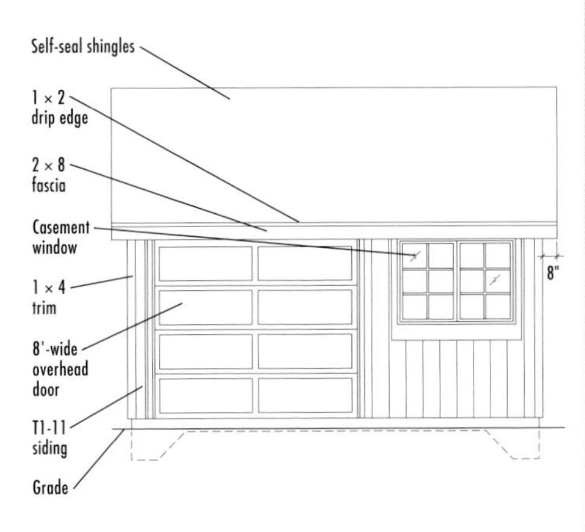

Building Section

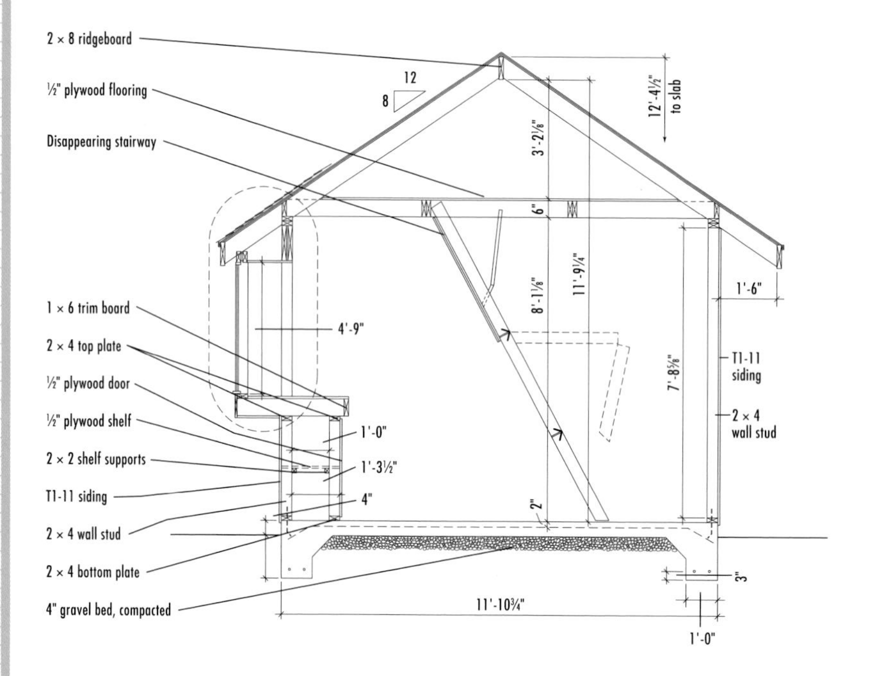

Right Side Elevation

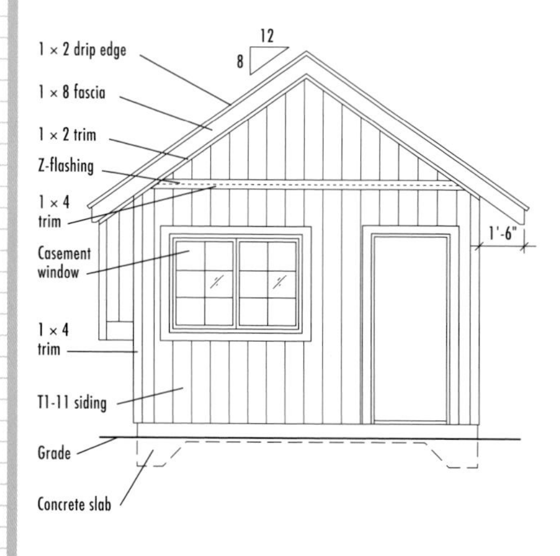

Rear Elevation

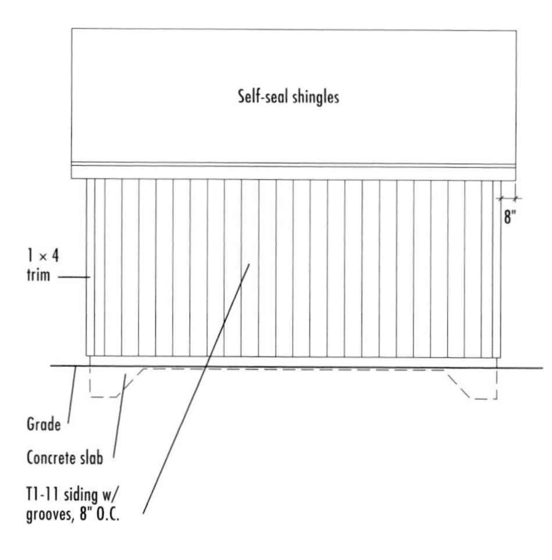

Wall Framing Plan

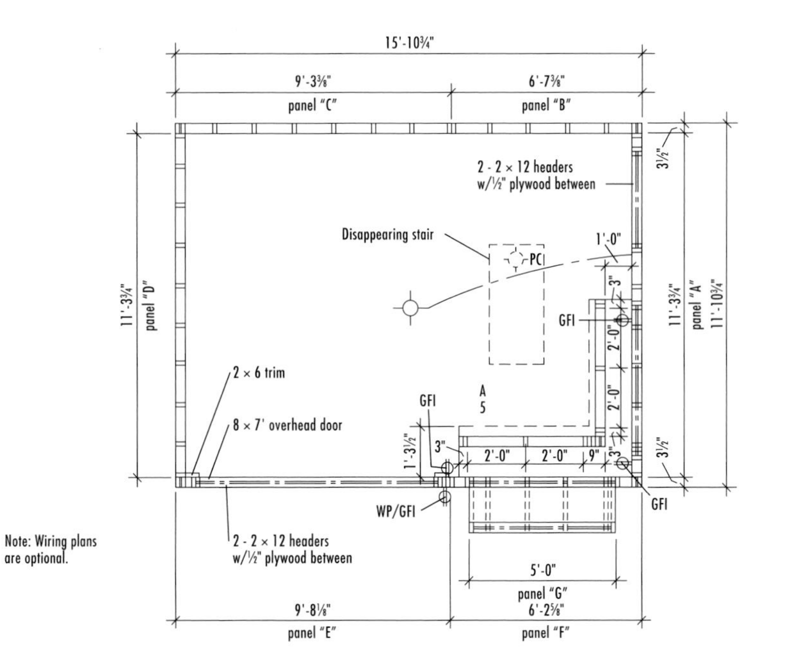

Back Side Framing

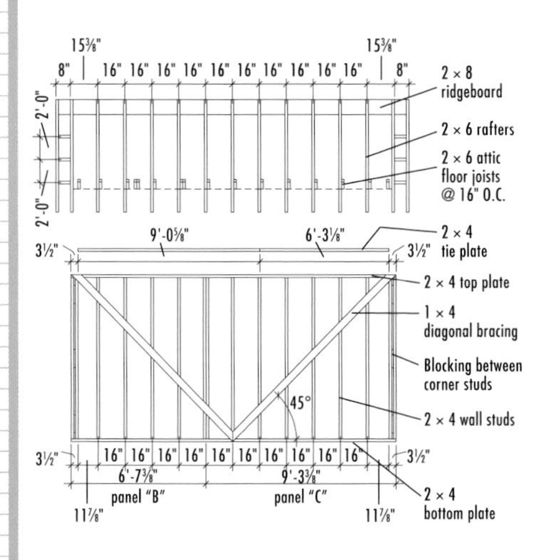

Side Framing

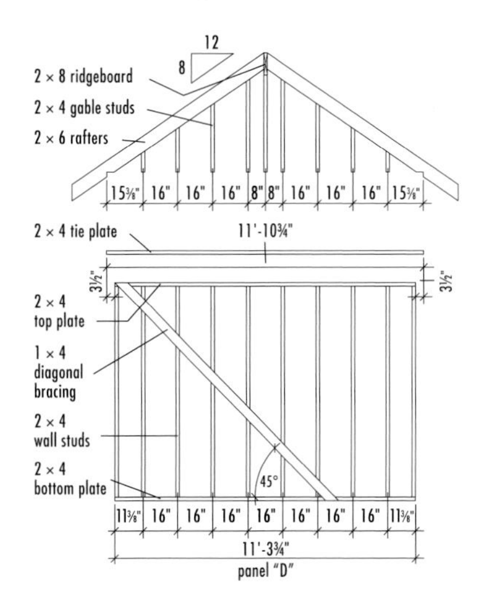

Front Side Framing

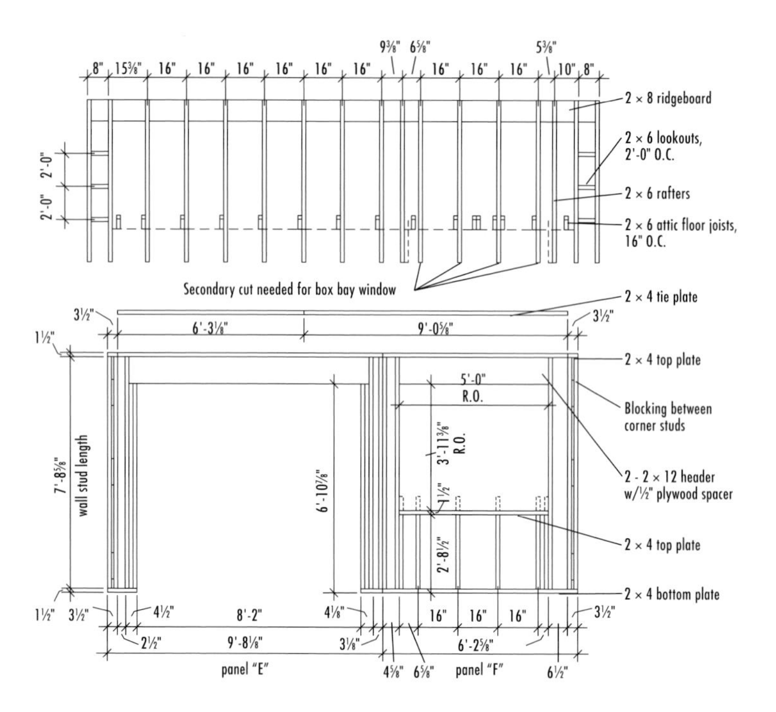

Attic Floor Joist Framing

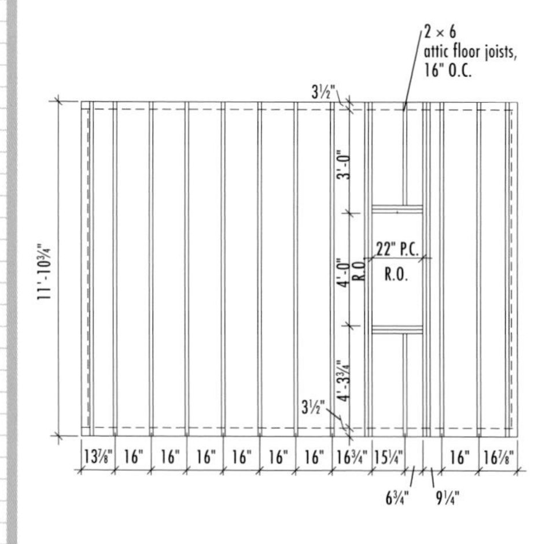

Box Bay Window Framing

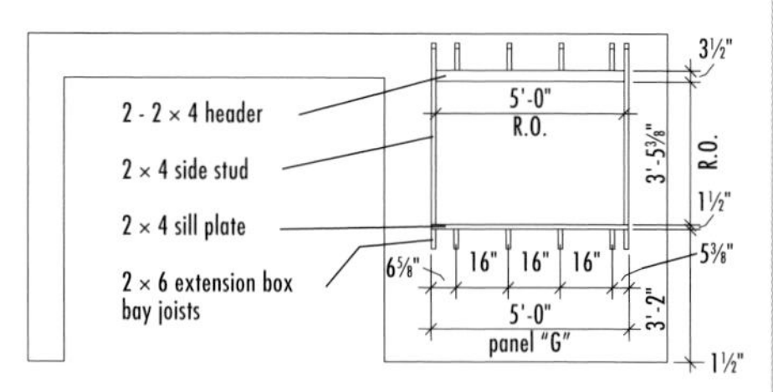

Overhead Door Header Detail

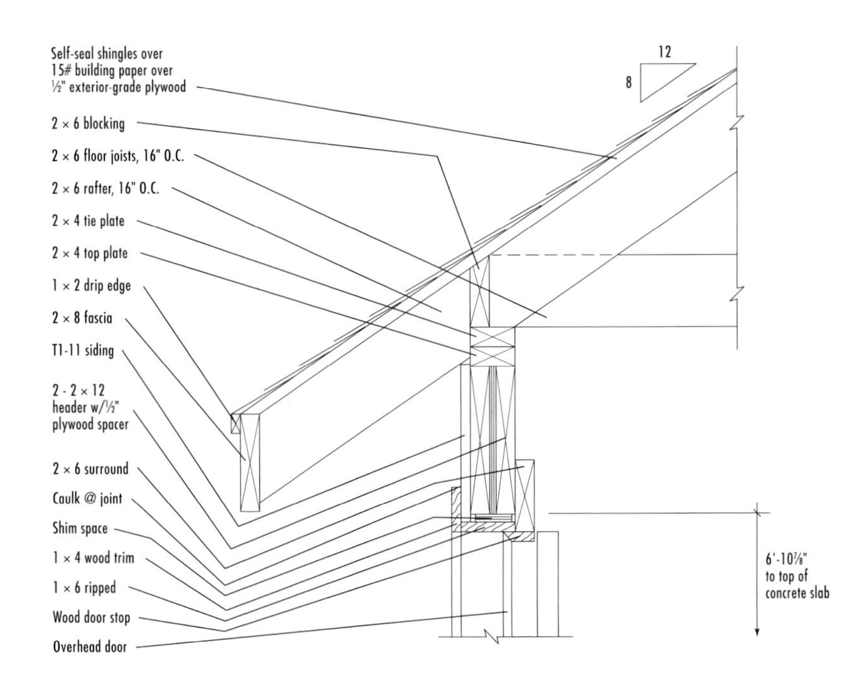

Overhead Door Jamb Detail

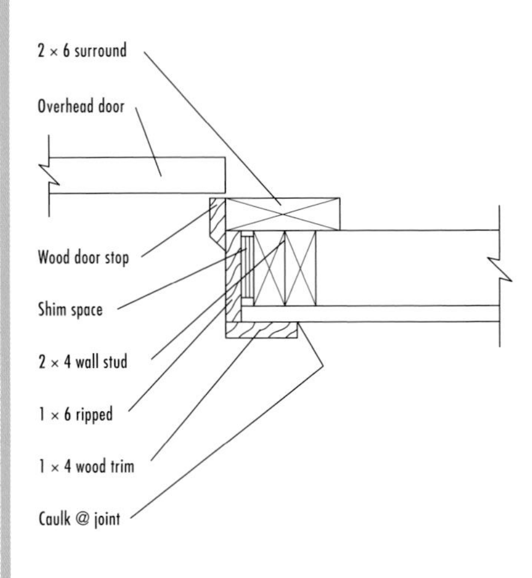

Service Door Header/Jamb Detail

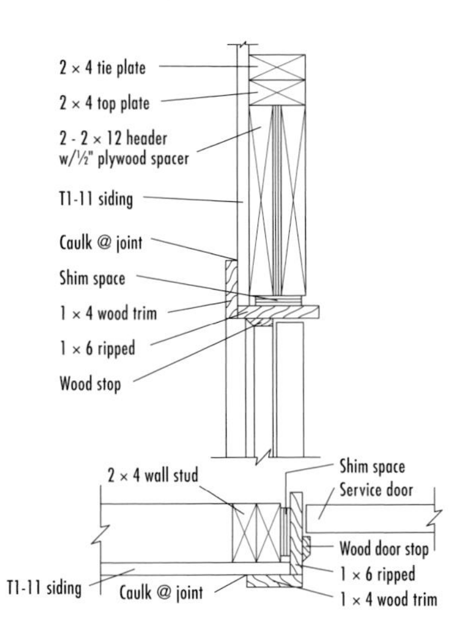

Rafter Template

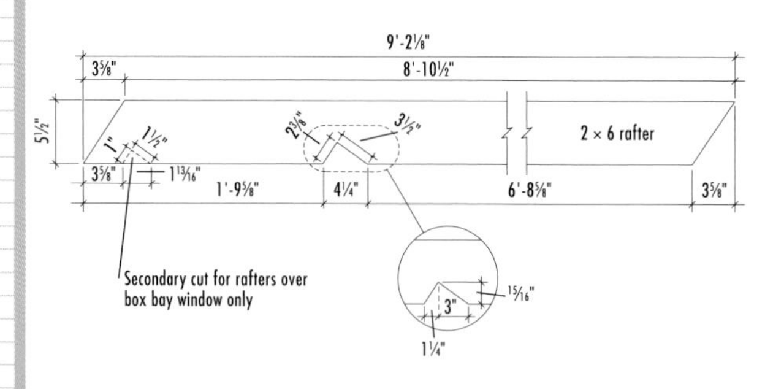

Corner Detail

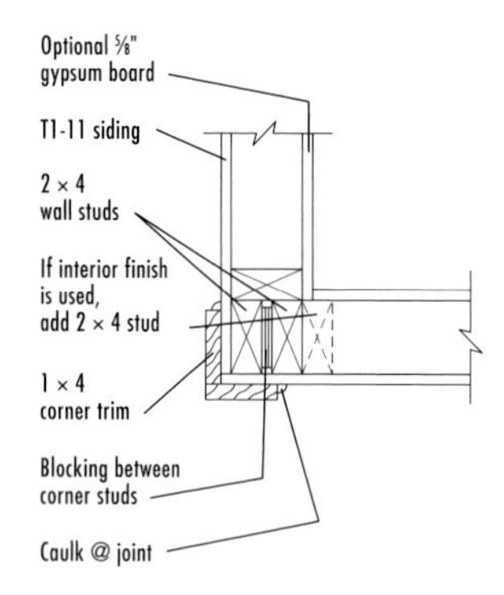

Box Bay Window Detail

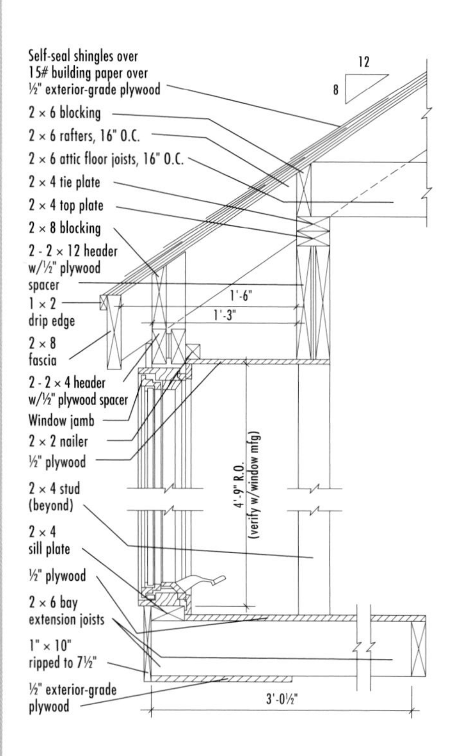

Isometric

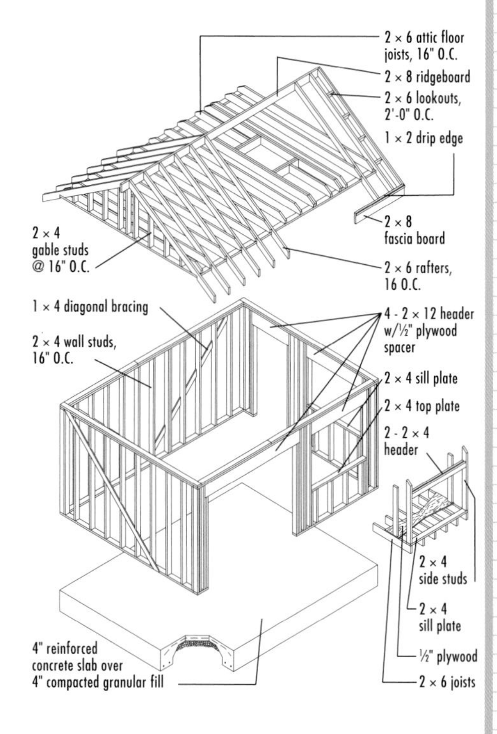

How to Build a Compact Garage

Build the concrete foundation following the FOUNDATION PLAN and FOUNDATION DETAIL (page 407). The slab should measure 190¾ × 142¾". Set the 14 J-bolts into the concrete as shown in the FOUNDATION PLAN. *Note: All foundation specifications must comply with local building codes.*

Snap chalk lines for the bottom plates so they will be flush with the outside edges of the foundation. You can frame the walls in four continuous panels or break them up into panels A through F, as shown in the WALL FRAMING PLAN (page 409). We completely assembled and squared all four walls before raising and anchoring them.

Frame the back wall(s) following BACK SIDE FRAMING (page 409). Use pressure-treated lumber for the bottom plate, and nail it to the studs with galvanized 16d common nails. All of the standard studs are 92⅝" long. Square the wall, then add 1 × 4 let-in bracing.

Raise the back wall and anchor it to the foundation J-bolts with washers and nuts. Brace the wall upright. Frame and raise the remaining walls one at a time, then tie all of the walls together with double top plates. Cover the outside of the walls with T1-11 siding.

(continued)

Cut fifteen 2 × 6 attic floor joists at 142¾". Cut the top corner at both ends of each joist: Mark 1⅞" along the top edge and 15⁄16" down the end; connect the marks, then cut along the line. Clipping the corner prevents the joist from extending above the rafters.

Mark the joist layout onto the wall plates following ATTIC FLOOR JOIST FRAMING (page 410). Leave 3½" between the outsides of the end walls and the outer joists. Toenail the joists to the plates with three 8d common nails at each end. Frame the rough opening for the staircase with doubled side joists and doubled headers; fasten doubled members together with pairs of 10d nails every 16". Install the drop-down staircase unit following the manufacturer's instructions.

Cover the attic floor with ½" plywood, fastening it to the joists with 8d nails.

Use the RAFTER TEMPLATE (page 412) to mark and cut two pattern rafters. Test-fit the rafters and adjust the cuts as needed. Cut all (24) standard rafters. Cut four special rafters with an extra birdsmouth cut for the box bay. Cut four gable overhang rafters—these have no birdsmouth cuts.

Cut the 2 × 8 ridgeboard at 206¾". Mark the rafter layout on the ridge and wall plates as shown in FRONT SIDE FRAMING (page 410) and BACK SIDE FRAMING (page 408). Frame the roof following the basic steps on pages 44 to 49. Install 6½"-long lookouts 24" on center, then attach the overhang rafters. Fasten the attic joists to the rafters with three 10d nails at each end.

Mark the stud layout for the gable end walls onto the end wall plates following SIDE FRAMING (page 409). Transfer the layout to the rafters using a level. Cut each of the 2 × 4 studs to fit, mitering the top ends at 33.5°. Install the studs flush with the end walls.

Construct the 2 × 4 half wall for the interior apron beneath the box bay. Cut two plates at 60" (pressure-treated lumber for bottom plate); cut five studs at 32½". Fasten one stud at each end, and space the remaining studs evenly in between. Mark a layout line 12" from the inside of the shed's front wall (see the BUILDING SECTION, page 408). Anchor the half wall to the slab using masonry screws or a powder-actuated nailer.

Cut six 2 × 6 joists at 36½". Toenail the joists to the inner and outer half walls following the layout in BOX BAY WINDOW FRAMING (page 410); the joists should extend 15" past the outer shed wall. Add a 60"-long 2 × 4 sill plate at the ends of the joists. Cut two 2 × 4 side studs to extend from the sill plate to the top edges of the rafters (angle top ends at 33.5°), and install them. Install a built-up 2 × 4 header between the side studs 41⅜" above the sill plate.

(continued)

Install a 2 × 2 nailer ½" up from the bottom of the 2 × 4 bay header. Cover the top and bottom of the bay with ½" plywood as shown in the BOX BAY WINDOW DETAIL on page 412. Cut a 2 × 4 stud to fit between the plywood panels at each end of the 2 × 4 shed wall header. Fasten these to the studs supporting the studs and the header.

Bevel the top edge of the 2 × 6 blocking stock at 33.5°. Cut individual blocks to fit between the rafters and attic joists, and install them to seal off the rafter bays. See the OVERHEAD DOOR HEADER DETAIL (page 411). The blocks should be flush with the tops of the rafters. Custom-cut 2 × 8 blocking to enclose the rafter bays above the box bay header; see the BOX BAY WINDOW DETAIL on page 412.

Add 2 × 8 fascia to the ends of the rafters along each eave so the top outer edge will be flush with the top of the roof sheathing. Cover the gable overhang rafters with 1 × 8 fascia. Add 1 × 2 trim to serve as a drip edge along the eaves and gable ends so it will be flush with the top of the roof sheathing. Install the ½" roof sheathing.

Add Z-flashing above the first row of siding, then cut and fit T1-11 siding for the gable ends. Cover the flashed seam with 1 × 4 trim.

To complete the trim details, add 1 × 2 along the gable ends and sides of the box bay. Use 1 × 4 on all vertical corners and around the windows, service door, and overhead door. Rip down 1 × 10s for horizontal trim along the bottom of the box bay. Also cover underneath the bay joists with ½" exterior-grade plywood.

Rip cut 1 × 6 boards to 4⅛" wide for the overhead door jambs. Install the jambs using the door manufacturer's dimensions for the opening. Shim behind the jambs if necessary. Make sure the jambs are flush with the inside of the wall framing and are flush with the face of the siding. Install the 2 × 6 surround as shown in the OVERHEAD DOOR HEADER DETAIL and OVERHEAD DOOR JAMB DETAIL on page 411.

Install the two windows and the service door following the manufacturer's instructions. Position the jambs of the units so they will be flush with the siding, if applicable. Install the overhead door, then add stop molding along the top and side jambs. See the SERVICE DOOR HEADER/JAMB DETAIL on page 411.

Add building paper and asphalt shingles following the steps on pages 62 to 67.

Carport

A carport provides a low-cost alternative to a garage, protecting your vehicle from direct rain, snow, and sunlight. Because it is not an enclosed structure, a carport is not held to the same building restrictions as a garage. This carport plan provides a 12 × 16-foot coverage area that is large enough to accommodate most full-size vehicles. To simplify the construction process, premanufactured trusses are used. When ordering trusses, specify the roof pitch, the distance being spanned, and the amount of overhang of the rafter tails. Also, place your order a few weeks in advance of your project start date. Many lumberyard suppliers carry in-stock trusses in standard dimensions and roof pitches, such as a 10-foot span with a 6-in-12 pitch—the dimensions used in this project.

This project also features metal roofing panels, an attractive and easy-to-install roofing material that does not require conventional roof sheathing. The trusses are tied together with 2 × 4 purlins, which also provide nailers for the metal roof panels. The panels are fastened with self-tapping metal roofing screws with rubber washers to prevent water leakage. Because of the scale of this project, recruit the help of at least one other person.

A carport is faster, easier, and cheaper to build than a full garage and they don't block the view from the house, like an enclosed building would.

Tools, Materials & Cutting List

Eye and ear protection
Batter boards
Mason's string
Line level
Wood stakes
Tape measure
Power auger or clamshell digger
Shovel
Level
Reciprocating saw
Circular saw
Clamps
Hammer
Drill with bits
Ratchet wrench
Wood shims
Rafter square
Handsaw
Caulk gun
Rubber sealer strips
Framing square
Finishing tools and materials

Description	Quantity/Size	Material
Foundation		
Batter boards/braces	10 @ 8'-0"	2 × 4
Drainage material	1⅔ cu. ft.	Compactable gravel
Concrete tube forms	6 @ 14"-dia.	
Concrete	Field measure	3,000 psi concrete
Beam Framing		
Posts (6)	6 @ 12'	6 × 6 rough-sawn cedar
Side beams (4)	4 @ 16'	2 × 8 pressure treated
End beams (2)	2 @ 12'	2 × 8 pressure treated
Lateral beams (4)	4 @ 10'	2 × 8 pressure treated
Diagonal supports (8)	4 @ 8'	4 × 4 cedar
Roof Framing		
Gable braces (8)	4 @ 10', 2 @ 8'	2 × 4
Trusses, 2 end and 11 common (13)	13 @ 10' span	2 × 4 with 6-in-12 pitch
Purlins (10)	20 @ 8'	2 × 4
Metal hurricane ties	22, with nails	Simpson H-1
Metal hurricane ties	4, with nails	Simpson H-2.5
Roofing		
Metal roofing panels	8 @ 4' × 8'	With ridge cap and sealer strip

Description	Quantity/Size	Material
Gable Finishes		
Blocking (8)	5 @10'	1 × 6
Gable sheathing (4)	2 @ 4 × 8'	¾" CDX plywood
Gable end fascia (4)	4 @ 8'	1 × 6 cedar
Side fascia (2)	4 @ 10'	1 × 8 cedar
Siding (14)	14 @ 8'	Cedar siding with 6" reveal
Fasteners		
1½" deck screws		
2½" deck screws		
6d galvanized common nails		
8d galvanized common nails		
8d joist hanger nails		
10d galvanized common nails		
⅜ × 4" galvanized lag screws	48, with washers	
⅜ × 5" galvanized lag screws	12, with washers	
10d galvanized ringshank nails		
6d galvanized casing nails		
6d siding nails		
1" self-tapping metal roofing screws with rubber washers (as specified by metal roofing manufacturer)		
2½" self-tapping metal roofing screws with rubber washers (as specified by metal roofing manufacturer)		

Front Elevation

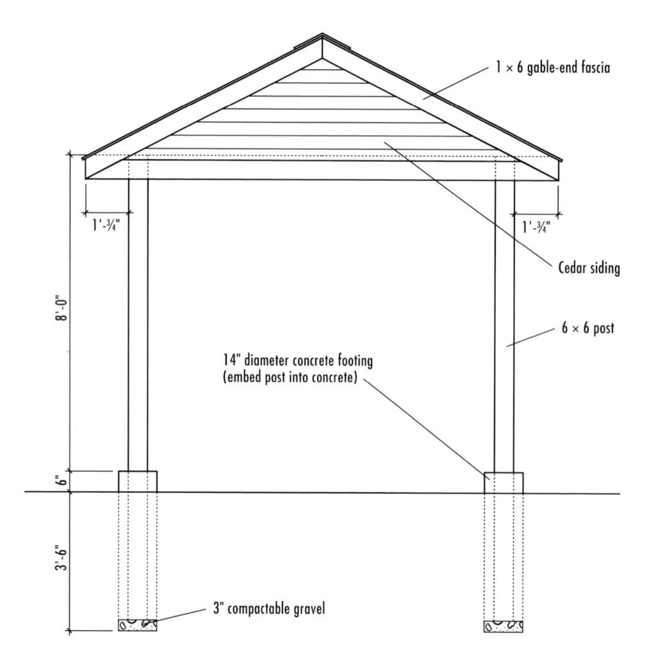

Side Elevation

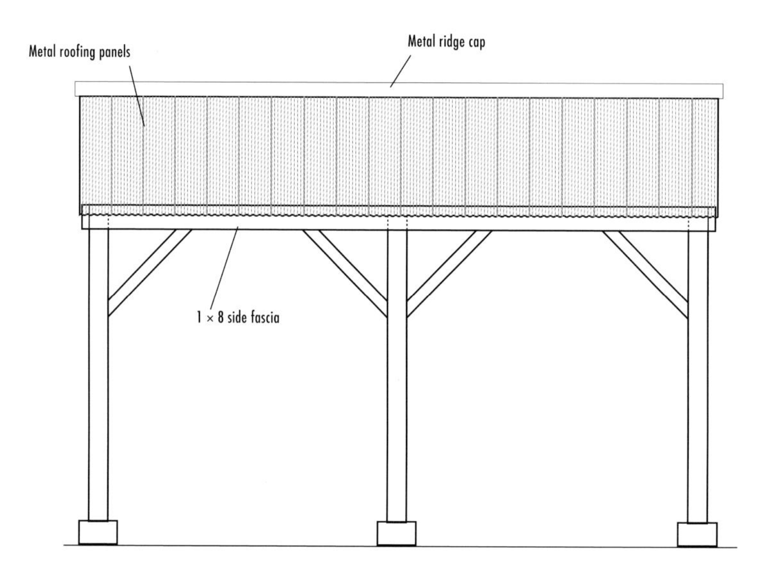

Front Section

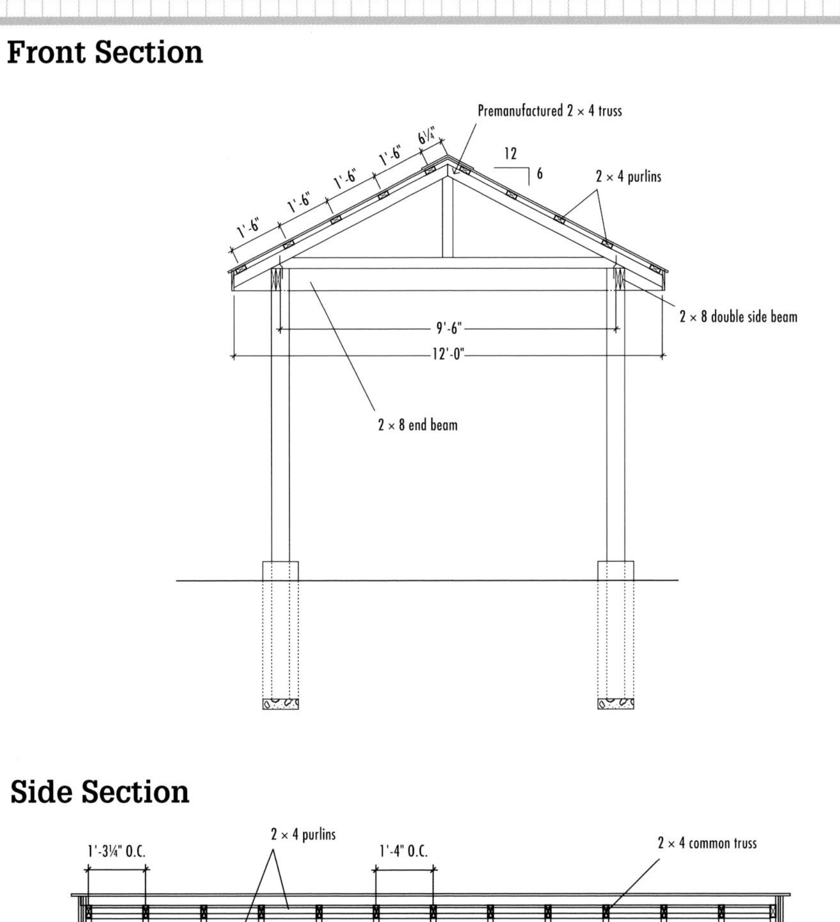

Side Section

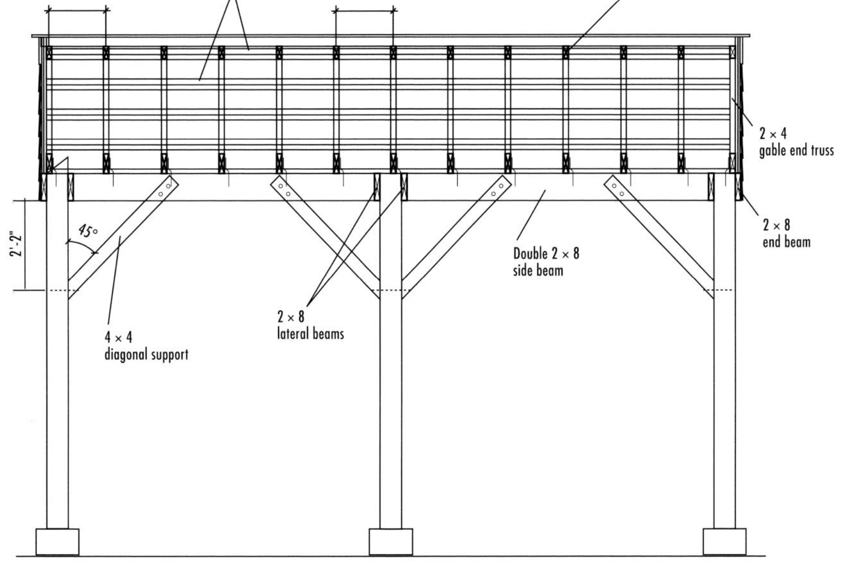

Beam Framing Plan

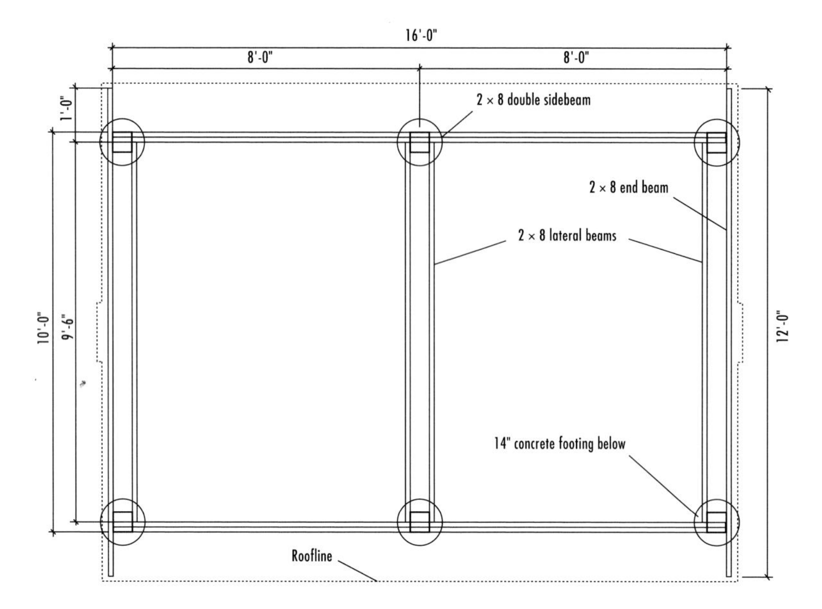

Diagonal Support Detail

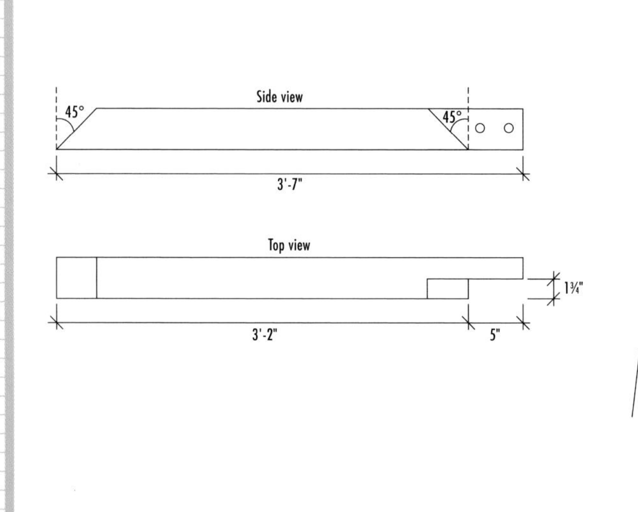

Gable End Detail

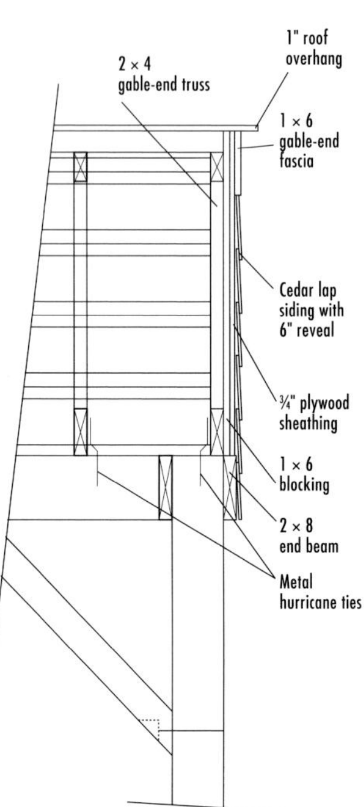

How to Build a Carport

Lay out the rough location of the carport with stakes and string, creating an area 10 feet wide and 16 feet long. Install ten 2 × 4 batter board sets with crosspieces about 2" below the tops of their support stakes. Run level mason's strings between the batter boards at the planned post locations. Measure and mark the exact post locations on the layout strings according to your plan, and then drive wooden stakes to mark their locations on the ground.

Dig post footing holes for 14"-dia. footings at least 6" deeper than your local frost line. Use a power auger or clamshell digger. Make sure the holes are centered around the stake locations. (Some building codes require bell-shaped flares at the footing bases.) Pour 3" of compactable gravel into each footing hole and tamp. Set a concrete tube form into each hole, then insert a 6 × 6 post that extends slightly higher than the final post height. Brace and plumb the posts and secure the forms with tamped soil. Fill the forms with concrete.

Trim the post tops. Mark the finished height (102") onto one post and draw a cutting line. Transfer the cutting line to all other posts using a mason's string and line level or a laser level as a reference. Trim the posts to height (see page 369). Mark a 3"-wide × 7¼"-deep notch on the outside face of each post, and then cut the notches with a reciprocating saw.

Install the side beams. Cut four 2 × 8s at 192" using a circular saw. Then clamp the boards together in pairs and facenail with 10d common nails to make the side beams. Lift the beams into the notches and clamp them into position so the ends of the beams are flush with the edges of the posts. Fasten each beam with two ⅜ × 5" galvanized lag screws and washers.

(continued)

Install the end beams. Cut two 2 × 8 end beams at 144" using a circular saw. Then lift and position the end beams against the ends of the posts with the top edges flush with the post tops. The beams should extend 12" past each post on each end. Securely clamp the beams in position and fasten with two 3/8 × 4" lag screws with washers per joint.

Install the lateral beams. Cut four 2 × 8 lateral beams to size and lift each beam into position between the side beams. Make sure the top edges of the beams are flush with the top of the posts and clamp them in place. Drill a pair of 1/2"-deep counterbore holes using a 1" spade bit, then drill 3 1/2"-deep, 1/4" pilot holes at each location. Fasten the lateral beams with 3/8 × 4" lag screws with washers.

Install the diagonal supports. Cut eight 4 × 4 diagonal supports to size, beveling one end and notching the other (see the DIAGONAL SUPPORT DETAIL, page 422). At each post, measure down from the side beam and mark at 26". Position the beveled end of the support against the post aligned with the mark and the notched end against the bottom edge of the inner member of the side beam. Clamp the support to the side beam and attach with 3/8 × 4" lag screws with washers driven through counterbored pilot holes. Mark the truss layout onto the tops of the side beams, following SIDE SECTION on page 421.

Install the first truss. Place a gable truss on the ends of the side beams. Extend a pair of long 2 × 4s to the ground and clamp them to the truss so the truss is held plumb. Use wood shims at the braces to keep the truss plumb, if necessary. Align each truss with the reference marks on the side beams, and then measure the overhang of each rafter tail to ensure proper placement. Toenail the truss in place using 10d galvanized common nails. Install the truss at the other end of the carport. *Tip: String a taut mason's string between the rafter tails at each carport end so it spans the length. Use these as references for installing the common trusses.*

Install the common trusses. Lift each truss up so its ends rest on the side beams—it's easier to do this if the truss is upside down (with the peak facing downward). When you are ready to install it, flip the truss right-side-up and position it on the beams. Toenail the trusses to the beams with 10d nails. Install the trusses in order, tacking a 1 × 4 brace to the top chord to maintain the correct spacing and alignment. If you will be installing purlins for a panel roof, install one now in lieu of a brace (see step 10, below). Anchor the trusses to the side beams at each end with the appropriate framing connectors.

Install the purlins. Metal and fiberglass roof panels don't require a deck, but they normally need to have evenly spaced sets of wood strips beneath them for reinforcement. Called purlins, these strips are mounted perpendicular to the trusses. Often, they are used to secure a profiled filler strip that fits underneath the roofing panel to support the profiled shape and create a seal. To install the purlins, snap chalk lines across the trusses, following the spacing shown in FRONT SECTION on page 421. Fasten the purlins to the rafter chords with 10d galvanized ringshank nails.

(continued)

Enclose the gable ends. Install 1 × 6 blocking to fur out the chords and struts of the gable-end trusses. Measure the triangular shape of the gable-end wall from the top edge of the end beam to the top edge of the blocking. Divide the area into two equal-sized triangular areas, and cut ¾" plywood sheathing to fit. Attach the sheathing with 1½" deck screws.

Cut 1 × 6 fascia boards—two for each gable end—long enough to extend from the peak to several inches past the ends of the rafter tails. Use a rafter square to mark the peak ends of the boards for the roof pitch, and then cut the angles. Fasten the gable-end fascia boards to the gable sheathing using 6d galvanized casing nails. Cut 1 × 8s to size for the side fascia boards and fasten with 6d galvanized casing nails driven into the ends of the rafter tails. Make sure the top edge of the fascia boards do not protrude above the top of the last row of purlins. Trim the ends of the gable-end fascia flush with the side fascia using a handsaw.

Install the roofing panels. Lay the first metal roofing panel across the purlins and position it so the finished edge of the panel extends approximately 1" beyond the gable-end fascia and 1" past the side fascia. Drive 1" metal roofing screws (these are sometimes called pole barn screws) with rubber washers through the roof panel into the purlins. Space the fasteners according to the manufacturer's directions. Install all panels, overlapping each preceding panel according to the manufacturer's directions. Work from one gable end to the other. Install the final panel so the finished edge overhangs the gable-end fascia by 1".

Install the ridge cap. To seal the roof ridge, a metal cap piece that matches the roof panels is screwed over the open seam. Mark the location for the rubber sealer strip on the starter purlin 6¼" from the peak of the roof. Run a bead of caulk along the reference line, and then install sealer strips on both sides of the peak. Apply a caulk bead to the tops of the sealer strips, and then center the preformed metal ridge cap over the peak so it overhangs the finished edges of the gable-end roof panels by 1". At each ridge of the metal roof panels, drive 2½" metal roofing screws with rubber washers through the ridge cap and sealer strip.

Ridge Caps ▸

A cutaway view of the ridge cap shows how the cap fits over the sealer strip. The caulk and the rubber sealer strip form a barrier to water and pests.

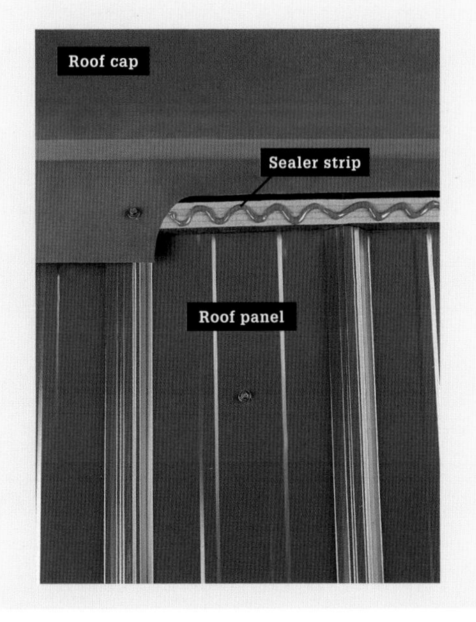

Install the siding. Choose a style and color of siding that matches or blends with your house siding and install it in the gable area. Here, cedar lap siding is being installed. Use a framing square or rafter square to mark cutting lines on the ends of each piece to match the roof pitch. Install a 2"-wide starter strip at the bottom, and then work your way up toward the gable peak. Maintain a consistent reveal, and nail the siding as specified by the siding manufacturer. Stain or seal any exposed cedar, such as the gable ends, side fascia, and posts.

Additional Garage Plans

Garage with Covered Porch

- Size: 24 × 22 ft.
- Building height: 13 ft.
- Roof pitch: 5/12
- Ceiling height: 8 ft.
- Overhead door: 9 × 7 ft.
- Roomy garage has space for storage
- Distinctive covered porch provides perfect area for entertaining

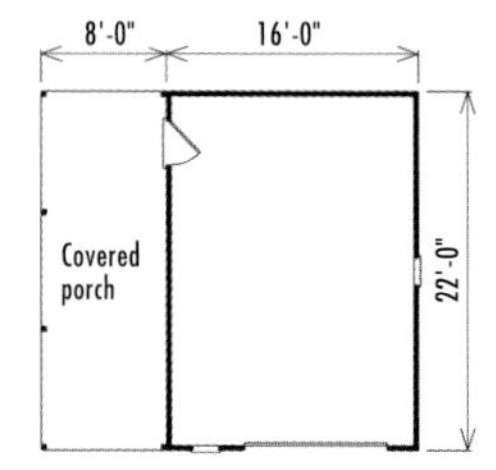

Design #002D-6010

Three-car Detached Garage

- Size: 32 × 22 ft.
- Building height: 12 ft., 2"
- Roof pitch: 4/12
- Ceiling height: 8 ft.
- Overhead doors: 9 × 7 ft., 16 × 7 ft.
- Side entry for easy access
- Perfect style with many types of homes

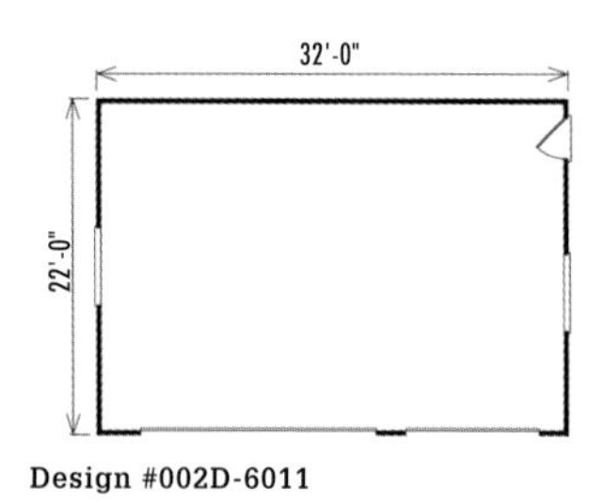

Design #002D-6011

Visit www.projectplans.com to order and view garage plans.

Two-car Detached Garage

- Size: 24 × 22 ft.
- Building height: 14 ft.
- Roof pitch: 5/12
- Ceiling height: 8 ft.
- Overhead door: 16 × 7 ft.
- Design with wonderful versatility

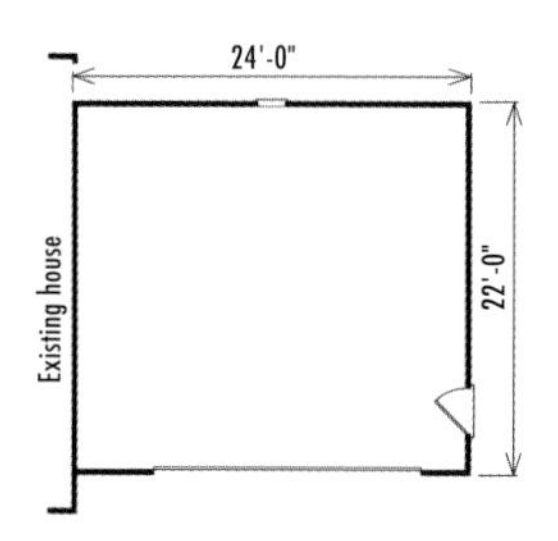

Design #002D-6014

Victorian Garage

- Size: 24 × 24 ft.
- Building height: 16 ft., 7"
- Roof pitch: 8/12
- Ceiling height: 8 ft.
- Overhead doors: (2) 9 × 7 ft.
- Accented with Victorian details
- Functional side entry

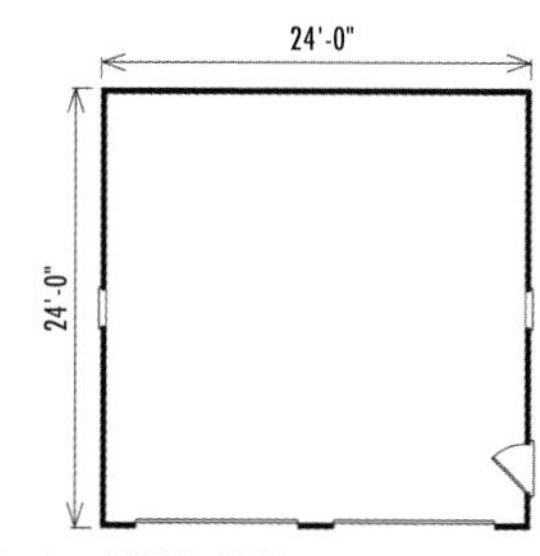

Design #002D-6018

Visit www.projectplans.com to order and view garage plans.

Reverse Gable Garage

- Size: 24 × 22 ft.
- Building height: 14 ft., 8"
- Roof pitch: $\frac{5}{12}$, $\frac{8.5}{12}$
- Ceiling height: 8 ft.
- Overhead doors: (2) 9 × 7 ft.
- Roof overhang above garage doors adds custom look
- Handy side door

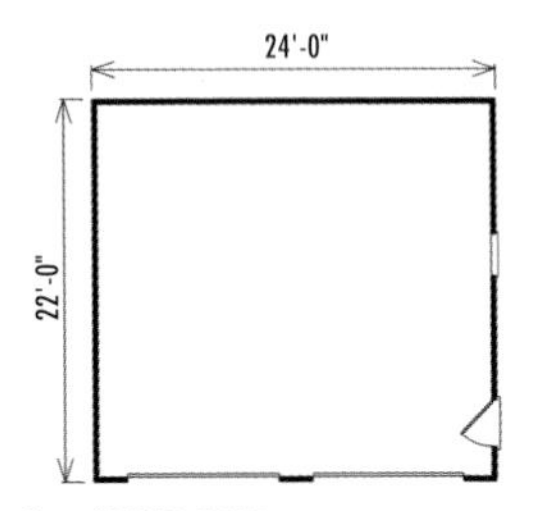

Design #002D-6040

Three-car Garage/Workshop

- Size: 24 × 36 ft.
- Building height: 14 ft., 6"
- Roof pitch: $\frac{4}{12}$
- Ceiling height: 10 ft.
- Overhead doors: (2) 9 × 8 ft.
- Oversized for storage
- Ideal size for workshop or maintenance building

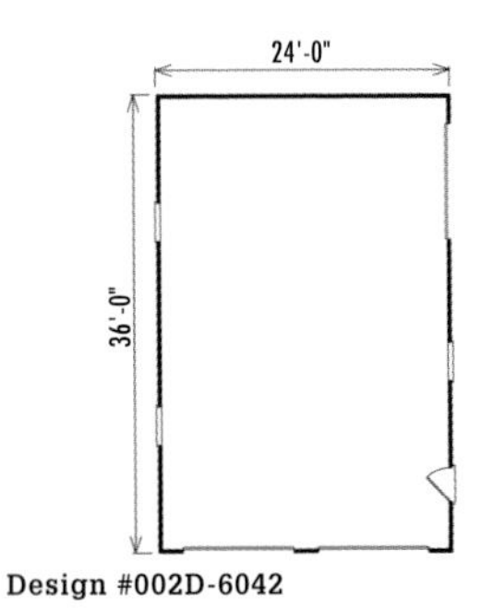

Design #002D-6042

Visit www.projectplans.com to order and view garage plans.

Three-stall Garage

- Size: 40 × 24 ft.
- Building height: 15 ft., 6"
- Roof pitch: 6/12
- Ceiling height: 9 ft.
- Overhead doors: (3) 9 × 7 ft.
- Oversized with plenty of room for storage
- Side door for easy access

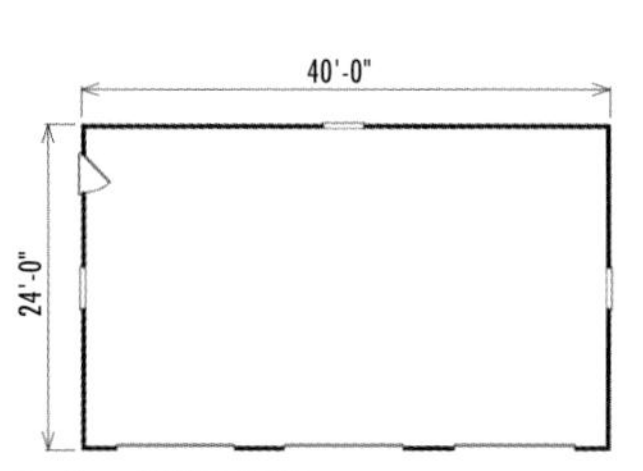

Design #002D-6046

Garage with Loft

- Size: 22 × 25 ft. 4"
- Building height: 20 ft., 6"
- Roof pitch: 7/12 (main), 3/12 (roof dormer)
- Ceiling height: 8 ft.
- Overhead door: 18 × 7 ft.
- Slab foundation

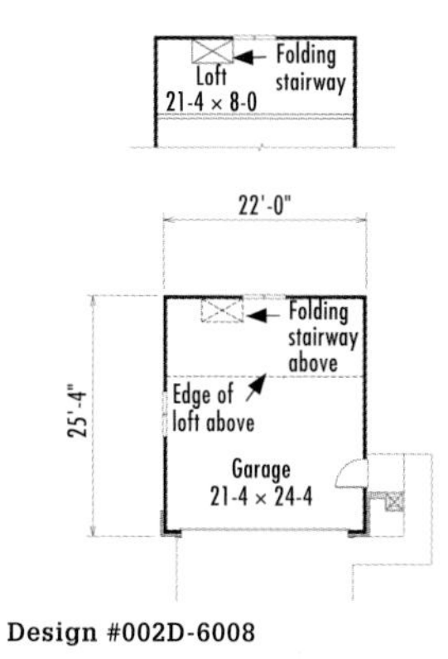

Design #002D-6008

Visit www.projectplans.com to order and view garage plans.

Appendix A: Foundations

Your outbuilding's foundation provides a level, stable structure to build upon and protects the building from moisture and erosion. In this brief appendix, you'll learn how to build three additional foundation types that are utilized in this book. The wooden skid foundation is an "on-grade" design, meaning that it is built on top of the ground and can be subject to rising and lowering a few inches during seasonal freezing and thawing of the underlying soil. This usually isn't a problem when an outbuilding is a small, freestanding structure. However, it can adversely affect some interior finishes (wallboard, for example).

When choosing a foundation type for your building, consider the specific site and the performance qualities of all systems in your climate; then check with the local building department to learn what's allowed in your area. Some foundations, such as concrete slabs, may classify your building as a permanent structure (see pages 28 to 33 to learn how to pour a concrete slab). *Note: Information for forming, reinforcing, and bracing deeper foundation walls is not included here. A safe rule of thumb is that the depth required to get below the frost line in cold climates is four feet, though colder places like Canada and Alaska can have frost depths up to eight feet. Check with your local building authority.*

Wooden Skid Foundation

A skid foundation couldn't be simpler: two or more treated wood beams or landscape timbers (typically 4 × 4, 4 × 6, or 6 × 6) set on a bed of gravel. The gravel provides a flat, stable surface that drains well to help keep the timbers dry. Once the skids are set, the floor frame is built on top of them and is nailed to the skids to keep everything in place.

Building a skid foundation is merely a matter of preparing the gravel base, then cutting, setting, and leveling the timbers. The timbers you use must be rated for ground contact. It is customary, but purely optional, to make angled cuts on the ends of the skids—these add a minor decorative touch and make it easier to "skid" the shed to a new location, if necessary.

Because a skid foundation sits on the ground, it is subject to slight shifting due to frost in cold-weather climates. Often a shed that has risen out of level will correct itself with the spring thaw, but if it doesn't, you can lift the shed with jacks on the low side and add gravel beneath the skids to level it.

Tools & Materials ▸

Excavation tools	Square
Rake	Treated wood timbers
4-ft. level	Compactable gravel
Straight, 8-ft. 4 × 4	Wood sealer-preservative
Hand tamper	Plate compactor
Circular saw	Eye and ear protection

How to Build a Wooden Skid Foundation

STEP 1: PREPARE THE GRAVEL BASE

A. Remove 4" of soil in an area about 12" wider and longer than the dimensions of the building.

B. Fill the excavated area with a 4" layer of compactable gravel. Rake the gravel smooth, then check it for level using a 4-ft. level and a straight, 8-ft.-long 2 × 4. Rake the gravel until it is fairly level.

C. Tamp the gravel thoroughly using a hand tamper or a rented plate compactor. As you work, check the surface with the board and level, and add or remove gravel as needed until the surface is flat and level.

STEP 2: CUT & SET THE SKIDS

A. Cut the skids to length using a circular saw (see page 369). (Skids typically run parallel to the length of the building and are cut to the same dimension as the floor frame.)

B. To angle-cut the ends, measure down 1½" to 2" from the top edge of each skid. Use a square to mark a

45° cutting line down to the bottom edge, then make the cuts.

C. Coat the cut ends of the skids with a wood sealer-preservative and let them dry.

D. Set the skids on the gravel so they are parallel and their ends are even. Make sure the outer skids are spaced according to the width of the building.

STEP 3: LEVEL THE SKIDS

A. Level one of the outside skids, adding or removing gravel from underneath as needed. Set the level parallel and level the skid along its length, then set the level perpendicular and level the skid along its width.

B. Place the straight 2 × 4 and level across the first and second skids, then adjust the second skid until it's level with the first. Make sure the second skid is also level along its width.

C. Level the remaining skids in the same fashion, then set the board and level across all of the skids to make sure they are level with one another.

Excavate the building site and add a 4" layer of compactable gravel. Level, then tamp the gravel with a hand tamper or rented plate compactor (inset).

If desired, mark and clip the bottom corners of the skid ends. Use a square to mark a 45° angle cut.

Using a board and a level, make sure each skid is level along its width and length and is level with the other skids.

Concrete Pier Foundation

Foundation piers are poured concrete cylinders that you form using cardboard tubes. The tubes come in several diameters and are commonly available from building materials suppliers. For a standard 8 × 10-ft. shed, a suitable foundation consists of one row of three 8"-diameter piers running down each long side of the shed.

You can anchor the structure's frame to the piers using a variety of methods. The simplest method (shown here) is to bolt a wood block to the top of each pier, then fasten the floor frame to the blocks. Other anchoring options involve metal post bases and various framing connectors that either set into the wet concrete or fasten to the piers after the concrete has cured. Be sure to consult your local building department for the recommended or required anchoring specifications.

Piers that extend below the frost line will keep your shed from shifting during annual freeze-thaw cycles. This is a standard requirement for major structures, like houses, but not typically for freestanding sheds (check with your building department). Another advantage of the pier foundation is that you can extend the piers well above the ground to accommodate a sloping site. *Note: Be sure to add a layer of compacted gravel underneath each footing to provide a stable base and promote water drainage.*

Concrete pier foundations can be used for sheds, gazebos, shelters, and other backyard structures. Before building, check with your local building authority to find out how deep the piers should be set into the ground in your area.

How to Build a Concrete Pier Foundation

Tools & Materials ▸

Circular saw
Drill
Mason's line
Sledgehammer
Line level
Framing square
Plumb bob
Shovel
Posthole digger
Reciprocating saw or handsaw
Utility knife
Ratchet wrench
2 × 4 lumber
2½" screws
Stakes
Nails
Masking tape
Cardboard concrete forms
Paper
Concrete mix
J-bolts with washers and nuts
2 × 10 pressure-treated lumber (rated for ground contact)
Eye, ear, and glove protection

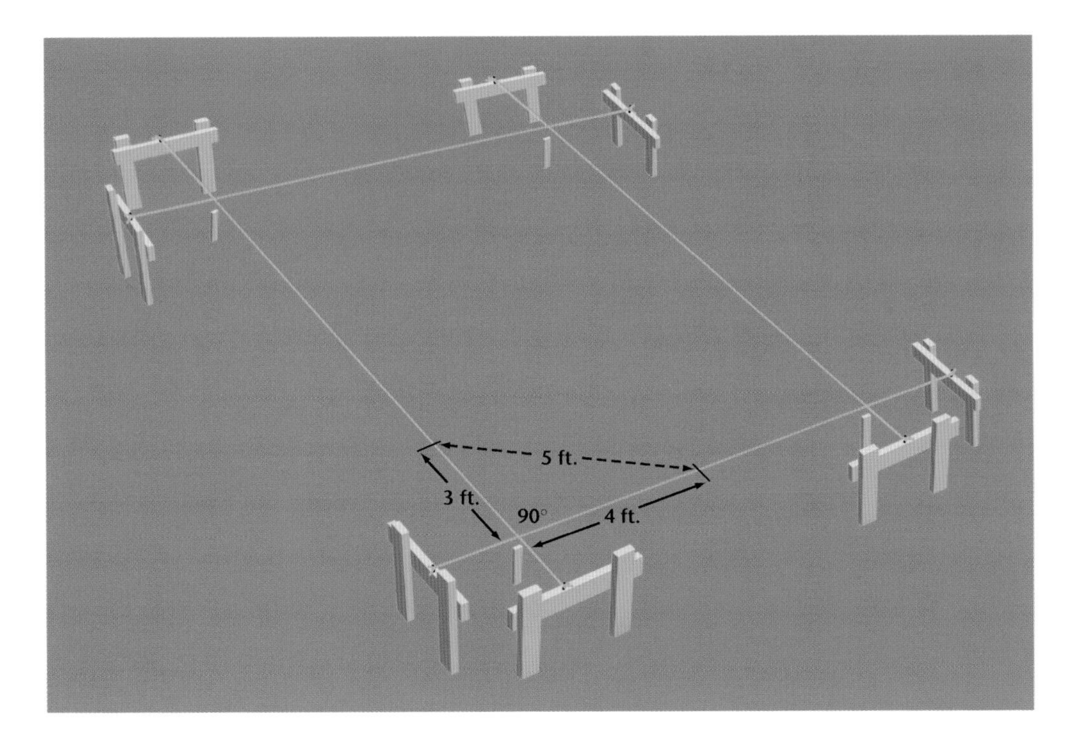

STEP 1: CONSTRUCT THE BATTER BOARDS

A. Cut two 24"- long 2 × 4 legs for each batter board (for most projects you'll need eight batter boards total). Cut one end square and cut the other end to a sharp point using a circular saw. Cut one 2 × 4 crosspiece for each batter board at about 18".

B. Assemble each batter board using 2½" screws. Fasten the crosspiece about 2" from the square ends of the legs. Make sure the legs are parallel and the crosspiece is perpendicular to the legs.

STEP 2: SET THE BATTER BOARDS & ESTABLISH PERPENDICULAR MASON'S LINES

A. Measure and mark the locations of the four corner piers with stakes, following your project plan.

B. Set two batter boards to form a corner about 18" behind each stake. Drive the batter boards into the ground until they are secure, keeping the crosspieces roughly level with one another.

(continued)

Cut the batter board pieces from 2 × 4 lumber and assemble them with screws.

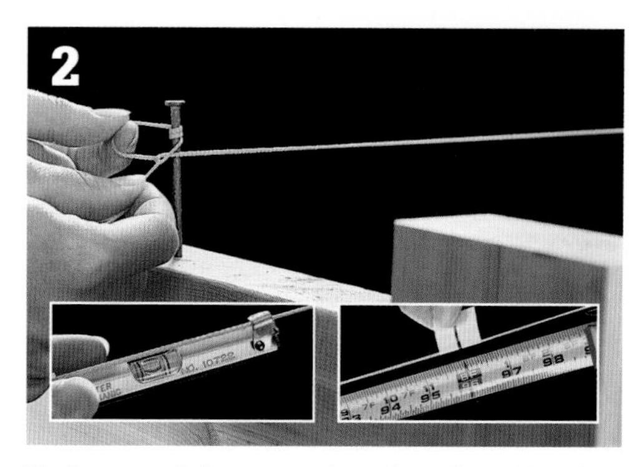

Tie the mason's lines securely to the nails, and level the lines with a line level (inset, left). Use tape to mark points on the lines (inset, right).

C. Stretch a mason's line between two batter boards at opposing corners (not diagonally) and tie the ends to nails driven into the top edge of the crosspieces; align the nails and line with the stakes. Attach a line level to the line, and pull the line very taut, making sure it's level before tying it.

D. Run a second level line perpendicular to the first: Tie off the end that's closest to the first string, then stretch the line to the opposing batter board while a helper holds a framing square at the intersection of the lines. When the lines are perpendicular, drive a nail and tie off the far end.

E. Confirm that the lines are exactly perpendicular using the 3-4-5 method: starting at the intersection, measure 3 ft. along one string and make a mark onto a piece of masking tape. Mark the other string 4 ft. from the intersection. Measure diagonally between the two marks; the distance should equal 5 ft. Reposition the second string, if necessary, until the diagonal measurement is 5 ft.

STEP 3: MARK THE FOOTING LOCATIONS

A. Following your plan, measure from the existing lines and use the 3-4-5 method to add two more perpendicular lines to form a layout with four 90° corners. Use the line level to make sure the mason's lines are level. The intersections of the lines should mark the centers of the corner piers, not necessarily the outside edge of floor framing.

B. Check the squareness of your line layout by measuring diagonally from corner to corner: when the measurements are equal, the frame is square. Make any necessary adjustments.

C. Plumb down with a plumb bob and place a stake directly under each line intersection. Mark the locations of intermediate piers onto the layout strings, then plumb down and drive stakes at those locations.

D. Untie each line at one end only, then coil the line and place it out of the way. Leaving one end tied will make it easier to restring the lines later.

STEP 4: SET THE FORMS

A. Dig holes for the forms, centering them around the stakes. The holes should be a few inches larger in diameter than the cardboard forms. The hole depth must meet the local building code requirements—add 4" to the depth to allow for a layer of gravel. For deep holes, use a posthole digger or a rented power auger. Add 4" of gravel to the bottom of each hole and tamp it with a 2 × 4.

B. Cut each cardboard form so it will extend at least 3" above the ground. The tops of all piers/forms should be level with each other. Also, the top ends of the forms must be straight, so place the factory-cut end up, whenever possible. Otherwise, mark a straight cutting line using a large piece of paper with at least one straight edge: Wrap the paper completely around the form so that it overlaps itself a few inches. Position the straight edge of the paper on the cutting mark, and align the overlapping edges of the paper with each other. Mark around the tube along the edge of the paper. Cut the tube with a reciprocating saw or handsaw.

C. Set the tubes in the holes and fill in around them with dirt. Set a level across the top of each tube to make sure the top is level as you secure the tube with dirt. Pack the dirt firmly, using a shovel handle or a stick.

Use a plumb bob to mark the pier locations. Drive a stake into the ground directly below the plumb bob pointer.

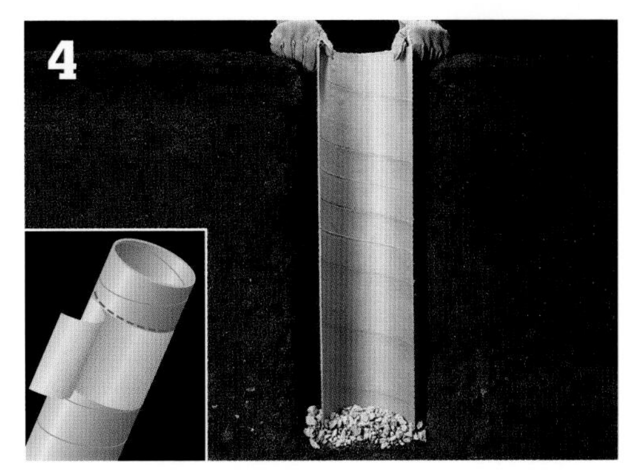

Wrap paper around the form to mark a straight cutting line (inset). Set the forms in the holes on top of a 4" gravel layer.

Fill the forms with concrete, then set the J-bolts. Check with a plumb bob to make sure the bolts are centered.

Anchor a block to each pier with a washer and nut. If desired, countersink the hardware (inset).

STEP 5: POUR THE CONCRETE

A. Restring the mason's lines and confirm that the forms are positioned accurately.

B. Mix the concrete following the manufacturer's directions; prepare only as much as you can easily work with before the concrete sets. Fill each form with concrete, using a long stick to tamp it down and eliminate air pockets in the concrete. Overfill the form slightly.

C. Level the concrete by pulling a 2 × 4 on edge across the top of the form using a side-to-side sawing motion. Fill low spots with concrete so that the top is perfectly flat.

D. Set a J-bolt into the wet concrete in the center of the form. Lower the bolt slowly, wiggling it slightly to eliminate air pockets. Use a plumb bob to make sure the bolt is aligned exactly with the mark on the mason's line.

Note: You can set the bolt at 1½" above the concrete so it will be flush with the top of the block, or extend it about 2½" so the washer and nut will sit on top of the block; doing the latter means you won't have to countersink the washer and nut. Make sure the bolt is plumb, then smooth the concrete around the bolt and let the concrete cure.

STEP 6: INSTALL THE WOOD BLOCKS

A. Cut 8 × 8" square blocks from 2 × 10 pressure-treated lumber that's rated for ground contact.

B. Drill a hole for the J-bolt through the exact center of each block; if you're countersinking the hardware, first drill a counterbore for the washer and nut.

C. Position each block on a pier, then add a galvanized washer and nut. Use the layout strings to align the blocks, then tighten the nuts to secure the blocks.

Setting Posts in Concrete

Burying posts in the ground with concrete provides strength and lateral stability, making it a good foundation system for small- to mid-scale projects. The concrete also helps protect the posts from ground moisture, so they last longer than they would if buried directly in the soil. However, it's a good idea to treat the bottom ends of posts before burying them, as an added measure against rot.

When digging postholes, make them 6" deeper than the post footing depth specified by the local building code. This leaves room for a layer of gravel that keeps water from collecting at the base of the post. Also follow the building code specs for posthole diameter; as a minimum, the holes should be several inches larger in diameter than the post size—about 8" for 4 × 4 posts, and 12"–14" for 6 × 6 posts.

Tools & Materials ▸

- Plumb bob
- Stakes & string
- Hand maul
- Power auger or posthole digger
- Shovel
- Coarse gravel
- Carpenter's level
- Drill screws
- Concrete
- Mason's trowel
- Pressure-treated, cedar, or redwood 4 × 4 posts
- Scrap lengths of 2 × 4
- Screw gun
- Screws
- Eye, ear, and glove protection

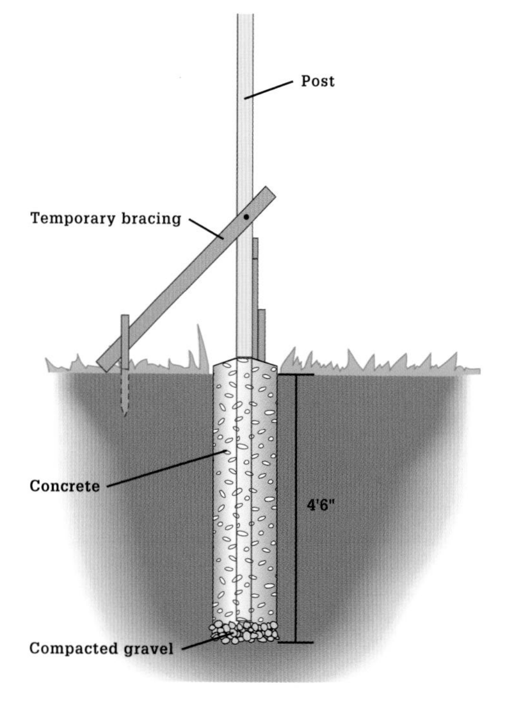

Protecting Buried Posts ▸

The most vulnerable parts of a buried post are the bottom end, which soaks up water when untreated, and the point where the post emerges from the ground or surrounding concrete. Here are a couple of popular methods for protecting posts against rot from moisture contact.

Stand posts in a pan of wood preservative and let them soak overnight. This protects the porous end grain from moisture.

Coat the buried portion of posts with roofing tar, covering the ends and all sides up to several inches above the points where they will emerge from the ground.

How to Set Posts in Concrete

STEP 1: MARK POST LOCATIONS

A. Transfer the marks from your layout string to the ground using a plumb bob to pinpoint the post locations.

B. Mark each post location with a stake, and remove the string.

STEP 2: DIG POSTHOLES

A. Dig postholes, using a power auger (available at rental centers) or posthole digger. Make each hole 6" deeper than the post footing depth specified by local building code or 12" past the frost line in cold climates. Keep the holes about twice the width of the post (check local regulations).

B. Pour a 6" layer of gravel into each hole for improved drainage. Tamp the gravel with a 2 × 4.

STEP 3: POSITION THE POSTS

A. Position each post in its hole. Check posts for plumb with a level. Unless you will trim them later, adjust posts to the correct height by adding or removing gravel until each post is at the same height.

B. Brace each post with scrap 2 × 4s secured to adjacent sides making sure the post is plumb.

C. If you're setting more than one post, make sure they're properly aligned with one another using mason's string. Adjust if necessary.

STEP 4: FILL THE POSTHOLES

A. Mix concrete and fill each posthole, overfilling it slightly.

B. Check to make sure each post is still plumb, then use a mason's trowel to shape the concrete around the bottom of the post to form a rounded crown that will shed water.

C. Let the concrete cure for 2 days before removing the braces.

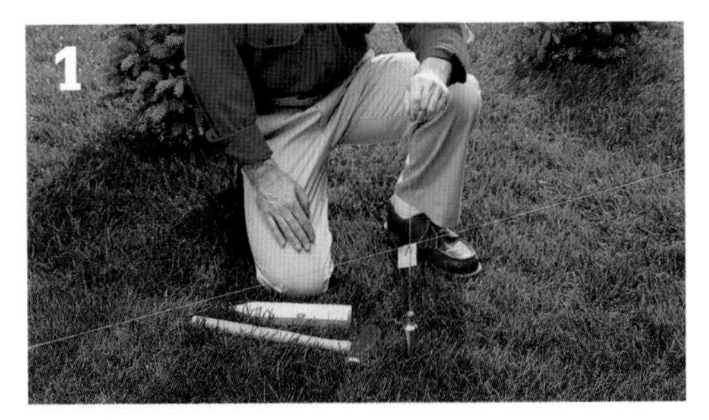

Drop a plumb bob from each post reference mark on the string to pinpoint the post centers on the ground.

Dig postholes 6" deeper than specified by local building code. Pour 6" of gravel into each hole to improve drainage.

Position each post in its hole. Brace the post with scrap pieces of 2 × 4 and stakes on adjacent sides.

Fill the postholes with premixed concrete, overfilling each slightly. Recheck the post for plumb and shape the concrete into a crown to shed water.

Appendix B: Cedar Shingle Roofing

Cedar shingles come in 16", 18", and 24" lengths and in random widths, generally between 3" and 10" wide. The exposure of the shingles depends on the slope of the roof and the length and quality of the shingles (check with the manufacturer). Because they're sold in a few different grades, make sure the shingles you get are good enough to be used as roofing. Also, be aware that galvanized nails may cause some staining or streaking on the shingles; if you can't accept that, use stainless-steel nails.

The project shown here includes 18" shingles with a 5½" exposure installed on a gable roof. At the ridge, the shingles are covered with a 1× cedar ridge cap, which is easier to install than cap shingles.

Tools & Materials ▸

Utility knife	T-bevel	15# or 30# building paper	Caulk gun
Chalk line	Stapler	Cedar roofing shingles	Caulk
Circular saw	Staples	6d galvanized roofing nails (3d, 6d)	Hammer
Table saw	2 × 4 lumber	1 × 4 and 1 × 6 cedar lumber	Eye, ear, and glove protection

How to Install Cedar Shingles

STEP 1: INSTALL THE STARTER COURSE

A. Apply building paper to the entire roof, overhanging the eaves by ⅜" (see pages 62 to 63).

B. Position the first shingle in the starter course so it overhangs the gable edge by 1" and the eave edge by 1½". Tack or clamp a 2 × 4 spacer to the fascia to help set the overhang. Make sure the butt (thick) end of the shingle is pointing down. Fasten the shingle with two 3d roofing nails, driven 4" up from the butt end and at least 1" from the side edges. Drive the nails just flush with the surface—countersinking creates a cavity that collects water.

C. Install the remaining shingles in the starter course, maintaining a ¼" to ⅜" gap between shingles. If necessary, trim the last shingle to width.

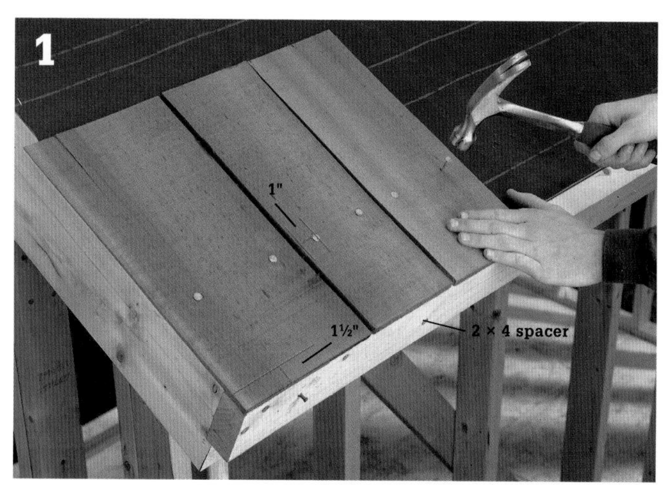

Install the starter row of shingles, overhanging the gable end by 1" and the eave by 1½".

STEP 2: INSTALL THE REMAINING COURSES

A. Set the first shingle in the first course so its butt and outside edges are flush with the shingles in the starter course and it overlaps the shingle gap below by 1½". Fasten the shingle 1" to 2" above the exposure line and 1" from the side edges.

B. Install the remaining shingles in the first course, maintaining a ¼" to ⅜" gap between shingles.

C. Snap a chalk line across the shingles at the exposure line (5½" in this example). Install the second course, aligning the butt ends with the chalk line. Make sure shingle gaps are offset with the gaps in the first course by 1½".

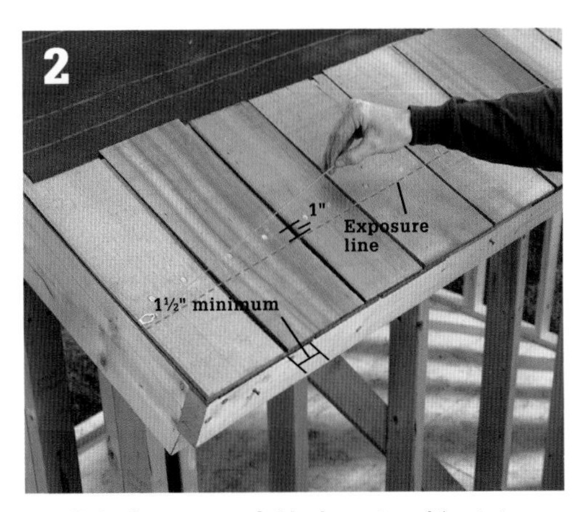

Install the first course of shingles on top of the starter course, offsetting the shingle gaps 1½" between the courses.

Cover the ridge with 24" of building paper, then a course of trimmed shingles. Repeat with 12" of paper and shingles.

D. Install the remaining courses using chalk lines to set the exposure. Measure from the ridge periodically to make sure the courses are parallel to the ridge. Offset the shingle gaps by 1½" with the gaps in the preceding three courses—that is, any gaps that are aligned must be four courses apart. Add courses until the top (thin) ends of the shingles are within a few inches of the ridge.

E. Shingle the opposite side of the roof.

STEP 3: SHINGLE THE RIDGE

A. Cut a strip of building paper to 24" wide and as long as the ridge. Fold the paper in half and lay it over the ridge so it overlaps the shingles on both sides of the roof; tack it in place with staples.

B. Install another course of shingles on each side, trimming the top edges so they are flush with the ridge. Cut another strip of building paper 12" wide, fold it, and lay it over these shingles.

C. Install the final course on each side, trimming the ends flush with the ridge. Nail the shingles about 2½" from the ridge.

STEP 4: INSTALL THE RIDGE CAP

A. Find the angle of the ridge using a T-bevel and two scraps of 1× board. Position the boards along the ridge with their edges butted together. Set the T-bevel to match the angle.

B. Transfer the angle to a table saw or circular saw and rip test pieces of 1×. Test-fit the pieces on the ridge, and adjust the angles as needed.

C. Cut the 1 × 6 and 1 × 4 cap boards to run the length of the ridge. Join the boards with caulk and 6d galvanized box nails. Attach the cap to the ridge with 6d roofing nails driven every 12".

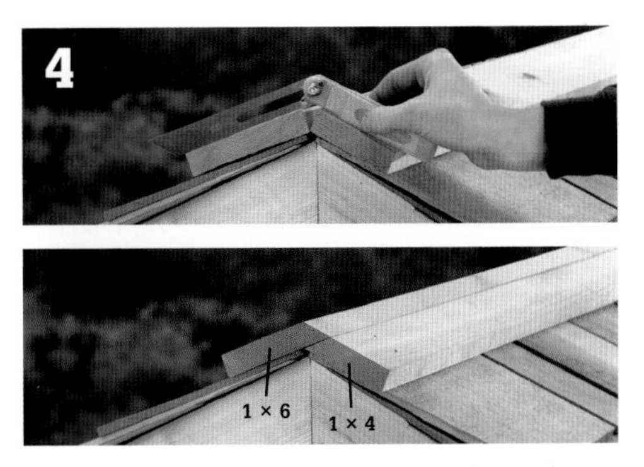

Use a T-bevel and scrap boards to find the ridge angle (above), then cut the 1 × 4 and 1 × 6 for the ridge cap.

Resources

Black & Decker
Portable Power Tools & More
www.blackanddecker.com

HDA Inc.
p. 160-173, 211-212, 248-261, 404-417, 428-431
800-373-2646
www.projectplans.com

Jamaica Cottage Shop
p. 6-9, 90-91, 210-211, 262-263, 264-281
866-297-3760
www.jamaicacottageshop.com

Palram Americas
p. 347, 349
800-999-9928
www.palramamericas.com

Quikrete Cos.
p. 33
800-282-5828
www.quikrete.com

Red Wing Shoes Co.
Workshoes and boots shown throughout book
800-733-9464
www.redwingshoes.com

Simpson Strong-Tie
p. 17
800-999-5099
www.strongtie.com

Sherman Pole Buildings
p. 4, 35, 94-113
800-464-6220
www.shermanpolebuildings.com

Photography Credits

Janet Backhaus
p. 11

Betty Mills Company
www.bettymills.com
p. 144 (top)

Cocinero Pty, Ltd T/A Sydney Sheds and Garages
www.sydneysheds.com.au
p. 418

DaVinci
www.davinciroofscapes.com
p. 22 (lower right)

EcoStar Carlisle / 800-211-7170
p. 22 (right)

Finley Products, Inc.
www.2×4basics.com
p. 145

iStock Photo
p. 23 (both), 50 (top), 225 (lower left)

Spirit Elements
www.spiritelements.com
p. 143

Sturdi-built
www.sturdi-built.com
p. 239 (top)

Summerwood Outdoors, Inc.
www.summerwood.com
p. 144 (lower)

Susan Teare
p. 6–7, 8–9, 90–91, 262–263, 264

TrestleWood.com
p. 22 (lower left)

Conversion Charts

Converting Measurements

To Convert:	To:	Multiply by:
Inches	Millimeters	25.4
Inches	Centimeters	2.54
Feet	Meters	0.305
Yards	Meters	0.914
Square inches	Square centimeters	6.45
Square feet	Square meters	0.093
Square yards	Square meters	0.836
Cubic inches	Cubic centimeters	16.4
Cubic feet	Cubic meters	0.0283
Cubic yards	Cubic meters	0.765
Pounds	Kilograms	0.454

To Convert:	To:	Multiply by:
Millimeters	Inches	0.039
Centimeters	Inches	0.394
Meters	Feet	3.28
Meters	Yards	1.09
Square centimeters	Square inches	0.155
Square meters	Square feet	10.8
Square meters	Square yards	1.2
Cubic centimeters	Cubic inches	0.061
Cubic meters	Cubic feet	35.3
Cubic meters	Cubic yards	1.31
Kilograms	Pounds	2.2

Lumber Dimensions

Nominal - U.S.	Actual - U.S. (in inches)	Metric
1 × 2	¾ × 1½	19 × 38 mm
1 × 3	¾ × 2½	19 × 64 mm
1 × 4	¾ × 3½	19 × 89 mm
1 × 5	¾ × 4½	19 × 114 mm
1 × 6	¾ × 5½	19 × 140 mm
1 × 7	¾ × 6¼	19 × 159 mm
1 × 8	¾ × 7¼	19 × 184 mm
1 × 10	¾ × 9¼	19 × 235 mm
1 × 12	¾ × 11¼	19 × 286 mm
2 × 2	1½ × 1½	38 × 38 mm

Nominal - U.S.	Actual - U.S. (in inches)	Metric
2 × 3	1½ × 2½	38 × 64 mm
2 × 4	1½ × 3½	38 × 89 mm
2 × 6	1½ × 5½	38 × 140 mm
2 × 8	1½ × 7¼	38 × 184 mm
2 × 10	1½ × 9¼	38 × 235 mm
2 × 12	1½ × 11¼	38 × 286 mm
4 × 4	3½ × 3½	89 × 89 mm
4 × 6	3½ × 5½	89 × 140 mm
6 × 6	5½ × 5½	140 × 140 mm
8 × 8	7¼ × 7¼	184 × 184 mm

Metric Plywood

Standard Sheathing Grade	Sanded Grade
7.5 mm (5/16")	6 mm (4/17")
9.5 mm (3/8")	8 mm (5/16")
12.5 mm (1/2")	11 mm (7/16")
15.5 mm (5/8")	14 mm (9/16")
18.5 mm (3/4")	17 mm (2/3")
20.5 mm (13/16")	19 mm (3/4")
22.5 mm (7/8")	21 mm (13/16")
25.5 mm (1")	24 mm (15/16")

Counterbore, Shank & Pilot Hole Diameters

Screw Size	Counterbore Diameter for Screw Head	Clearance Hole for Screw Shank	Pilot Hole Diameter	
			Hard Wood	Soft Wood
#1	.146 (9/64)	5/64	3/64	1/32
#2	1/4	3/32	3/64	1/32
#3	1/4	7/64	1/16	3/64
#4	1/4	1/8	1/16	3/64
#5	1/4	1/8	5/64	1/16
#6	5/16	9/64	3/32	5/64
#7	5/16	5/32	3/32	5/64
#8	3/8	11/64	1/8	3/32
#9	3/8	11/64	1/8	3/32
#10	3/8	3/16	1/8	7/64
#11	1/2	3/16	5/32	9/64
#12	1/2	7/32	9/64	1/8

Index

A

A-frame greenhouses, building
  about, 226
  step-by-step, 228-231
  tools & materials list, 226-227
American Gothic style, 186
Arbor retreats, building
  about, 360
  arch detail drawing, 367
  elevation drawings, 362-363
  post layout drawing, 363
  roof framing diagrams, 364, 366
  roof/slat plan, 365
  screen layout drawing, 367
  seat framing diagram, 364
  seat level roof framing diagram, 366
  seat section drawing, 366
  seat slat layout drawing, 367
  slat plan at seating, 365
  step-by-step, 368-373
  tools & materials list, 361
Asian-inspired gazebos, building
  about, 310
  corner detail drawing, 314
  door arch drawing, 312
  elevation drawings, 312, 313, 314
  floor plan, 313
  roof plan, 314
  roof truss template, 315
  slat section drawing, 315
  step-by-step, 316-319
  tools & materials list, 311
  truss top chord template, 315
  window bracket drawing, 312
Asphalt shingles
  about, 61
  installing, 63-66
  installing & preparing roof decks for, 62-63
Attics
  adding to simple storage sheds, 140
  See also Mini gambrel barns, building

B

Barns, building mini gambrel
  about, 114-115
  building section drawing, 117
  doors' details drawings, 122
  eave detail drawing, 121
  elevation drawings, 120, 122, 123
  floor plan, 118
  foundation detail drawing, 121
  materials list, 116
  overhangs drawings, 121
  rafter templates, 119
  step-by-step, 124-129
  window detail drawing, 122
Barn/shed plans
  barn storage, 212
  barn storage with loft, 212
  combined firewood & storage, 211
  gable storage with cupola, 213
  large gable storage, 213
  post & beam barn, 210
  run-in shed, 211
Barn storage shed plan, 212
Barn storage shed with loft plan, 212
Barns with partial lean-to overhang, building pole
  about, 94
  elevation drawings, 96-97
  floor plan, 96
  framing plans, 99
  roof detail drawings, 99, 100
  step-by-step, 101-113
  tools & materials list, 95
Birdsmouths
  gable end rafters and, 49
  template for, 45
Board lumber, about, 18
Building codes, 14

C

Carpenter Gothic style, 186
Carports, building
  about, 418
  beam framing plan, 422
  diagonal support detail drawing, 422
  elevation drawings, 420
  front section drawing, 421
  gable end detail drawing, 422
  side section drawing, 421
  step-by-step, 423-427
  tools & materials list, 419
Cast veneer stones (siding)
  about, 83-84
  applying, 84-85
Cedar lumber, 18
Cedar shingle roofing, installing, 440-441
Classical pergolas, building
  about, 374
  beam end templates, 378
  column connection drawing, 378
  elevation drawings, 376
  foundation plan, 377
  framing plans, 377, 378
  step-by-step, 379-383
  tools & materials list, 375
Clerestory studios, building
  about, 160-161
  building section drawing, 164
  door detail drawing, 166
  elevation drawings, 163, 167
  floor plan, 167
  framing drawings, 165-166
  jamb/corner detail drawing, 166
  materials list, 162
  rafter templates, 168
  rake detail drawing, 166
  step-by-step, 169-173
Closure strips, using, 357
Columns
  structural fiberglass, 374
  timeless, 380
Combined firewood & storage shed plan, 211
Compact detached garages, building
  about, 404-405
  attic floor joist framing drawing, 410
  box bay window detail drawing, 412
  box bay window framing drawing, 410
  building section drawing, 408
  corner detail drawing, 412
  elevation drawings, 407, 408
  foundation detail drawing, 407
  foundation plan, 407
  framing drawings, 409, 410
  isometric drawing, 412
  materials list, 407
  overhead door header detail drawing, 411
  overhead door jamb detail drawing, 411
  rafter template, 412
  service door header/jamb detail drawing, 411
  step-by-step, 413-417
Concrete
  piers, pouring, 290
  posts, setting in, 438-439
  slab-on-grade foundations, pouring, 30-33
  working with, 20
Concrete pier foundations, building, 434-437
Conservation greenhouses, described, 219
Construction (CONST) lumber, 18
Construction, overview of general outbuilding, 26-27
Corner lounges, building
  about, 384
  elevation drawing, 386
  fullum roof framing plan, 388
  post layout plan, 386
  roof framing plans, 388
  roof slat plan, 387
  screen layout, 387
  seat framing plan, 389
  seat slat layout, 389
  step-by-step, 390-395
  tools & materials list, 385
Corrugated polycarbonate materials, about, 349
Counterbored pilot holes, drilling, 159

D

Detached garages, building compact
  about, 404-405
  attic floor joist framing drawing, 410
  box bay window detail drawing, 412
  box bay window framing drawing, 410
  building section drawing, 408
  corner detail drawing, 412
  elevation drawings, 407, 408
  foundation detail drawing, 407
  foundation plan, 407
  framing drawings, 409, 410

isometric drawing, 412
materials list, 407
overhead door header detail drawing, 411
overhead door jamb detail drawing, 411
rafter template, 412
service door header/jamb detail drawing, 411
step-by-step, 413-417
Detached garages, building single
about, 398
building section drawing, 402
corner detail drawing, 401
elevation drawings, 400, 403
floor plan, 401
framing drawings, 403
materials list, 399
overhead door jamb detail drawing, 401
service door jamb detail drawing, 401
Dimensional lumber, about, 19
Doors
buying tips, 73
framing, 39-40, 42, 43, 69
installing, 68, 73-74
overhead, installing, 76-80
security, 75
sizing, 69
Drawings, working with, 15
Dutch kick, 115

E

8-sided gazebos, building
about, 320
building section drawing, 325
center pier detail drawing, 323
corner detail at roof beam line drawing, 328
deck edge detail drawing, 328
decking plan, 327
elevation drawing, 323
floor beam support detail drawing, 327
floor framing plan, 324
rafter hub detail drawing, 328
rafter templates, 326
roof edge detail drawing, 328
roof framing plan, 326
stair detail drawing, 327
step-by-step, 329-335
stringer template, 327
tools & materials list, 320, 321-322
Electricity, conduit installation, 30
Energy conservation, 23
Environmentally friendly materials, 22-23

F

Fascia, installing, 56-59
Fasteners, selecting, 17
Fiber-cement (lap) siding
about, 83
cutting, 87
installing, 86-89
working with, 240
Finished boards, about, 19
Finish lumber, about, 18
Firewood & storage shed plan, 211
Foundations, building
concrete pier, 434-437
slab-on-grade, 28-33
wooden skid, 432-433
Framing elements near ground, 94
Framing lumber, about, 18
Framing & raising walls, 34-43
Framing roofs, 44, 46-49
Furring strips, about, 19

G

Gable roofs, framing
about, 44
step-by-step, 46-49
Gable storage shed with cupola plan, 213
Gambrel barns, building mini
about, 114-115
building section drawing, 117
doors' details drawings, 122
eave detail drawing, 121
elevation drawings, 120, 122, 123
floor plan, 118
foundation detail drawing, 121
materials list, 116
overhangs drawings, 121
rafter templates, 119
step-by-step, 124-129
window detail drawing, 122
Garage doors, installing
about, 76
opener safety tips, 81
step-by-step, 77-80
Garage plans
with covered porch, 428
with loft, 431
reverse gable, 430
three-car detached, 428
three-car workshop and, 430
three-stall, 431
two-car detached, 429
Victorian, 429
Garages, building compact detached
about, 404-405
attic floor joist framing drawing, 410
box bay window detail drawing, 412
box bay window framing drawing, 410
building section drawing, 408
corner detail drawing, 412
elevation drawings, 407, 408
foundation detail drawing, 407
foundation plan, 407
framing drawings, 409, 410
isometric drawing, 412
materials list, 407
overhead door header detail drawing, 411
overhead door jamb detail drawing, 411
rafter template, 412
service door header/jamb detail drawing, 411
step-by-step, 413-417
Garages, building single detached
about, 398
building section drawing, 402
corner detail drawing, 401
elevation drawings, 400, 403
floor plan, 401
framing drawings, 403
materials list, 399
overhead door jamb detail drawing, 401
service door jamb detail drawing, 401
Garage with covered porch plan, 428
Garage with loft plan, 431
Garden sheds, building mini
about, 240-241
drawings of views, 243
step-by-step, 244-247
Garden sheds, building sunlight
about, 248-249
building section drawing, 251
door construction drawings, 255
elevation drawings, 254
framing plans, 252-253
header & window drawing, 256
materials list, 250
rafter templates, 256
rake board detail drawing, 256
soffit detail drawing, 254
step-by-step, 257-261
table & lower window drawing, 256
window drawings, 256
Gazebos, building lattice
about, 310
corner detail drawing, 314
door arch drawing, 312
elevation drawings, 312, 313, 314
floor plan, 313
framing plan, 314
roof plan, 314
roof truss template, 315
slat section drawing, 315
step-by-step, 316-319
truss top chord template, 315
window bracket drawing, 312
Gazebos, building 3-season
about, 283
building section drawing, 287
door frame detail drawing, 287
eave detail drawing, 289
elevation drawing, 285
floor framing plan, 286
floor plans, 286
foundation drawing, 285
rafter hub/corner detail drawing, 289
rafter templates, 288
roof framing plan, 288
stair detail drawing, 289
step-by-step, 290-296
tools & materials list, 283-284
window frame detail drawing, 287

Gazebos, building 8-sided
about, 320
building section drawing, 325
center pier detail drawing, 323
corner detail at roof beam line drawing, 328
deck edge detail drawing, 328
decking plan, 327
elevation drawing, 323
floor beam support detail drawing, 327
floor framing plan, 324
framing plans, 324, 326
rafter hub detail drawing, 328
rafter templates, 326
roof edge detail drawing, 328
roof framing plan, 326
stair detail drawing, 327
step-by-step, 329-335
stringer template, 327
tools & materials list, 320, 321-322
Gazebos, 6-sided. *See* Gazebos, building lattice
Gothic playhouses, building
about, 186-187
board & batten detail drawing, 191
deck railing detail drawing, 190
door architecture template, 191
door detail drawing, 191
drawing of entire, 189
floor plan, 190, 193
framing drawings, 192-193
materials list, 188
rafter template, 194
spire detail drawing, 190
step-by-step, 195-199
verge board template, 190
window box detail drawing, 194
Gothic Revival style, 186
Greek columns, 380
Greenhouses
conservation described, 219
lean-to described, 217
mansard described, 218
mini, described, 217
software to design, 218
three-quarter span, described, 217
Traditional span, described, 216
Greenhouses, building A-frame
about, 226
step-by-step, 228-231
tools & materials list, 226-227
Greenhouses, building kit
about, 232
preparation, 233, 237
step-by-step, 233-238
ventilation, 239

H

Hard-sided greenhouses, building
about, 232
preparation, 233, 237
step-by-step, 233-238
ventilation, 239
Hardware, selecting, 17
Holes, drilling counterbored pilot, 159
Hoophouses, building
about, 219, 220
anatomy of, 221
building tips, 222
materials list, 221
step-by-step, 222-224
variations, 225
Housewrap, 54-55

J

Japanese-inspired gazebos, building
about, 310
corner detail drawing, 314
door arch drawing, 312
elevation drawings, 312, 313, 314
floor plan, 313
roof plan, 314
roof truss template, 315
slat section drawing, 315
step-by-step, 316-319
tools & materials list, 311
truss top chord template, 315
window bracket drawing, 312

K

Kit greenhouses, building
about, 232
preparation, 233, 237
step-by-step, 233-238
ventilation, 239
Kit sheds, building
about, 142-143
metal, anchoring, 150
metal, maintaining, 150
metal, step-by-step, 146-150
shopping for, 144-145
wood, delivery, 152
wood, step-by-step, 152-159

L

Lap siding, fiber-cement
about, 83
cutting, 87
installing, 86-89
working with, 240
Large gable storage shed plan, 213
Lattice gazebos, building
about, 310
corner detail drawing, 314
door arch drawing, 312
elevation drawings, 312, 313, 314
floor plan, 313
roof plan, 314
roof truss template, 315
slat section drawing, 315
step-by-step, 316-319
tools & materials list, 311
truss top chord template, 315
window bracket drawing, 312
Lean-to greenhouses, described, 217
Lean-to overhang, building pole barns with partial
about, 94
elevation drawings, 96-97
floor plan, 96
framing plans, 99
roof detail drawings, 99, 100
step-by-step, 101-113
tools & materials list, 95
Lean-to tool bins, building
about, 200-201
building section drawing, 205
door jamb detail drawing, 206
elevation drawings, 204, 205, 206
floor plan, 206
framing plans, 203-204
materials list, 202
overhang detail drawing, 206
rafter template, 206
roof drawings, 203, 206
step-by-step, 207-209
Location, choosing
general considerations, 12
for greenhouses, 218
sunlight and seasonal considerations, 13
for sunlight garden shed, 249
Locksets, installing, 75
Lounge areas, building corner
about, 384
elevation drawing, 386
fullum roof framing plan, 388
post layout plan, 386
roof framing plans, 388
roof slat plan, 387
screen layout, 387
seat framing plan, 389
seat slat layout, 389
step-by-step, 390-395
tools & materials list, 385
Lumber posts, cutting, 369
Lumber, selecting, 18-19

M

Mansard greenhouses, described, 218
Metal kit sheds, building
about, 142
anchoring, 150
maintaining, 150
shopping for, 144-145
step-by-step assembly, 146-150
Metal roofing panels, 418
Micro-lam, about, 19
Mini gambrel barns, building
about, 114-115
building section drawing, 117
doors' details drawings, 122
eave detail drawing, 121
elevation drawings, 120, 122, 123
floor plan, 118
foundation detail drawing, 121
materials list, 116
overhangs drawings, 121
rafter templates, 119
step-by-step, 124-129
window detail drawing, 122
Mini garden sheds, building
about, 240-241
drawings of views, 243
step-by-step, 244-247
tools & materials list, 242
Mini (lean-to) greenhouses, described, 217
Movable structures. *See* Hoophouses, building; Mini garden sheds, building
Multitab shingles, about, 61

N

Nailing techniques, 17

O

Overhead doors, installing
about, 76
opener safety tips, 81
step-by-step, 77-80

P

Patio shelters, building
about, 346-347
plan, 348
step-by-step, 350-359
tools & materials list, 348-349
Pavilions, building. *See* Pool pavilions, building
Pergolas, building classical
about, 374
beam end templates, 378
column connection drawing, 378
elevation drawings, 376
foundation plan, 377
framing plans, 377, 378
step-by-step, 379-383
tools & materials list, 375
Pilot holes, drilling counterbored, 159
Plans, working with, 15
Playhouses, building Gothic
about, 186-187
board & batten detail drawing, 191
deck railing detail drawing, 190
door architecture template, 191
door detail drawing, 191
drawing of entire, 189
floor plan, 190, 193
framing drawings, 192-193
materials list, 188
rafter template, 194
spire detail drawing, 190
step-by-step, 195-199
verge board template, 190
window box detail drawing, 194
Pole barns with partial lean-to overhang, building
about, 94
elevation drawings, 96-97
floor plan, 96
framing plans, 99
roof detail drawings, 99, 100
step-by-step, 101-113
tools & materials list, 95
Pole framing, 35
Polycarbonate materials, about, 349
Pool pavilions, building
about, 336
bracket hub detail drawing, 340
building section drawing, 339
elevation drawing, 338
floor plan, 340
hip rafter template, 341
intermediate rafter template, 341
rim beam template, 341
roof framing plan, 341
roof hub detail drawing, 340
step-by-step, 342-345
tools & materials list, 337
Post & beam barn/shed plan, 210
Posts
cutting lumber, 369
setting in concrete, 438-439
Power tools, 16
Prefabricated structures. *See* Kit greenhouses, building; Kit sheds, building
Pressure-treated lumber, 18

Q

Quonset houses. *See* Hoophouses, building

R

Rafters, 44
Rafter ties, 44
Rental tools, 16
Reverse gable garage plan, 430
Ridge caps
cutting, 65
fitting over sealer strips, 427
Ridge vents
installing, 66-67
ventilation and, 60
Rocks, working around, 103
Roman columns, 380
Roofs
about completing, 60
asphalt shingles, installing, 61, 63-66
cedar shingles, installing, 440-441
corrugated panels for, installing, 245-246
corrugated polycarbonate materials for, 349
deck, installing & preparing, 62-63
fascia & soffits, installing, 56-59
framing, 44, 46-49
gambrel, about, 115
greenhouse ventilation, 239
metal panels for, 418
safety considerations, 64
shade and slatted, 395
trusses, using, 50-51
windows for ventilation, 231
Run-in shed plan, 211
Rustic summerhouses, building
about, 264-265
common rafter template, 271
dormer common rafter template, 271
elevation drawings, 268-269
floor plan, 267
framing drawings, 269-270
jack rafter template, 271
roof drawings, 270-271
roof template for cutting corrugated metal, 271
step-by-step, 272-281
tools & materials list, 266-267
valley rafter template, 271

S

Safety considerations
fiber-cement lap siding, cutting, 87
garage door openers, installing, 81
general, 21
on roofs, 64
Screened porches, building. *See* Summer pavilions, building
Select Structural (SEL STR) lumber, 18
Shade and slatted roof projects, 395
Shakes & shingles, wood, 89
Sheathing, about spaced, 174
Sheathing walls, 52-55
Shed plans
barn storage, 212
barn storage with loft, 212
combined firewood & storage, 211
gable storage with cupola, 213
large gable storage, 213
post & beam barn, 210
run-in shed plan, 211
Sheds, building metal kit
about, 142
anchoring, 150
maintaining, 150
shopping for, 144-145
step-by-step assembly, 146-150
Sheds, building mini garden
about, 240-241
drawings of views, 243
step-by-step, 244-247
tools & materials list, 242
Sheds, building simple storage
about, 130-131
door detail drawing, 137
elevation drawings, 133-134
floor plan, 136
framing drawings, 134-135
materials list, 132
rafter template, 137
roof plan, 136
step-by-step, 138-141
Sheds, building sunlight garden
about, 248-249
building section drawing, 251
door construction drawings, 255
elevation drawings, 254
framing plans, 252-253
header & window drawing, 256
materials list, 250
rafter templates, 256
rake board detail drawing, 256
soffit detail drawing, 254
step-by-step, 257-261
table & window drawing, 256
window drawings, 256
Sheds, building timber-frame
about, 174-175
building section drawing, 178
door detail drawing, 180
door jamb detail drawing, 180
eave detail drawing, 180
elevation drawings, 177, 179
floor plan, 178
framing drawings, 177, 178
gable overhand detail drawing, 180
materials list, 176
rafter template, 178
roof drawings, 180
step-by-step, 181
Sheds, building wood kit
about, 142, 143
delivery, 152
shopping for, 144-145
step-by-step assembly, 152-159

Shingle roofing, installing cedar, 440-441
Siding
about, 82
applying cast veneer stones, 84-85
types, 83-84, 89
Simple storage sheds, building
about, 130-131
door detail drawing, 137
elevation drawings, 133-134
floor plan, 136
framing drawings, 134-135
materials list, 132
rafter template, 137
roof plan, 136
step-by-step, 138-141
Single detached garages, building
about, 398
building section drawing, 402
corner detail drawing, 401
elevation drawings, 400, 403
floor plan, 401
framing drawings, 403
materials list, 399
overhead door jamb detail drawing, 401
service door jamb detail drawing, 401
Site, choosing
general considerations, 12
for greenhouses, 218
sunlight and seasonal considerations, 13
for sunlight garden shed, 249
6-sided gazebos. *See* 3-season gazebos, building
Slab-on-grade foundations
about, 28
plan views, 29
pouring, 30-33
Soffits, installing, 56-59
Spaced sheathing, about, 174
Speed square, using, 45
Stairs, building, 296
Standard (STAND) lumber, 18
Stick framing walls, 34-43
Storage & firewood shed plan, 211
Storage sheds, building. *See* Entries beginning with *shed* or *sheds;* Storage sheds, building simple
Storage sheds, building simple
about, 130-131
door detail drawing, 137
elevation drawings, 133-134
floor plan, 136
framing drawings, 134-135
materials list, 132
rafter template, 137
roof plan, 136
step-by-step, 138-141
Studios, building clerestory
about, 160-161
building section drawing, 164
door detail drawing, 166
elevation drawings, 163, 167
floor plan, 167
framing drawings, 165-166
jamb/corner detail drawing, 166
materials list, 162
rafter templates, 168
rake detail drawing, 166
step-by-step, 169-173
Summer pavilions, building
about, 297
eave details drawing, 303
elevation drawings, 300-301
foundation plan, 301
framing drawings, 302
roof beam detail drawing, 303
roof beam template, 301
roof plan, 303
step-by-step, 304-309
tools & materials list, 298-299
Sunlight
site selection and, 13
slatted roof projects and, 395
Sunlight garden sheds, building
about, 248-249
building section drawing, 251
door construction drawings, 255
elevation drawings, 254
framing plans, 252-253
header & window drawing, 256
materials list, 250
rafter templates, 256
rake board detail drawing, 256
soffit detail drawing, 254
step-by-step, 257-261
table & window drawing, 256
window drawings, 256

T

Three-car detached garage plan, 428
Three-car garage/workshop plan, 430
Three-quarter span greenhouses, described, 217
3-season gazebos, building
about, 283
building section drawing, 287
door frame detail drawing, 287
eave detail drawing, 289
elevation drawing, 285
floor plans, 286
foundation drawing, 285
framing drawings, 286, 288
rafter hub/corner detail drawing, 289
rafter templates, 288
roof framing plan, 288
stair detail drawing, 289
step-by-step, 290-296
tools & materials list, 283-284
window frame detail drawing, 287
Three-stall garage plan, 431
Timber-frame sheds, building
about, 174-175
building section drawing, 178
door detail drawing, 180
door jamb detail drawing, 180
eave detail drawing, 180
elevation drawings, 177, 179
floor plan, 178
framing drawings, 177, 178
gable overhand detail drawing, 180
materials list, 176
rafter template, 178
roof drawings, 180
step-by-step, 181
Tongue-and-groove paneling, about, 19
Tool bins, building lean-to
about, 200-201
building section drawing, 205
door jamb detail drawing, 206
elevation drawings, 204, 205, 206
floor plan, 206
framing plans, 203-204
materials list, 202
overhang detail drawing, 206
rafter template, 206
roof drawings, 203, 206
step-by-step, 207-209
Tools, power & rental, 16
Traditional span greenhouses, described, 216
Trusses
about, 50
plan for lateral in pole barns, 99
premanufactured, 418
working with, 51
Tunnel houses. *See* Hoophouses, building
Two-car detached garage plan, 429

U

Utility (UTIL) grade lumber, 18
Utility lines, 14

V

Ventilation
for kit greenhouses, 239
ridge vents for, 60
windows for, 231
Victorian garage plan, 429
Vinyl lap siding, 83
Virtual Grower (software), 218

W

Walls
framing & raising, 34-43
sheathing, 52-55
White oak, using, 358
Windows
framing, 39-40, 69
installing, 68, 70-72
sizing, 69
for ventilation, 231
Wood A-frame greenhouses, building
about, 226
step-by-step, 228-231
tools & materials list, 226-227
Wooden skid foundations, building, 432-433
Wood kit sheds
about, 142, 143
delivery, 152
shopping for, 144-145
step-by-step assembly, 152-159
Wood lap siding, 83
Wood shakes & shingles, 89

Z

Zoning laws, 14